OXFORD MATHEMATICAL MONOGRAPHS

Series Editors

J. M. BALL   W. T. GOWERS
N. J. HITCHIN   L. NIRENBERG
R. PENROSE   A. WILES

For a full list of titles please visit

http://ukcatalogue.oup.com/category/academic/series/science/maths/omm.do

Cyclic Modules and the Structure of Rings

S. K. Jain
King Abdulaziz University, SA and Ohio University, USA

Ashish K. Srivastava
Saint Louis University, USA

Askar A. Tuganbaev
Russian State University of Trade and Economics, Moscow, Russia

OXFORD
UNIVERSITY PRESS

OXFORD
UNIVERSITY PRESS

Great Clarendon Street, Oxford, OX2 6DP,
United Kingdom

Oxford University Press is a department of the University of Oxford.
It furthers the University's objective of excellence in research, scholarship,
and education by publishing worldwide. Oxford is a registered trade mark of
Oxford University Press in the UK and in certain other countries

© S. K. Jain, Ashish K. Srivastava, Askar A. Tuganbaev 2012

The moral rights of the authors have been asserted

First Edition published in 2012

Impression: 1

British Library Cataloguing in Publication Data
Data available

Library of Congress Cataloging in Publication Data
Library of Congress Control Number: 2012942948

ISBN 978–0–19–966451–1

Printed and bound by
CPI Group (UK) Ltd, Croydon, CR0 4YY

Preface

The origin of this monograph is a survey article that we wrote for the proceedings of an international conference in Rings and Modules. We promised to write a monograph that provided, in one place, interested readers with an up-to-date account of the literature on the subject of determining structure of rings over which cyclic modules or proper cyclic modules have a finiteness condition or a homological property. The finiteness conditions and homological properties are closely interrelated in the sense that either hypothesis induces the other in some form. The main objective behind writing this volume was the absence of such a book containing most of the relevant material on the subject. This monograph will be the first of its kind in the literature.

Over the last 50 years, numerous authors including Armendariz, Baccella, Beidar, Boyle, Byrd, Camillo, Chatters, Clark, Cohen, Faith, Farkas, Fisher, Goodearl, Gómez Pardo, Guil Asensio, Hajarnavis, Huynh, Koehler, Jain, Levy, López-Permouth, Mohamed, Ornstein, Osofsky, Salce, Singh, Skornyakov, Smith, Tuganbaev, and Wisbauer have investigated rings whose factor rings or factor modules have finiteness condition or a homological property. Fundamental contributions on this subject have continually opened new directions. The bibliography has more than 200 references and does not claim to be exhaustive.

The monograph assumes knowledge of basic noncommutative ring theory and homological algebra, equivalent to about a one-year graduate course in rings and modules.

Chapter 1, however, presents basic definitions and results in ring theory and module theory with plenty of examples. These concepts and results are needed throughout this monograph. Most of these results are stated without proofs.

Chapter 2 considers rings characterized by their proper factor rings. The first section studies rings each of whose proper factor rings is artinian. The next section deals with rings each of whose proper factor rings is perfect, a property that relates to the existence of strong flat covers for commutative rings. The third section discusses nonprime rings each of whose proper factor rings is von Neumann regular. We conclude this chapter by considering results due to Cohen and Levy for commutative rings for which every proper factor ring is self-injective—a property shared by Dedekind domains.

Chapter 3 deals with rings over which each proper cyclic module has a chain condition, considered by various authors including Camillo, Huynh, Jain, and Krause. The first section of this chapter studies rings each of whose proper cyclic modules is artinian. In the second section we study rings with the restricted minimum condition as considered by Chatters and Ornstein. We close this chapter discussing rings each of whose proper cyclic modules is perfect, as studied by Salce, Jain, and others.

Chapter 4 begins with Osofsky's celebrated theorem that gives the structure of rings over which every cyclic module is injective, which was also considered by Skornyakov. In the next section, we discuss a powerful result of Osofsky–Smith for rings over which each cyclic module satisfies the property that complement submodules are direct summands.

Chapter 5 considers rings each of whose proper cyclic modules is injective and prove theorems of Faith, Cozzens, and Damiano. Chapter 6 opens with Villamayor rings (V-rings), that is, rings over which each simple module is injective. In the second section, we discuss weakly-V rings, that is, rings over which every simple module is injective relative to proper cyclic modules. The next section deals with rings each of whose simple modules is Σ-injective and gives recent results due to Srivastava. Further generalizations are given in the fourth section that studies rings over which every cyclic module has the property that its injective hull is Σ-injective.

In Chapter 7, the first section deals with rings each of whose cyclic modules is quasi-injective as given in the Ph.D. dissertations of Ahsan and Koehler. This is followed by Section 7.2 in which we discuss rings over which each proper cyclic module is quasi-injective, as studied in the Ph.D. dissertation of Symonds. In Chapter 8, the first section deals with rings each of whose cyclic modules is continuous and we prove the structure theorem of Jain and Mohamed. In the next section, we discuss semiperfect rings over which each proper cyclic module is continuous as given by Jain and Mueller.

In Chapter 9, the first section deals with rings for which each cyclic module is π-injective. In the next section, we discuss rings each of whose proper cyclic modules is π-injective.

In Chapter 10, we begin with results due to Tuganbaev for rings each of whose cyclic modules is $\aleph_0$-injective. In Section 10.2, we discuss rings over which each cyclic module is weakly injective. The third section contains Koehler's theorem on the structure of rings each of whose cyclic modules is quasi-projective. The first section of Chapter 11 studies hypercyclic rings due to Osofsky and Caldwell. The second and third sections deal with q-hypercyclic and π-hypercyclic rings as studied by Malik in his dissertation and other authors.

Chapter 12 considers rings each of whose cyclic modules is essentially embeddable in a free module (or a projective module) and gives results from Gómez-Pardo, Guil Asensio, Jain, López-Permouth, and others. In this chapter, we present a survey of the work done on both the FGF problem and CF problem. In Chapter 13, we study the structure of rings whose modules are serial or distributive. Next, in Chapter 14, we consider rings characterized by properties of decomposition of cyclic modules that pertains to the works of Chatters, Huynh, Jain, Leroy, and Rizvi, among others. Chapter 15 contains selected results of Cohen, Kaplansky, Koethe, and others on the structure of rings over which each module is a direct sum of cyclic modules. In Chapter 16, we consider rings where every module is an I_0-module. The results of Chapter 16 are based on the papers of Abyzov and Tuganbaev. In Chapters 17 and 18, we present a brief study of completely integrally closed modules and rings that pertain to results mostly due to Tuganbaev.

The last chapter of the monograph deals with the dual notion of structure of rings determined by properties of its one-sided ideals. The first section deals with rings in which each right ideal is quasi-injective, equivalently, rings which are right self-injective and whose each essential right ideal is two-sided. Structure theorems from Beidar, Byrd, Jain, Ivanov, Mohamed, and others, are included in this section. The second section deals with rings in which each right ideal is a finite direct sum of quasi-injective right ideals, a property shared by artinian serial rings as shown by Nakayama. The last four sections deal with rings in which each one-sided ideal is π-injective, weakly injective, quasi-projective, or their direct sums.

At the end of most chapters is a list of classical open problems. It is expected that this monograph will be valuable reading material and a reference volume for students working in Algebra and for working mathematicians. It should inspire young researchers to work on the open problems discussed in the text.

We would like to thank John Clark, J. L. Gómez Pardo, Greg Marks, and Surjeet Singh for their valuable and helpful comments.

S. K. Jain would like to dedicate this monograph to his wife, children, and grandchildren: Parvesh, Nisha, Chirag, Steve, Priya, Sahanna, Eshan, Lakshmi, and Arjun.

S. K. Jain would also like to dedicate this monograph to his friend Professor Abdulrahman O. Alyoubi, Vice President of Educational Affairs, King Abdulaziz University, Saudi Arabia, who has been instrumental in encouraging both strong and highest quality academic programs at the university, and with whom Jain has been associated as an Advisor to KAU for the past three years.

Ashish K. Srivastava would like to dedicate this monograph to his parents Suresh Chand, Giriza, wife Sweta, and his whole family: Abhijeet, Amitabh, Amrita, Anamika, Armaan, Ashutosh, Asheesh, Divyanshi, Mansi, Prem Shankar, Rekha, Ragini, and last but not the least his newborn son Anupam who has brought immense joy in his life.

Askar Tuganbaev would like to dedicate this monograph to his wife, sister, children, and grandchildren: Natalie, Saule, Saule Jr, Askar Jr, Diar, Timur, Arman, Maxim, Ivar, Polina, Anvar, and Dina.

Askar Tuganbaev would also like to dedicate this monograph to his friends Professors Victor N. Latyshev, Alexander V. Mikhalev, and Mikhail V. Zaicev of Moscow State University.

Contents

1 Preliminaries

All our rings are rings with identity. Let R be a ring. Let M be an additive abelian group. Let $\operatorname{End}(M)$ be the ring of all group endomorphisms of M. We will call M a right R-module written usually as M_R if there is a ring homomorphism from R to $\operatorname{End}(M)$. Equivalently, there is a mapping $f : M \times R \to M$ such that the following conditions (a)–(d) hold. Write for convenience, $f(m, r) = mr$.

(a) $(m_1 + m_2)r = m_1 r + m_2 r$,
(b) $m(r_1 + r_2) = mr_1 + mr_2$,
(c) $m_1(r_1 r_2) = (m_1 r_1)r_2$, and
(d) $m1 = m$
for all $m, m_1, m_2 \in M$ and $r, r_1, r_2 \in R$.

Left R-modules are defined analogously using a mapping $g : R \times M \to M$ and usually denoted by ${}_R M$. If M is a right R-module and also a left S-module, then we say M is an $(S; R)$-bimodule if $(sm)r = s(mr)$ for all $s \in S$ and $r \in R$, denoted as ${}_S M_R$. Informally, we also say R acts on M on the right and S acts on the left.

Any additive abelian group is a module over the ring of integers $\mathbb{Z}$. Over a commutative ring R any module M can be made into an $(R; R)$-bimodule by defining $rm = mr$.

Let M be a right R-module. Let $f : M \to M$ be an additive group homomorphism such that $f(mr) = f(m)r$ for all $m \in M$ and $r \in R$. Then f is called an R-homomorphism of M into M or an R-endomorphism of M. Any R-module M can be made into an $(S; R)$-bimodule in a natural way, where S is the ring of all R-endomomorphisms of the right R-module M.

We call a family $\{N_i : i \in \mathcal{I}\}$ of submodules of an R-module M an *independent family* of submodules if $\Sigma_{i \in \mathcal{I}} N_i$ is a direct sum. A set $\{m_i\}_{i \in \mathcal{I}}$ of elements of M is called a *generating set* of M if $M = \Sigma_{i \in \mathcal{I}} m_i R$. If $\mathcal{I}$ is a finite set then M is called *finitely generated*. If $\mathcal{I}$ is a singleton set then M is called a *cyclic module*.

Let $\{M_i\}_{i \in \mathcal{I}}$ be a family of right R-modules. The cartesian product $\prod_{i \in \mathcal{I}} M_i$ of the family can be made into an R-module in a natural way. If the cardinality of $\mathcal{I}$ is n, then $\prod_{i \in \mathcal{I}} M_i$ is expressed as $M_1 \times M_2 \times \ldots \times M_n$. If M_R is a right R-module then it is trivially generated by the set $\{x_i\}$, $x_i \in M$ and the cardinality of the generating set is same as that of M. For any module S and any set $\mathcal{I}$, we denote by $S^{(\mathcal{I})}$ the submodule of the module of the cartesian product $S^{\mathcal{I}}$ of $\mathcal{I}$ copies of S that consists of those elements (s_i) of $S^{\mathcal{I}}$ for which all but finitely many s_i are 0. Then by considering the canonical R-epimorphism, $f : R^{(M)} \to M = \Sigma_{i \in M} x_i R$, given by $(r_i)_{i \in M} \mapsto \Sigma_{i \in M} x_i r_i$, we have $M \cong R^{(M)}/K$, where $K = \{(r_i) \in R^{(M)} : \Sigma_{i \in M} x_i r_i = 0\}$. In particular, if M is cyclic, then $M \cong R/A$ for some right ideal A. Conversely, the right R-module R/A is a cyclic module generated by $1 + I$. It follows from above discussion that every R-module is a homomorphic image of $R^{(\mathcal{I})}$ for some set $\mathcal{I}$.

A nonzero module M is called a *simple module* if it has no proper submodules. The *socle* of a module M is defined as the sum of all simple submodules of M and is denoted by $\mathrm{Soc}(M)$; we set $\mathrm{Soc}(M) = 0$ if M does not have simple submodules. A module M is called *semisimple* (or *completely reducible*) if $\mathrm{Soc}(M) = M$. It is not difficult to see that a module M is semisimple if and only if each submodule of M is a direct summand of M. A ring R is called a *semisimple ring* if R_R (or equivalently, $_R R$) is semisimple.

If M is a module, then any module that is isomorphic to a submodule of some homomorphic image of a direct sum of copies of M, is called an *M-subgenerated* module. The full subcategory of the category of all right R-modules whose objects are all M-subgenerated modules is denoted by $\sigma[M]$ and called the *Wisbauer category* of the module M.

1.1 Artinian and noetherian modules

A family of subsets $\{S_i : i \in \mathcal{I}\}$ in a set S is said to satisfy the *Ascending Chain Condition* (*ACC* in short) if for any ascending chain $S_{i_1} \subseteq S_{i_2} \subseteq \ldots$ in the family, there exists an integer n such that $S_{i_n} = S_{i_{n+k}}$ for each $k \in \mathbb{N}$.

A family of subsets $\{S_i : i \in \mathcal{I}\}$ in a set S is said to satisfy the *Descending Chain Condition* (*DCC* in short) if for any descending chain $S_{i_1} \supseteq S_{i_2} \supseteq \ldots$ in the family, there exists an integer n such that $S_{i_n} = S_{i_{n+k}}$ for each $k \in \mathbb{N}$.

A module M is called *artinian* (resp., *noetherian*) if the family of all submodules of M satisfies *DCC* (resp., *ACC*). A ring R is called *right artinian* (resp., *noetherian*) if R_R is artinian (resp., noetherian).

For any module M, the *Jacobson radical* of M is defined as the intersection of all maximal submodules of M and is denoted by $J(M)$. For a ring R, it can be proved that $J(R_R)$, $J(_R R)$ are equal, and we denote both of them by $J(R)$. A ring R is called a *semiprimitive ring* if $J(R) = 0$.

Theorem 1.1 *(Hopkins–Levitzki Theorem) If R is a right artinian ring, then R is also right noetherian and $J(R)$ is nilpotent.*

This was proved by Hopkins [108] and Levitzki [165]. Note that an artinian module need not be noetherian. The ring of p-adic integers as a module over $\mathbb{Z}$ is artinian but not noetherian. Examples are known of finitely generated artinian modules which are not noetherian (see [98], [198]).

Example Let R be the ring of 2×2 matrices over $\mathbb{Q}$ of the form $\begin{bmatrix} a & b \\ 0 & c \end{bmatrix}$, where $a, b \in \mathbb{Q}$ and $c \in \mathbb{Z}$. Then R is left noetherian but not right noetherian. It may also be verified that R is neither left nor right artinian. $\square$

Example Let R be the ring of 2×2 matrices over $\mathbb{Q}$ of the form $\begin{bmatrix} a & b \\ 0 & c \end{bmatrix}$, where $a \in \mathbb{Z}$ and $b, c \in \mathbb{Q}$. Then R is right noetherian but not left noetherian. $\square$

Example Let $\mathbb{Q}(x)$ be the field of fractions of $\mathbb{Q}[x]$. Let R be the ring of 2×2 matrices over $\mathbb{Q}(x)$ of the form $\begin{bmatrix} a & b \\ 0 & c \end{bmatrix}$, where $a \in \mathbb{Q}$ and $b, c \in \mathbb{Q}(x)$. Then R is right artinian but not left artinian. $\square$

Any endomorphism of an artinian (or a noetherian) module is an automorphism if it is a monomorphism (or an epimorphism). The following lemma, known as the Fitting Lemma is valuable in a number of applications. The reader can find this in [12] or [160].

Lemma 1.2 *Let M be a module such that M is both artinian and noetherian. If $f \in \mathrm{End}(M)$, then for some positive integer n, $M = \mathrm{Im}(f^n) \oplus \mathrm{Ker}(f^n)$.*

A *composition series* for a module M is a chain of submodules

$$0 = M_0 \subset M_1 \subset M_2 \subset \ldots \subset M_{n-1} \subset M_n = M$$

such that each of the factors M_i/M_{i-1} is a simple module. The number of gaps (namely n) is called the *length* of the composition series. One can show that any two composition series of M have the same length. If a module M has a composition series of length n, then M is said to *have length n* and we denote this by $l(M) = n$. It is not difficult to see that a module M has finite length if and only if M is both noetherian and artinian.

Lemma 1.3 *(Nakayama's Lemma) If M is a finitely generated right R-module then $MJ(R) = M$ implies $M = 0$.*

If R is right (left) artinian and $J(R) = 0$, then R is semisimple artinian and then by the Wedderburn–Artin Structure Theorem, R is isomorphic to a finite direct product of matrix rings over division rings.

For a ring R, the intersection of all prime ideals of R is denoted by $N(R)$; it is called the *prime radical* or the *lower nil-radical* of R. The ideal $N(R)$ contains all nilpotent right or left ideals of R. A ring R is called *semiprime* if the prime radical $N(R) = 0$. An ideal I of a ring R is called a *semiprime ideal* if R/I is a semiprime ring. A ring R is called a *prime ring* if the zero ideal is a prime ideal of R.

Next, we state the Krull–Schmidt–Azumaya Theorem.

Theorem 1.4 *Suppose an R-module M has two decompositions, $M = \oplus_{i \in \mathcal{I}} A_i = \oplus_{j \in \mathcal{J}} B_j$ such that each A_i, B_j is indecomposable and their endomorphism rings are local. Then $|\mathcal{I}| = |\mathcal{J}|$ and there exists a bijection $\sigma : \mathcal{I} \to \mathcal{J}$ such that $A_i \cong B_{\sigma(i)}$, $i \in \mathcal{I}$.*

The reader is referred to Lambek ([162], Proposition 10, page 24) if both $\mathcal{I}$ and $\mathcal{J}$ are assumed to be finite. For the general case, one may look up Faith ([75]).

A module M is said to be *uniserial* if any two submodules of M are comparable with respect to inclusion. A ring R is called a *right* (resp. *left*) *uniserial ring* if R_R (resp. $_RR$) is a uniserial module. Any direct sum of uniserial modules is called a *serial* module. A ring R is said to be a *right* (resp. *left*) *serial ring* if the module R_R (resp. $_RR$) is serial. A ring R is called a *serial ring* if R is both left as well as right serial. If R is a right or left serial ring, then $R/J(R)$ is a semisimple artinian ring.

We give below an example of a right serial ring which is not left serial.

Example Let R be the ring of 2×2 matrices of the form $\begin{bmatrix} a & b \\ 0 & c \end{bmatrix}$, where $a \in \mathbb{R}$ and $b, c \in \mathbb{C}$. Then R is a right serial ring but not a left serial ring. $\qquad\square$

The *Krull dimension* of a right R-module M, which will be denoted by $\mathcal{K}(M)$, is defined by transfinite recursion as follows: if $M = 0$, $\mathcal{K}(M) = -1$; if α is an ordinal and $\mathcal{K}(M) \not< \alpha$ then $\mathcal{K}(M) = \alpha$ provided there is no infinite descending chain $M = M_0 \supset M_1 \supset \cdots$ of submodules M_i such that for $i = 1, 2, 3, \ldots$

$$\mathcal{K}(M_{i-1}/M_i) \not< \alpha.$$

If there is no ordinal α such that $\mathcal{K}(M) = \alpha$, then we say that M has no Krull dimension. The Krull dimension of a ring R, $\mathcal{K}(R)$ is defined to be Krull dimension of R_R.

1.2 Free modules, projective modules, and injective modules

A module F is called a *free module* if it has an independent set of generators. For example, if R is a ring with identity, then R is a free R-module and so is $R^{(\mathcal{I})}$ for any set $\mathcal{I}$. It is a simple exercise to show that every free R-module is isomorphic to $R^{(\mathcal{I})}$ for some set $\mathcal{I}$. We have already seen that every R-module is a homomorphic image of $R^{(\mathcal{I})}$ for some set $\mathcal{I}$. Thus every module is a homomorphic image of a free module. Free modules F have the property that every homomorphism $f : F \to X/K$ can be lifted to $g : F \to X$, but modules with this property need not be free.

A right R-module M is called *finitely presented* if for some finitely generated free module F and some finitely generated submodule K of F, $M = F/K$.

An R-module P is called a *projective module* if for any R-epimorphism $f : X \to Y$ and any R-homomorphism $g : P \to Y$, there exists an

R-homomorphism $h : P \to X$ such that $fh = g$. A module is projective if and only if it is a direct summand of a free module.

Dual Basis Lemma A right R-module P is projective if and only if there exist a family of elements $\{a_i : i \in \mathcal{I}\} \subseteq P$ and linear functionals $\{f_i : i \in \mathcal{I}\} \subseteq \text{Hom}_R(P, R)$ such that, for each $a \in P$, $f_i(a) = 0$ for all but finitely many $i \in \mathcal{I}$, and $a = \Sigma_{i \in \mathcal{I}} a_i f_i(a)$.

A module N is called a *small submodule* of M if for any submodule K of M, $K + N = M$ implies $K = M$. A *projective cover* of a module M_R is an epimorphism $f : P_R \to M_R$ from a projective module P_R onto M_R such that $\text{Ker}(f)$ is a small submodule of P. It is not difficult to see that the Jacobson radical $J(M)$ of a module M is the sum of all small submodules of M.

A ring R is called a *left perfect ring* if every left R-module has a projective cover.

Theorem 1.5 *For a ring R, the following are equivalent:*

1. *R is left perfect.*
2. *R has DCC on principal right ideals.*
3. *$R/J(R)$ is semisimple artinian and $J(R)$ is left T-nilpotent (that is, for every sequence $a_1, a_2, \ldots$ in $J(R)$, $a_1 a_2 \ldots a_n = 0$ for some positive integer n).*
4. *$R/J(R)$ is semisimple artinian and every nonzero left R-module contains a maximal submodule.*
5. *R contains no infinite set of orthogonal idempotents and every nonzero right R-module contains a minimal submodule.*

The following example from Bass [20] is an example of a left perfect ring which is not a right perfect ring.

Example Let K be a field and let K_ω be the K-algebra of all matrices with entries in K, countably many rows and columns, and in which each row has only finitely many nonzero entries. Let N be the set of all strictly lower triangular matrices in K_ω with only finitely many nonzero entries and let R be the subalgebra $K + N$ of K_ω. The Jacobson radical $J(R) = N$. Now as $R/J(R) \cong K$ and $J(R) (= N)$ is left T-nilpotent but not right T-nilpotent, it follows that R is left perfect but not right perfect. $\qquad\square$

A ring R is called *semiperfect* if every finitely generated right R-module has a projective cover. A ring R is called a *local ring* if R has a unique maximal right (left) ideal. If A is a left (or right) ideal of R, we say that *idempotents lift* modulo A if, given any $r \in R$ with $r^2 - r \in A$, there exists $e^2 = e \in R$ with $e - r \in A$.

Theorem 1.6 *For a ring R, the following are equivalent:*

1. *R is semiperfect.*
2. *$R/J(R)$ is semisimple artinian and idempotents lift modulo $J(R)$.*

3. $R = e_1 R \oplus \cdots \oplus e_n R$ with $e_1 + \cdots + e_n = 1$, where e_i are orthogonal idempotents of R and each $e_i R e_i$ is a local ring.

A ring R is called a *semilocal ring* if $R/J(R)$ is semisimple artinian. A ring R is called a *primary ring* if $R/J(R)$ is simple artinian and $J(R)$ is nilpotent. An ideal I of a ring R is called a *primary ideal* if R/I is a primary ring. A ring R is called a *semiprimary ring* if $R/J(R)$ is semisimple artinian and $J(R)$ is nilpotent.

A right R-module M is called *semi-artinian* if for every submodule $N \neq M$, $\operatorname{Soc}(M/N) \neq 0$. A ring R is called right semi-artinian if R_R is semi-artinian. Bass [20] proved the following.

Theorem 1.7 *A ring R is left perfect if and only if R is semiperfect and right semi-artinian.*

The *socle series* of a module M is the ascending chain

$$0 \subset \operatorname{Soc}_1(M) = \operatorname{Soc}(M) \subset \ldots \subset \operatorname{Soc}_\alpha(M) \subset \operatorname{Soc}_{\alpha+1}(M) \subset \ldots,$$

where $\operatorname{Soc}_\alpha(M)/\operatorname{Soc}_{\alpha-1}(M) = \operatorname{Soc}(M/\operatorname{Soc}_{\alpha-1}(M))$ for every nonlimit ordinal number α and $\operatorname{Soc}_\alpha(M) = \bigcup_{\beta < \alpha} \operatorname{Soc}_\beta(M)$ for every limit ordinal number α. We denote by $L(M)$ the submodule of the form $\operatorname{Soc}_\xi(M)$, where ξ is the least ordinal number such that the relation $\operatorname{Soc}_\xi(M) = \operatorname{Soc}_{\xi+1}(M)$ holds.

It follows that a module M is semi-artinian if $M = L(M)$. In that case the least ordinal ξ such that $M = \operatorname{Soc}_\xi(M)$, is called the *socle length of M*. For an arbitrary ring R, we denote the ideals $L(R_R)$ and $\operatorname{Soc}(R_R)$ by $L(R)$ and $\operatorname{Soc}(R)$ respectively.

Dual to projective modules is the concept known as injective modules. This concept was initially studied by R. Baer in 1940 for abelian groups, although he used the word "complete" instead of "injective." Every module can be embedded in an injective module which is unique up to isomorphism.

An R-module M is called an *injective module* if for any given R-module A, a submodule B of A, and an R-homomorphism $f : B \to M$, there exists an R-homomorphism $g : A \to M$ that extends f.

In particular, if R_R is injective, then any R-homomorphism f from a right ideal of R to R can be extended to an endomorphism g of R. Then, setting $g(1) = c$, we have for all $a \in A$, $f(a) = g(a) = g(1a) = g(1)a = ca$, that is, the image of every element of a right ideal is a multiple of a fixed element of the module. The following theorem shows that this condition is also sufficient for a module to be injective.

Theorem 1.8 *Let M be an R-module. A necessary and sufficient condition that M is injective is that for each right ideal I of R and each R-homomorphism $f : I \to R$, there exists an element $m \in M$ such that $f(a) = ma$ holds for all $a \in I$.*

Proof Necessity is clear from above. To prove sufficiency, consider any R-module A, its submodule B, and an R-homomorphism $g : B \to M$. Consider the family $F = \{(C, h) : B \subseteq C \subseteq A,$ where $h : C \to M$ extends $g\}$. This family is nonempty and can be partially ordered in a natural way. It can be shown that F is an inductive set and hence by Zorn's Lemma, F has a maximal element, say (C, h). We claim $C = A$. Assume to the contrary that $C \neq A$. Consider $x \in A, x \notin C$. Set $I = x^{-1}C = \{t \in R : xt \in C\}$. Clearly I is a right ideal. Define $\theta : I \to M$ by $\theta(t) = h(xt)$. Note that θ is an R-homomorphism and, by hypothesis, there exists an element $m \in M$ such that $\theta(t) = mt$ for all $t \in I$. We define an R-homomorphism $k : C + xR \to M$ by $k(c + xr) = h(c) + mr$. But then $(C, h) \subset (C + xR, k)$, a contradiction to the maximality of (C, h). Hence $C = A$. This proves sufficiency. $\qquad\qquad\square$

Theorem 1.9 *Every module is a submodule of an injective module.*

For a proof, the reader is referred to Sharpe and Vamos ([202], Theorem 2.11, page 37) or Lambek ([162], Lemma 2, Lemma 3, and Proposition 4, pp. 89–90).

Theorem 1.10 *An R-module M is injective if and only if for each module N containing M, there exists a submodule K of N such that $N = M \oplus K$.*

For a proof, see Sharpe and Vamos ([202], Theorem 2.15, p. 39).

Thus, from the above theorem, it follows that a module M is semisimple if each submodule of M is injective.

An R-module B is called an *essential extension* of an R-module A (or, in other words, A is called an *essential submodule* of B) if for every nonzero submodule C of B, $C \cap A \neq 0$. Equivalently, for every nonzero element $b \in B$, there exists an element $r \in R$ such that $0 \neq br \in A$. If an R-module B is an essential extension of an R-module A, we generally denote this by $A \subseteq_e B$.

A submodule N of a module M is called *closed* (in M) if N has no proper essential extension in M. For a given submodule X of a module M, the family of essential extensions of X inside of M contains a maximal member, say C (by Zorn's Lemma). Clearly, C is closed in M. This module C is called an essential closure of X.

It may be shown using Zorn's Lemma that every module M has a maximal essential extension E, which can also be shown to be a minimal injective extension of M. This minimal injective extension E of M is unique up to isomorphism and is called the *injective hull* of M, denoted as $E(M)$. We state below a sequence of results that leads to the preceding statement.

Theorem 1.11 *Let M be an R-module. Then M is injective if and only if it has no proper essential extensions.*

Theorem 1.12 *Let M be an R-module and let N be an R-module containing M. Then the following statements are equivalent:*

1. N is a maximal essential extension of M.
2. N is a minimal injective extension of M.

For a proof, see Sharpe and Vamos ([202], Theorem 2.17, Propositions 2.18–2.20, and Theorem 2.21, pp. 41–43).

If R_R is injective, then R is called a *right self-injective* ring.

Example Let D be a division ring and V be an infinite-dimensional right D-space. Let $R = \text{End}(V_D)$ be the ring of all left linear transformations on V_D. It may be verified that R is a right self-injective ring. Now, it is well known that a ring which is both right as well as left self-injective must be directly-finite. Since the above ring R is not directly-finite, it follows that R is not a left self-injective ring.

For additional properties of self-injective rings and injective modules, the reader is referred to ([161], 3E–3G, pp. 81–99).

A nonzero module M is called *uniform* if the intersection of any two nonzero submodules of M is nonzero, or, equivalently, every nonzero submodule of M is essential in M. A module M is said to have *Goldie dimension* (or *uniform dimension*) n, denoted as $\text{Gdim}(M) = n$, if $E(M)$ is a direct sum of n uniform submodules, equivalently if M has an essential submodule which is a direct sum of n uniform submodules. A ring R is called a *right q.f.d. ring* if each cyclic right R-module has finite Goldie dimension.

For any subset X of R, the set $ann_r(X) = \{r \in R : Xr = 0\}$ is called the right annihilator of X. A *right (left) annihilator* in a ring R is any right (left) ideal of R which equals the right (left) annihilator of some subset of R. A ring R is called a *right Goldie ring* if R_R has finite Goldie dimension and R has the ACC on right annihilators.

For a ring Q, a unitary subring R in Q is called a *right order* in Q if every nonzero-divisor of the ring R is invertible in Q and every element of the ring Q can be represented in the form as^{-1}, where $a, s \in R$ and s is a nonzero-divisor in R. In this case, the ring Q is called the *classical right ring of quotients* of the ring R.

By a well-known theorem of Goldie ([74], 9.13), every semiprime right Goldie ring has a semisimple artinian classical right ring of quotients.

1.3 Hereditary and semihereditary rings

A ring R is called *right hereditary* if each right ideal of R is projective. A ring R is called *right semihereditary* if each finitely generated right ideal of R is projective.

For example, Dedekind domains and Prufer domains are hereditary and semihereditary rings, respectively.

The following fundamental theorem for hereditary rings is due to Kaplansky [151].

Theorem 1.13 *If R is a right hereditary ring then every submodule of a free right R-module is the direct sum of modules each of which is isomorphic to a right ideal of the ring R.*

For details of the proof, the reader may refer to ([41], Theorem 5.3). The corresponding result for semihereditary rings is as below.

Theorem 1.14 *If R is a right semihereditary ring then every finitely generated submodule of a free right R-module is the direct sum of a finite number of modules each of which is isomorphic to a finitely generated right ideal of the ring R.*

Another fundamental result on the characterization of hereditary and semihereditary rings is contained in the following theorem. But first we define a *f.g. injective* module M_R as one which has the property that for any homomorphism $f : A \to M$, where A is a finitely generated right ideal of R, there exists an element m of M such that $f(a) = ma$ for all $a \in A$.

Theorem 1.15 *For a ring R, the following are equivalent:*

1. *R is right hereditary (semihereditary).*
2. *Every submodule (finitely generated submodule) of a projective module is projective.*
3. *Every factor module of an injective module is injective (f.g. injective).*

For a proof, see ([41], Theorem 5.4, Proposition 6.2) and [104].

A ring R is called a *von Neumann regular ring* if for each $x \in R$, there exists $y \in R$ such that $xyx = x$.

Theorem 1.16 *For a ring R, the following are equivalent:*

1. *R is von Neumann regular.*
2. *Each principal right (left) ideal of R is generated by an idempotent.*
3. *Each finitely generated right (left) ideal of R is generated by an idempotent.*

For a proof, see [97].

A ring R is called *semiregular* if $R/J(R)$ is von Neumann regular and idempotents lift modulo $J(R)$. A von Neumann regular ring R is called *abelian regular* (or *strongly regular*) if all the idempotents of R are central. A ring R is called a *unit regular ring* if for each $x \in R$, there exists a unit element $u \in R$ such that $x = xux$.

For any module M, we define $Z(M) = \{x \in M : ann_r(x) \subseteq_e R_R\}$. It can be easily checked that $Z(M)$ is a submodule of M. It is called the *singular submodule* of M. If $Z(M) = M$, then M is called a *singular module*. If $Z(M) = 0$, then M is called a *nonsingular module*. In particular, if we take $M = R_R$, then $Z(R_R)$ is an ideal of the ring R; it is called the *right singular ideal*. If $Z(R_R) = 0$, then R is called a *right nonsingular ring*.

A module M_R is called a *rational extension* of a submodule N if for any elements $m, n \in M$ with $m \neq 0$, there exists an element $a \in R$ such that $ma \neq 0$ and $na \in N$.

Theorem 1.17 *Any ring R has a maximal rational extension $Q^r_{max}(R)$ contained in $E = E(R_R)$. Moreover, $Q^r_{max}(R) = \mathrm{End}_S(E)$, where $S = \mathrm{End}_R(E)$. This ring $Q^r_{max}(R)$ is called the maximal right quotient ring of R.*

If R is right nonsingular, then $Q^r_{max}(R)$ is the injective hull $E(R_R)$ of R, and is a von Neumann regular right self-injective ring.

1.4 Generalizations of injectivity

A module M is said to be *injective with respect to the module N* (or *N-injective*) if for every submodule N_1 of the module N, all homomorphisms $N_1 \to M$ can be extended to homomorphisms $N \to M$.

Thus, a right R-module M is *injective* if M is N-injective for every $N \in$ Mod-R. A module M is said to be *quasi-injective* if M is M-injective.

Consider the following three conditions;

C1: Every submodule of M is essential in a direct summand of M.

C2: Every submodule of M isomorphic to a direct summand of M is itself a direct summand of M.

C3: If N_1 and N_2 are direct summands of M with $N_1 \cap N_2 = 0$ then $N_1 \oplus N_2$ is also a direct summand of M.

A module M is called a *continuous module* if it satisfies conditions C1 and C2. A module M is called *π-injective* (or *quasi-continuous*) if it satisfies conditions C1 and C3. A module M is called a CS *module* (or *extending module*) if it satisfies condition C1.

In general, we have the following implications.

$$\text{Injective} \implies \text{Quasi-injective} \implies \text{Continuous} \implies \pi\text{-injective} \implies \text{CS}$$

In general, reverse implications do not hold in the above hierarchy.

Example

(i) For $n > 1$, $\mathbb{Z}/n\mathbb{Z}$ is a quasi-injective $\mathbb{Z}$-module which is not injective.

(ii) Let R denote the ring of all sequences of real numbers which are eventually rational. Then R as an R-module is continuous but not quasi-injective.

(iii) $\mathbb{Z}$ is a π-injective $\mathbb{Z}$-module which is not continuous.

(iv) $(\mathbb{Z}/2\mathbb{Z})\oplus(\mathbb{Z}/4\mathbb{Z})$ is a CS $\mathbb{Z}$-module which is not π-injective.

A ring R is called a right continuous (right CS) ring if R_R is continuous (CS). We give below some interesting examples of such rings.

Example (See [132].) Group algebra $K[D_\infty]$ of infinite dihedral group is right CS if and only if $char(K) \neq 2$. $\qquad\qquad\square$

Example (See [8].) Let p be a prime number and $G = P_\infty = \cup_{n=1}^\infty P_n$ where for each n, P_n is a Sylow p-subgroup of S_{p^n} and $P_n \subset P_{n+1}$. Then G is a locally finite p-group and $\Delta(G) = 1$. Let K be a field of characteristic p. Then $K[G]$ is a continuous group algebra. $\qquad\qquad\square$

Let X be a π-injective module. It follows from C3 that a direct sum $A_1 \oplus A_2 \oplus \cdots \oplus A_n$ in X is a summand of X if and only if each A_i is a summand of X. It is also evident that a complement submodule of a π-injective module X is a summand.

A module M is said to be *skew-injective* if for every submodule X of M, every homomorphism $X \to X$ can be extended to a homomorphism $M \to M$ (i.e., every endomorphism of any submodule of M can be extended to an endomorphism of M).

Every skew-injective module is π-injective (see Proposition 17.4). For any noncommutative division ring D, the polynomial ring $D[x]$ is right and left π-injective, which is not right or left skew-injective (see [223]).

Lemma 1.18 *Any direct summand of an injective (resp., quasi-injective, skew-injective, π-injective) module is injective (resp., quasi-injective, skew-injective, π-injective).*

We denote by $\mathrm{End}(M)$ the endomorphism ring of the module M. If $f\colon X \to M$ is a module homomorphism and N is a submodule of M, then we denote the submodule $\{x \in X : f(x) \in N\}$ of X by $f^{-1}(N)$.

Set $\Delta(M) = \{f \in \mathrm{End}(M) : \mathrm{Ker}(f) \subseteq_e M\}$. It is easy to check that $\Delta(M)$ is an ideal of the endomorphism ring $\mathrm{End}(M)$. If M is a right module over the ring R and X, Y are two subsets in M, then $(X : Y)$ denotes the subset $\{a \in R : Xa \subseteq Y\}$ of the ring R.

Lemma 1.19 *Let R be a ring, M be a right R-module, and X be an essential submodule of M.*

1. *If N is a module, $f\colon N \to M$ is a homomorphism and there exists a homomorphism $g\colon N \to X$ such that f coincides with g on $f^{-1}(X)$, then $f(N) \subseteq X$.*

2. *If N is a module and X is N-injective, then $f(N) \subseteq X$ for every homomorphism $f\colon N \to M$.*

3. *If the module X is quasi-injective, then $f(X) \subseteq X$ for every homomorphism $f\colon X \to M$. In particular, X is a fully invariant submodule in M.*

4. *A module X is quasi-injective $\Leftrightarrow X$ is a fully invariant submodule in the injective hull of M.*

5. *If $f \in \mathrm{End}(M)$ and there exists an endomorphism g of X such that f coincides with g on $X \cap f^{-1}(X)$, then $f(X) \subseteq X$.*

6. *If $Y_1 \oplus \cdots \oplus Y_n$ is a submodule in X, then there exists a direct decomposition $M' = Z' \oplus M_1 \oplus \cdots \oplus M_n$ such that X is an essential extension of the module $(Z' \cap X) \oplus Y_1 \oplus \cdots \oplus Y_n$ and $X \cap M_i$ is an essential extension of the module Y_i for every i.*

7. *If $f \in \mathrm{End}(M)$ is an idempotent and X is π-injective, then $f(X) \subseteq X$. Consequently, $X = \oplus_{i \in I}(X \cap M_i)$ for every direct decomposition $M = \oplus_{i \in I} M_i$.*

   8. *If the module X is π-injective and $Y_1 \oplus \cdots \oplus Y_n$ is a submodule in X, then there exists a direct decomposition $X = Z \oplus X_1 \oplus \cdots \oplus X_n$ such that X_i is an essential extension of the module Y_i for every i.*

   9. *If the module X is π-injective and $X = X_1 \oplus X_2$, then modules X_1 and X_2 are injective with respect to each other.*

Proof

(1) We assume that $y = (f - g)(x) \in X \cap (f - g)(N)$, where $x \in N$. Then $f(x) = (f - g)(x) + g(x) = y + g(x) \in X, x \in f^{-1}(N)$. Thus $y \in (f - g)$ $(f^{-1}(X)) = 0$ and so $X \cap (f - g)(N) = 0$. Since M is an essential extension X, we have $(f - g)(N) = 0$. Therefore $f(N) = g(N) \subseteq X$.

(2) Since $f(f^{-1}(X)) \subseteq X$ and X is N-injective, there exists a homomorphism $g \colon N \to X$ such that g coincides with f on $f^{-1}(X)$. By (1), $f(N) \subseteq X$.

(3) The assertion follows from (2) with $N = X$.

(4) The implication $\Rightarrow$ follows from (3). We prove the implication $\Leftarrow$. Let Y be a submodule in X, $f \in \mathrm{Hom}(Y, X)$, and let M' be the injective hull of the module X. Since the module M' is injective, f can be extended to an endomorphism f' of the module M'. Since X is a fully invariant submodule in M', we have that $f'(X) \subseteq X$ and the restriction of f' to X is the required extension of the homomorphism f to an endomorphism of the module X.

(5) Let $h = f|_X \colon X \to M$. In (1) take $N = X$. Now $h^{-1}(X) = f^{-1}(X) \cap X$ and g coincide with $h^{-1}(X)$ by the hypothesis. Thus by (1), $h(X) \subseteq X$. Hence $f(X) \subseteq X$.

(6) The injective hull M' of the module X contains injective hulls $M_1, \ldots, M_n$ of the modules $Y_1, \ldots, Y_n$. It is directly verified that $M_1 + \cdots + M_n = M_1 \oplus \cdots \oplus M_n$. Then the module $M_1 \oplus \cdots \oplus M_n$ is injective and there exists a direct decomposition $M' = Z' \oplus M_1 \oplus \cdots \oplus M_n$. It is directly verified that X is an essential extension of the module $(Z' \cap X) \oplus Y_1 \oplus \cdots \oplus Y_n$ and $X \cap M_i$ is an essential extension of the module Y_i for every i.

(7) We set $Y = X \cap f^{-1}(X)$. Now $M = f(M) \oplus (1 - f)(M)$. Let $z \in Y$. Then $f(z) \in X$. But $z = f(z) + (1 - f)(z)$. Thus $Y = X \cap f(M) \oplus (1 - f)(M) \cap X$. But X is π-injective. So $X = A \oplus B$ with $X \cap f(M) \subset_e A$, and $X \cap (1 - f)(M) \subset_e B$. This gives an idempotent endomorphism g of X which is identity on A and zero on B. This g coincides with f on Y. Then by (5), $f(X) \subseteq X$.

(8) The assertion follows from (6) and (7).

(9) Let $u_1 \colon X_1 \to X$ be a natural embedding, $\pi_2 \colon X \to X_2$ be the projection with kernel X_1, Y_1 be a submodule in X_1, and let $f \colon Y_1 \to X_2$ be a homomorphism. We denote by g the endomorphism of the submodule $Y_1 \oplus X_2$ in X such that $g(y_1 + x_2) = y_1 + f(y_1) \in Y_1 \oplus X_2$ for all $y_1 \in Y_1$, $x_2 \in X_2$. Then $g^2 = g$. Since the module X is π-injective, g can be extended

to an endomorphism h of the module X. Then $\pi_2 h u_1$ is a homomorphism from X_1 into X_2; it extends the homomorphism f. Therefore, the module X_1 is X_2-injective. $\qquad\square$

Lemma 1.20 *Let R be a ring, M be a right R-module, and X be an essential submodule of M.*

1. *For any module homomorphism $f\colon N \to M$, the module $f^{-1}(X)$ is an essential submodule in N.*
2. *$M/X = Z(M/X)$, that is, the module M/X is singular.*
3. *If the module X is nonsingular, then the module M is nonsingular and the kernel of any nonzero module homomorphism $f\colon N \to M$ is not an essential submodule in N (in particular, $\Delta(M) = 0$).*

Proof

(1) Let N' be a nonzero submodule in N. We wish to show that $f^{-1}(X) \cap N' \neq 0$. If $f(N') = 0$, then $N' \subseteq f^{-1}(0) \subseteq f^{-1}(X)$. Thus we assume that $f(N') \neq 0$. Then $X \cap f(N') \neq 0$, whence $0 \neq f^{-1}(X) \cap N'$, as required.

(2) Let $m \in M$ and let $f\colon R_R \to M$ be the homomorphism such that $f(a) = ma$ for all $a \in R$. By (1), $f^{-1}(X)$ is an essential right ideal of the ring R. Since $f^{-1}(X)$ is the annihilator of the element $m + X \in M/X$, we have $M/X = Z(M/X)$.

(3) Since $X \cap Z(M) = 0$ and M is an essential extension of the module X, the module M is nonsingular. If $f \in \mathrm{Hom}(N, M)$ and N is an essential extension of $\mathrm{Ker}(f)$, then $Z(N/\mathrm{Ker}(f)) = N/\mathrm{Ker}(f) \cong f(N)$ by (2). Therefore, $f(N) = Z(f(N)) \subseteq Z(M) = 0$ and f is the zero homomorphism. $\qquad\square$

Theorem 1.21 *[233] Let M be a quasi-injective right R-module and let $S = \mathrm{End}(M)$. Then $J(S) = \{f \in S : \mathrm{Ker}(f) \subseteq_e M\}$ and $S/J(S)$ is a von Neumann regular right self-injective ring whose idempotents lift modulo $J(S)$.*

In particular, if R is a right self-injective ring then $R/J(R)$ is a von Neumann regular right self-injective ring.

2 Rings characterized by their proper factor rings

2.1 Restricted artinian rings

A ring R is called a *restricted right artinian ring* if each proper homomorphic image of R is a right artinian ring, that is, R/I is a right artinian ring for every nonzero two-sided ideal I of R.

Clearly every right artinian ring is a restricted right artinian ring but a restricted right artinian ring need not be right artinian. For example, $\mathbb{Z}$, the ring of integers is a restricted artinian ring but not an artinian ring.

It is well-known that a right (left) artinian ring is right (left) noetherian. This was proved by Hopkins [108] and Levitzki [165]. In the case of commutative rings this result was proved several years earlier by Akizuki [7]. So, it is natural to ask: what is the connection between restricted artinian rings and noetherian rings or restricted noetherian rings? Cohen studied this question for commutative rings [51].

We start with a classical result due to Cohen.

Theorem 2.1 *Let R be a commutative ring. Then R is artinian if and only if R is noetherian and every prime ideal $P\,(\neq R)$ of R is maximal.*

Proof Suppose R is artinian. Then by the Hopkins–Levitzki theorem, R is noetherian. Now let $P\,(\neq R)$ be a prime ideal of R. Then $\overline{R} = R/P$ is an artinian integral domain. Therefore, $\overline{R}$ is a field and hence P is a maximal ideal of R.

Conversely, suppose R is noetherian and every prime ideal $P\,(\neq R)$ of R is maximal. Then every prime ideal of R is both a maximal ideal and a minimal prime ideal. Since R is noetherian, it has finitely many minimal prime ideals. Thus, it follows that R has finitely many maximal ideals, say $M_1, M_2, \ldots, M_n$. Therefore, $J(R) = M_1 \cap M_2 \cap \ldots \cap M_n = M_1 M_2 \ldots M_n$. Since R is noetherian and $J(R)$ is nil, $J(R)$ is nilpotent. So there exists an integer $k > 0$ such that $M_1^k M_2^k \ldots M_n^k = 0$. But, as $M_1^k, M_2^k, \ldots, M_n^k$ are all pairwise relatively prime, the Chinese Remainder Theorem gives $R = R/M_1^k M_2^k \ldots M_n^k \cong R/M_1^k \times R/M_2^k \times \ldots \times R/M_n^k$. Each R/M_i^k is local noetherian with its unique maximal ideal nilpotent. Therefore each R/M_i^k is artinian. Hence R is artinian. $\square$

As a consequence, we have the following characterization of a commutative restricted artinian ring.

Corollary 2.2 *A commutative ring R is restricted artinian if and only if R is noetherian and every nonzero prime ideal of R is maximal.*

Corollary 2.3 *A commutative ring R is restricted artinian but not artinian if and only if R is a noetherian integral domain not a field, in which every nonzero prime ideal is maximal.*

Now we move on to noncommutative rings. Ornstein [184] studied restricted right artinian rings. We start with the following simple observation.

Lemma 2.4 *Let R be a ring such that R/M is a simple artinian ring for each maximal ideal M of R. If the zero ideal of R is a finite product of maximal ideals, then R is right (left) artinian if and only if R is right (left) noetherian.*

Proof If R is right artinian then, by the Hopkins–Levitzki theorem, R is right noetherian. Conversely, suppose R is a right noetherian ring such that R/M is a simple artinian ring for each maximal ideal M of R and the zero ideal of R is a finite product of maximal ideals. Consider the series $R = M_0 \supset M_1 \supset M_1 M_2 \supset \cdots \supset M_1 M_2 \ldots M_k = 0$, where each M_i, $1 \le i \le k$ is a maximal ideal of R. The R-modules $N_i = M_1 \ldots M_{i-1}/M_1 \ldots M_i$ are semisimple R/M_i-modules, and therefore have a composition series. But $N_i \subset R/M_1 \ldots M_i$ is noetherian. Thus the above series can be refined to a composition series of the module R_R. Hence R is right artinian. $\qquad\square$

The next result shows the relation between restricted right artinian rings and restricted right noetherian rings.

Theorem 2.5 *A ring R is a restricted right artinian ring if and only if R is a restricted right noetherian ring and, for each prime ideal $P \neq 0$ in R, the ring R/P is a simple artinian ring.*

Proof Suppose R is a right restricted artinian ring. Then clearly R is a restricted right noetherian ring and for each prime ideal $P \neq 0$ in R, the ring R/P is a simple artinian ring. Conversely, suppose R is a restricted right noetherian ring and for each prime ideal $P \neq 0$ in R, the ring R/P is a simple artinian ring. Let I be a nonzero ideal of R and let $\overline{R} = R/I$. If I is not prime then the zero ideal of $\overline{R}$ is a finite product of prime ideals. If $\overline{P}$ is any prime ideal in $\overline{R}$, its inverse image P is a prime ideal in R, and $\overline{R}/\overline{P} \cong R/P$ is simple artinian. By the above lemma, it follows that $\overline{R}$ is right artinian. This shows R is a restricted right artinian ring. $\qquad\square$

Corollary 2.6 *Let R be a nonsimple prime ring. Then the following are equivalent:*

1. *R is right restricted artinian but not right artinian.*
2. *R is right restricted noetherian and for each prime ideal $P \neq 0$, R/P is a simple artinian ring.*

Let R be a nonprime right restricted artinian ring. Then there exist a finite number of prime ideals $P_1, P_2, \ldots, P_n$ (they are necessarily maximal ideals) of R such that $P_1 P_2 \ldots P_n = 0$.

Theorem 2.7 *A nonprime restricted right artinian ring R has only a finite number of prime ideals $P_1, P_2, \ldots, P_n, n \geq 1$, which are also the only maximal ideals of R. Their intersection $\cap_{i=1}^{n} P_i$ is nilpotent and is precisely the Jacobson radical $J(R)$.*

Thus it follows that a nonprime restricted right artinian ring has a nilpotent Jacobson radical.

Proof Let $P \subset R$ be a prime ideal. Since $P_1 \ldots P_n = 0 \subset P$, it follows that $P_i \subset P$ for some i, and therefore $P = P_i$ by the maximality of P_i. Let $A = \cap_{i=1}^{n} P_i$. Then $A^n = (P_1 \cap P_2 \cap \ldots \cap P_n)^n \subset P_1 P_2 \ldots P_n = 0$. This shows that A is nilpotent and so $A \subset J(R)$. Since R/P_i is simple and right artinian, $(R/P_i)J(R) = 0$ and so $J(R) \subset P_i$ for each i. Therefore $J(R) \subset A$, and hence $J(R) = A$. This shows $J(R) = \cap_{i=1}^{n} P_i$ and $J(R)$ is nilpotent. $\qquad\square$

As a consequence, we have the following

Corollary 2.8 *A nil ideal in a nonprime restricted right artinian ring is nilpotent.*

Corollary 2.9 *A nonprime ring R with $J(R) \neq 0$ is restricted right artinian if and only if it is right artinian.*

Corollary 2.10 *A nonprime semiprime ring is restricted right artinian if and only if it is right artinian.*

Theorem 2.11 *Let R be a prime ring containing a minimal right ideal. Then the following are equivalent:*

1. *R is right restricted artinian but not right artinian.*
2. *The socle $\mathrm{Soc}(R)$ is an infinite direct sum of minimal right ideals, and $R/\mathrm{Soc}(R)$ is right artinian.*

Proof Assume (1). By assumption, $R/\mathrm{Soc}(R)$ is right artinian. Now if $\mathrm{Soc}(R)$ is a finite direct sum of minimal right ideals, then R is right artinian. But R is assumed to be nonartinian, hence $\mathrm{Soc}(R)$ is an infinite direct sum of minimal right ideals.

Conversely, assume (2). As R is a prime ring, any minimal right ideal of R is of the form fR for some idempotent $f \in R$. Further, $\mathrm{Soc}(R)$ is homogeneous. So there exists an indecomposable idempotent $e \in R$ such that eR is a minimal right ideal and $\mathrm{Soc}(R) = ReR$. Let A be any nonzero ideal of R. Then $eR \cap A \neq 0$. Therefore $eR \subseteq A$ and $ReR \subseteq A$. Thus $\mathrm{Soc}(R) \subseteq A$. Thus if $R/\mathrm{Soc}(R)$ is right artinian, then obviously R is right restricted artinian. Since $\mathrm{Soc}(R)$ is an infinite direct sum of minimal right ideals, R is not right artinian. $\qquad\square$

2.2   Restricted perfect rings

Recall that a ring R is left perfect if and only if $R/J(R)$ is semisimple artinian and each nonzero right R-module has nonzero socle (see Theorem 1.7).

A ring R is called a *restricted left perfect ring* if every proper homomorphic image of R is left perfect, that is, R/I is a left perfect ring for every proper nonzero two-sided ideal I of R. Similarly we define restricted right perfect rings. Commutative restricted perfect rings were studied by Bazzoni and Salce [21]. Noncommutative restricted perfect rings have been studied by Abuhlail, Jain, and Laradji in [5] and Facchini and Parolin in [69].

Simple rings or right perfect rings are clearly restricted right perfect. Other examples of restricted right perfect rings are given by the nearly simple chain rings due to Dubrovin [61]. A *nearly simple chain ring* R is a noncommutative right and left uniserial ring with exactly three two-sided ideals. These three ideals are 0, R, and the maximal ideal $J(R)$. As $R/J(R)$ is a division ring and hence a perfect ring, R is a restricted right (and left) perfect ring. We give below example of a nearly simple chain ring due to G. Puninski [196].

Example Let G be the group of affine linear functions on $\mathbb{Q}$;

$$G = \{f_{a,b} : \mathbb{Q} \to \mathbb{Q} : a, b \in \mathbb{Q}, a > 0\},$$

where $f_{a,b}(t) = at + b$ and the group operation on G is the composition of functions. Fix a positive irrational number ϵ in the field of real numbers. It is possible to define a right order on G with generalized positive cone $P := \{f_{a,b} \in G : \epsilon \le f_{a,b}(\epsilon)\}$. Let D be a division ring, and let $D[P]$ be the semigroup ring. Consider $M := \Sigma_{f \in P^+} f D[P]$ in $D[P]$, where $P^+ = \{f_{a,b} \in G : \epsilon < f_{a,b}(\epsilon)\}$. Then M is a maximal ideal in $D[P]$. The subset $D[P] \setminus M$ is a right and left Ore set in $D[P]$. Let R denote the localization of $D[P]$ with respect to $D[P] \setminus M$. The ring R is a nearly simple chain domain, hence it is a right and left almost perfect ring. $\qquad\square$

The following example shows that restricted perfect rings are not left–right symmetric.

Example Let K be a field and let K_ω be the K-algebra of all matrices with entries in K, countably many rows and columns, and in which each row has only finitely many nonzero entries. Let N be the set of all strictly lower triangular matrices in K_ω with only finitely many nonzero entries and let R be the subalgebra $K + N$ of K_ω. We have already seen that R is left perfect but not right perfect (see the example on p. 5). In particular, R is restricted left perfect. Now we will show that R is not restricted right perfect. First, note that $J(R) = N$. Let us denote the matrix units by $E_{i,j}$. Consider the principal two-sided ideal I of R generated by $E_{2,1}$. It may be seen that I is the vector space generated by all products $E_{i,j}E_{2,1}E_{k,l}$ with $i > j$ and $k > l$, that is, I is the vector space generated by all $E_{i,1}$ with $i \ge 2$. Then $I \subseteq J(R)$. We have $J(R/I) = J(R)/I = N/I$. In the sequence $\ldots, E_{5,4}, E_{4,3}, E_{3,2}$ all the elements are in N and the products $E_{n,n-1}E_{n-1,n-2} \ldots E_{3,2} = E_{n,2}$ are not in I. This proves that N/I is not right T-nilpotent, and so R/I is not right perfect. This shows that R is restricted left perfect but not restricted right perfect. $\qquad\square$

A restricted perfect ring need not be perfect. The next example From Facchini et al. [69] shows this.

Example Let K be a field and let V be an infinite dimensional left vector space over K. Let $S = \mathrm{End}_K(V)$ be the ring of right linear operators on V. Let I be the two-sided ideal of S consisting of linear transformations of finite rank. Consider the subring $R = K + I$ of S. This ring R has exactly three ideals: 0, I, and R. Trivially, R is restricted right (and left) perfect. Now, $J(R)$ is a proper two-sided ideal, so it can only be 0 or I. If $J(R) = I$, then as $E_{11} \in I$, $1 - E_{11}$ would be invertible. However, clearly the element $1 - E_{11}$ is not invertible. Thus $J(R) = 0$. Then obviously R is not semilocal because otherwise $R/J(R) \cong R$ would be semisimple artinian, which is not. Thus R is not right or left perfect. $\qquad\qquad\Box$

Proposition 2.12 *Let A and B be nonzero ideals of a restricted left perfect ring R. If R is not left perfect, then $A \cap B$ is not zero.*

Proof Since R is restricted left perfect, R/A and R/B are left perfect. Bjork [26] proved that if R/A and R/B are left perfect, then $R/A \cap B$ is also left perfect (although, Bjork stated this result for commutative rings but his proof remains valid for noncommutative rings). Since R is given to be not left perfect, $A \cap B \neq 0$. $\qquad\qquad\Box$

Lemma 2.13 *A ring R is a restricted left perfect ring if and only if for every nonzero ideal I, every descending chain $a_1 R + I \supset a_2 R + I \supset \cdots$, where $a_1, a_2, \ldots \in R$, is finite.*

Proof Suppose R is a restricted left perfect ring and I is a nonzero ideal of R. The chain $a_1 R + I \supset a_2 R + I \supset \cdots$ induces the chain $(a_1 R + I)/I \supseteq (a_2 R + I)/I \supseteq \cdots$ of principal right ideals of R/I. Since R/I is left perfect, by Theorem 1.5, there exists $n \in \mathbb{N}$ such that $a_n \in a_{n+k} R + I$ for all $k \in \mathbb{N}$. The converse is straightforward. $\qquad\qquad\Box$

Proposition 2.14 *Every nonzero prime ideal of a restricted left perfect ring R is maximal and hence primitive.*

Proof Let P be a nonzero prime ideal of a restricted left perfect ring R. Then R/P is a prime and left perfect ring. Hence R/P is simple artinian. Therefore P is maximal and hence primitive. $\qquad\qquad\Box$

Theorem 2.15 *If a restricted left perfect ring R is not left perfect, then R is a prime ring.*

Proof We first consider the case when $J(R) = 0$. Then the prime radical of R is zero, and so R is semiprime. Let A and B be ideals of R such that $AB = 0$. Then we get $A \cap B = 0$. Thus, by Proposition 2.12, $A = 0$ or $B = 0$, and hence R is prime.

Now we consider the case that $J(R) \neq 0$. Then, by assumption, $R/J(R)$ is left perfect and hence $R/J(R)$ is a semisimple artinian ring (see Theorem 1.5).

Now, since R is not left perfect, there is a nonzero right R-module M such that $\text{Soc}(M) = 0$. If B is a nonzero submodule of M such that $(0 : B) = \{x \in R : Bx = 0\} \neq 0$, then B is a right $R/(0 : B)$-module, and since $R/(0 : B)$ is left perfect, $\text{Soc}(B) \neq 0$ as an $R/(0 : B)$ module. Since $\text{Soc}(B)$ is also an R-module, we get $\text{Soc}(M) \supseteq \text{Soc}(B) \neq 0$, a contradiction. Thus, $(0 : B) = 0$ for each nonzero submodule B of M. If I and J are ideals of R with $IJ = 0$ and $I \neq 0$, then $MI \neq 0$, and so $J \subseteq (0 : MI) = 0$. This shows that R is a prime ring. $\qquad\square$

Corollary 2.16 *If R is a commutative restricted perfect ring and R is not an integral domain, then R is perfect.*

There is no relation between being a restricted perfect ring and being a semiperfect ring. For example, $\mathbb{Z}$ is restricted perfect but not semiperfect and a commutative uniserial domain of Krull domain two is semiperfect but not restricted perfect.

2.3 Restricted von Neumann regular rings

A ring R is called a *restricted von Neumann regular ring* if each proper homomorphic image of R is a von Neumann regular ring, that is, R/I is a von Neumann regular ring for every nonzero two-sided ideal I of R. Jain et al.[131] studied such rings. They proved the following.

Theorem 2.17 *Let R be a nonprime right noetherian ring. Then R is a restricted von Neumann regular ring if and only if*

1. *R is semisimple artinian; or*
2. *R has exactly one nontrivial ideal, namely, the Jacobson radical $J(R)$, and is isomorphic to an $n \times n$ matrix ring over a local ring; or*
3. *R has exactly three nontrivial ideals, namely, $J(R)$, $\text{ann}_l(J(R))$; and $\text{ann}_r(J(R))$ and is isomorphic to $\begin{pmatrix} A & M \\ 0 & B \end{pmatrix}$, where A, B are simple artinian rings and M is an irreducible $(A; B)$-bimodule.*

Proof Let R be a nonprime right noetherian restricted von Neumann regular ring. We first consider the case that $J(R) = 0$. Since R is right noetherian, each ideal of R contains a product of finitely many prime ideals of R. Since prime ideals are maximal, we have $M_1 M_2 \ldots M_n = 0$, where M_i are maximal ideals. Now $(M_1 \cap \ldots \cap M_n)^n \subset M_1 M_2 \ldots M_n = 0$ gives us $M_1 \cap M_2 \cap \ldots \cap M_n \subset J(R) = 0$. Therefore R is a finite direct sum of simple artinian rings R/M_i. This gives us a ring as described in (1).

Next we consider the case that $J(R) \neq 0$. Assume $J(R)^2 \neq 0$. Since we have assumed that every proper homomorphic image of R is von Neumann regular, we have that $R/J(R)^2$ is von Neumann regular. This yields $J(R) = J(R)^2$. But then, as R is a right noetherian ring, by Lemma 1.3, $J(R) = 0$, a contradiction. Thus $J(R)^2 = 0$. Since $R/J(R)$ is right noetherian von Neumann regular it is semisimple artinian. As $J(R)$ is a finitely generated right $R/J(R)$-module, we get $J(R)$ is an artinian right R-module. Hence R is right artinian. If $R/J(R)$ is a simple ring, then $J(R)$ is a maximal primary ideal. So R is isomorphic to a matrix ring over a local ring S such that $J(S)$ is a unique minimal ideal of S with $J(S)^2 = 0$. This gives us a ring as described in (2).

Suppose now that $R/J(R)$ is not a simple ring. Then $R/J(R)$ being semisimple artinian gives that the products of its maximal ideals commute. Suppose $R/J(R)$ is a direct sum of more than two maximal ideals. Let M_1, M_2 be any two maximal ideals of R. If both M_1M_2 and M_2M_1 are nonzero, then we get $M_1M_2 = M_2M_1$. Let $M_1, M_2, \ldots, M_k$ be all distinct maximal ideals of R. As their products commute, $J(R) = \cap_{i=1}^{k} M_i = M_1M_2 \ldots M_k$, $J(R)^2 = M_1^2 M_2^2 \ldots M_k^2 = M_1M_2 \ldots M_k \neq 0$, which is a contradiction. Hence $k = 2$ and one of M_1M_2, M_2M_1, say $M_1M_2 = 0$. Then $J(R) = M_2M_1$, $ann_l(J(R)) = M_1$, $ann_r(J(R)) = M_2$. We can find an idempotent $e \in R$ which is central modulo $J(R)$ such that $R/J(R) = M_1/J(R) \oplus M_2/J(R)$ with $M_1/J(R) = eR/J(R)$ and $M_2/J(R) = (1 - e)R/J(R)$. Then $M_1M_2 = 0$ gives $eR(1 - e) = 0$, $M_1 = eR + J(R) = Re + J(R)$, $M_2 = (1 - e)R + J(R)$. Now $J(R) = J(R)M_1 = J(R)(Re + J(R)) = J(R)e$. Similarly $J(R) = (1 - e)J(R)$. Hence $J(R) = (1 - e)J(R)e = (1 - e)Re$. Thus eRe, $(1 - e)R(1 - e)$ are simple rings isomorphic to R/M_2 and R/M_1 respectively. In addition as $J(R) = (1 - e)Re$ is a unique minimal ideal, it follows that $N = (1 - e)Re$ is a simple $((1 - e)R(1 - e); eRe)$-bimodule. Clearly $R \cong \begin{pmatrix} (1 - e)R(1 - e) & N \\ 0 & eRe \end{pmatrix}$. This gives us a ring as described in (3).

The converse is obvious when R is a ring of type (1) or (2). So suppose now that R is a ring of type (3), that is, $R = \begin{pmatrix} A & M \\ 0 & B \end{pmatrix}$, where A, B are simple artinian rings and M is an irreducible $(A; B)$-bimodule. Since $R/J(R)$ is von Neumann regular, it is enough to show that each nonzero ideal I of R contains $J(R)$. Let $p \in I$. Then $p = ae_{11} + me_{12} + be_{22}$, where $a \in A$, $b \in B$ and $m \in M$. We have $a'e_{11}pb'e_{22} = a'mb'e_{12} \in I$, for all $a' \in A$, $b' \in B$. Thus $AmBe_{12} \subset I$. Since M is an irreducible $(A; B)$-bimodule, $AmB = M$. Thus $Me_{12} \subset I$. Hence $J(R) \subset I$. This completes the proof. $\qquad \square$

Theorem 2.18 *If R is a prime right noetherian restricted von Neumann regular ring, then R is semiprimitive and each nontrivial ideal is a unique product of maximal ideals.*

Proof Assume that R is a prime right noetherian restricted von Neumann regular ring. Let A be any nonzero ideal in R. Then $A^2 \neq 0$. Thus, by assumption, R/A^2 is a von Neumann regular ring. Since A/A^2 is a nilpotent ideal in the von Neumann regular ring R/A^2, we must have $A = A^2$. In particular, $J(R) = J(R)^2$. Since R is right noetherian, Lemma 1.3 yields that $J(R) = 0$. Furthermore, since R/A is von Neumann regular, there exists a finite set of maximal ideals M_i such that $A = M_1 \cap M_2 \cap \ldots \cap M_n$. Then $A = A^2 = (M_1 \cap M_2 \cap \ldots \cap M_n)^2 \subset M_1M_2 \ldots M_n \subset A$ gives that $A = M_1M_2 \ldots M_n$. If $A = M_1' M_2' \ldots M_k'$, then $k = n$ and $M_i = M_{\sigma(i)}'$ where σ is a permutation of $\{1, 2, \ldots, n\}$ and $1 \leq i \leq n$. $\qquad \square$

A ring R is called a *right duo ring* if each right ideal of R is two-sided. A ring R is called a duo ring if it is both left as well as right duo. The next result characterizes right duo restricted von Neumann regular rings without chain conditions.

Theorem 2.19 *A right duo ring R is a restricted von Neumann regular ring if and only if R is strongly regular or R has exactly one proper ideal.*

Proof Let R be a right duo restricted von Neumann regular ring. First, assume $J(R) = 0$. Then $(xR)^2 \neq 0$ for any nonzero $x \in R$. Since R is right duo, xR is a two-sided ideal. This implies that $R/(xR)^2$ is von Neumann regular and so $(xR)^2 = xR$. Therefore, $x \in x^2 R$, proving that R is strongly regular.

Now let $J(R) \neq 0$. As R is a right duo ring, every idempotent in R is central. Now $J(R)$ being a minimal ideal is principal. Therefore $J(R)^2 = 0$. So idempotents modulo $J(R)$ can be lifted. Now R contains no non-trivial idempotents, for otherwise R becomes von Neumann regular. Also, $R/J(R)$ contains no nontrivial idempotents. This implies that $R/J(R)$ is a division ring. Thus $J(R)$ is the only proper ideal of R. The converse is trivial. $\qquad\square$

In particular, we have

Corollary 2.20 *A commutative ring R is a restricted von Neumann regular ring if and only if it is either strongly regular or it has exactly one proper ideal.*

In [152] Kaplansky conjectured that a ring R is von Neumann regular if and only if R is semiprime and each prime factor ring of R is von Neumann regular. It is well known that the conjecture holds for commutative rings (see [28], p. 173). But the conjecture fails to hold, in general. The example below shows this.

Example Let R consist of all sequences of 2×2 matrices over a field which are eventually strictly upper triangular. Then R is semiprime and, if P is a prime ideal of R, then by Posner's Theorem [195], we know that R/P has a simple artinian classical quotient ring. It can be shown that nonzero divisors in R/P are invertible in R/P. Thus R/P coincides with its quotient ring and hence R/P is von Neumann regular. Finally, R is not von Neumann regular since the constant sequence $a = (x, x, \ldots)$ where $x = \begin{bmatrix} 0 & 1 \\ 0 & 0 \end{bmatrix}$ is not a von Neumann regular element. $\qquad\square$

Fisher and Snider [83] showed the following.

Theorem 2.21 *A ring R is von Neumann regular if and only if R is semiprime, each prime factor ring of R is von Neumann regular, and R satisfies the property that the union of every chain of semiprime ideals in R is semiprime.*

Proof Suppose that R is semiprime, each prime factor ring of R is von Neumann regular and R satisfies the property that the union of every chain of semiprime ideals in R is semiprime. To prove that R is von Neumann regular, we need to show that for each $a \in R$, there exists an $r \in R$ such that $a = ara$. Assume, to the contrary, that there exists an $a \in R$ such that $a = axa$ has no solution in R. By Zorn's lemma, we choose a semiprime ideal A which is maximal with respect to the property that $\overline{a} = \overline{axa}$ has no solution in R/A, that is $a - axa$ is not in A for every $x \in A$. Clearly A is not prime. Hence by passing to R/A, we may assume that A is zero and R is semiprime but not prime. Therefore there exist nonzero semiprime ideals I and J in R such that $I \cap J = 0$. By the choice of A, there exist x and y in R with $a - axa \in I$ and $a - aya \in J$. Thus $a - a(x + y - xay)a = (a - axa) - aya + (axa)ya$ and so $a - a(x + y - xay)a \in I$. Similarly $a - a(x + y - xay)a \in J$. Therefore, $a - a(x + y - xay)a \in I \cap J$. But, since $I \cap J = 0$, we have $a - a(x + y - xay)a = 0$. This contradicts the choice of A. Consequently, R is von Neumann regular. The converse is obvious. $\qquad\square$

Corollary 2.22 *If R is two-sided noetherian and each prime factor ring of R is von Neumann regular, then $R/N(R)$ is von Neumann regular and $N(R)$ is nilpotent.*

Proof The above theorem applied to the ring $R/N(R)$ yields that $R/N(R)$ is von Neumann regular. For the nilpotency of $N(R)$, see ([99], Theorem 3.11). $\qquad\square$

Corollary 2.23 *A ring R is von Neumann regular if and only if each factor ring of R is semiprime and each prime factor ring of R is von Neumann regular.*

Corollary 2.24 *A restricted von Neumann regular ring which is semiprime but not prime is von Neumann regular.*

2.4   Restricted self-injective rings

A ring R is called a *restricted self-injective ring* if every proper homomorphic image of R is self-injective. These rings have also been called *pre-self-injective rings* in the literature.

Levy [167] studied commutative noetherian restricted self-injective rings and obtained a complete characterization for such rings. To present his characterization, we need the following useful lemma.

Lemma 2.25 *Let R be a commutative self-injective ring and let M be a maximal ideal of R such that $M^2 = 0$. Then $0 \subset M \subset R$ are the only ideals of R.*

Proof Let $x \,(\neq 0) \in M$. Since $M^2 = 0$, we have $M \subseteq ann(x)$. As M is a maximal ideal of R, we get $M = ann(x)$. Thus $Rx \cong R/ann(x) = R/M$ as R-modules. Therefore, for any two nonzero elements x and y of M, $Rx \cong Ry$. Let φ be an isomorphism of Rx onto Ry. Since R is self-injective, φ can be extended to an endomorphism f of R. Suppose $f(1) = u$. Then $Rxu = Ry$. So $Rx \supseteq Ry$ for every nonzero $x, y \in M$. Thus we conclude that M has no proper submodules. Let N be any nonzero ideal of R such that $N \neq M$. Then, as M is a maximal ideal of R, $R = N + M$. Since $M^2 = 0$, we get $M = MN \subseteq N$. Thus $N = R$. This shows that $0 \subset M \subset R$ are the only ideals of R. $\qquad\square$

Theorem 2.26 *Let R be a commutative noetherian ring. Then R is a restricted self-injective ring if and only if*

1. *R is a Dedekind domain; or*
2. *R is an artinian principal ideal ring; or*
3. *R is a local ring whose maximal ideal M has composition length 2 and satisfies $M^2 = 0$.*

Proof Suppose that R is a restricted self-injective ring. Assume first that R is a domain. Let M be a maximal ideal of R, then $M^2 \neq 0$. As R/M^2 is self-injective and indecomposable, it must be uniform. Therefore there are no ideals between M and M^2 and so R must be a Dedekind domain.

Next, we consider the case when R is not a domain. Then the zero ideal is not a prime ideal. For each prime ideal P, by hypothesis, R/P is an injective and hence a divisible R/P-module. Thus R/P is a field and so P is a maximal ideal. But a commutative noetherian ring in which every prime ideal is maximal must be an artinian ring (see Theorem 2.1). Therefore $R = R_1 \oplus \ldots \oplus R_n$ where each R_i is a local artinian ring. Let M_i be the maximal ideal of R_i. Again we consider two cases. Suppose first that either $n > 1$, or $n = 1$ but $M_1^2 \neq 0$. Then for each i, $R/(M_i^2 + \Sigma_{j \neq i} R_j) \cong R_i/M_i^2$ is a self-injective ring. Now R_i/M_i^2 is a commutative self-injective ring with a maximal ideal $N_i = M_i/M_i^2$ such that $N_i^2 = 0$. Thus by the above lemma, $0 \subseteq N_i \subseteq R_i/M_i^2$ are all of the ideals of R_i/M_i^2. Therefore for any $m_i \in M_i$ with $m_i \notin M_i^2$, we have $M_i = R_i m_i + M_i^2 = R_i m_i + (J(R_i))M_i$; and Nakayama's Lemma then shows that $M_i = R_i m_i$. Thus R_i is a commutative noetherian ring in which every maximal ideal is principal. Therefore by Kaplansky [150], R_i must be a principal ideal ring. Hence R is an artinian principal ideal ring. Finally, suppose that R is a local artinian ring whose maximal ideal M satisfies $M^2 = 0$. We may assume that M has composition length at least 2, since otherwise R would be a principal ideal ring. Since R is artinian, M contains a minimal ideal, say K, of R. Now R/K is a commutative self-injective ring with a maximal ideal M/K such that $(M/K)^2 = 0$. Therefore by the above lemma, $0 \subseteq M/K \subseteq R/K$ are all the ideals of R/K. Hence M has composition length 2.

Conversely, suppose that R is a ring of type $(1), (2)$, or (3). Note that the proper homomorphic images of all three types of rings are artinian principal ideal rings. We know that an artinian principal ideal ring is a direct sum of rings R_i where the ring R_i has exactly one composition series $R_i \supset R_i x \supset R_i x^2 \supset \ldots \supset R_i x^t = 0$ (see [150], Theorem 13.3). Therefore it suffices to show that this ring R_i is self-injective. Let f be a homomorphism of $R_i x^k$ into R_i. Then the composition length of $f(R_i x^k)$ cannot exceed that of $R_i x^k$. Since the above composition series contains all the ideals of R_i, we have $f(R_i x^k) \subseteq R_i x^k$. Consequently, $f(x^k) = x^k y$ for some $y \in R_i$. The map $r \mapsto ry$ $(r \in R_i)$ is then the extension of f to an endomorphism of R_i, showing that R_i is self-injective. Hence R is a restricted self-injective ring. $\qquad\square$

Levy gave an example of a non-noetherian restricted self-injective domain.

Example (Levy, [167]). Let F be a field and x an indeterminate; and let $\mathcal{I}$ be the family of all well-ordered sets Λ of nonnegative real numbers, the order relation being the natural order of the real numbers. Let

$$R = \{\Sigma_{j \in \Lambda} a_j x^j : a_j \in F, \Lambda \in \mathcal{I}\}.$$

Note that every element of R whose constant term is nonzero is invertible in R. It follows that every nonzero element of R has the form $x^b u$ where u is invertible in R. This implies that R has only two types of nonzero ideals: The principal ideals (x^b), and those of the form $(x^{>b}) = \{x^c u : c > b$ and u is invertible or zero$\}$.

Let $S = R/I$ where $I \neq 0$ and set $y = x + I$. Then S can be considered as the collection of formal power series $\Sigma_{j \in \Lambda} a_j y^j$ with $a_j \in F, \Lambda \in \mathcal{I}$ and

1. $y^b = 0$ if $I = (x^b)$; or
2. $y^c = 0$ for $c > b$ if $I = (x^{>b})$.

Observe that for $c \leq b$, we have

(A) If $I = (x^b)$, then $ann(y^c) = (y^{b-c})$ and $ann(y^{>c}) = (y^{b-c})$.
(B) If $I = (x^{>b})$, then $ann(y^c) = (y^{>(b-c)})$ and $ann(y^{>c}) = (y^{b-c})$.

From (A) and (B) it follows that whether S is of type (1) or (2), the principal ideals of S satisfy $ann(ann(y^c)) = (y^c)$.

To see that S is self-injective, let f be an S-homomorphism from an ideal K of S to S. If $K = (y^c)$ then $ann(K)f(K) = f(0) = 0$ so that $f(K) \subseteq ann(ann(K)) = K$. Hence $f(y^c) = y^c p$ for some $p \in S$. Thus f can be extended to the endomorphism $s \to sp$ of S.

Next, let $K = (y^{>c})$, and choose an infinite sequence $c(1) > c(2) > \cdots$ such that $\lim_{i \to \infty} c(i) = c$. Then $K = \cup_{i=1}^{\infty} (y^{c(i)})$. For each i the previous paragraph shows that we can choose a "power series" p_i such that $f(y^{c(i)}) = y^{c(i)} p_i$. If $j > i$ so that $y^{c(i)} = y^{c(i)-c(j)} y^{c(j)}$ the fact that f is an S-homomorphism shows that $f(y^{c(i)}) = y^{c(i)} p_j$ so that we have $p_j - p_i \in ann(y^{c(i)})$.

If (A) holds, this means that all the terms of p_i of degree $< b - c(i)$ are equal to the terms of the same degree of p_j while the terms of higher degree do not affect the products $y^{c(i)} p_i$ and $y^{c(i)} p_j$.

A similar statement is true if (B) holds. Thus we can assemble a single "power series" p such that $(p - p_i)y^{c(i)} = 0$ for all i. Note that the collection of exponents appearing in p is well-ordered which shows p is in S. Then the map $s \mapsto sp$ extends f to an endomorphism of S. This shows that S is self-injective.

Thus R is a non-noetherian restricted self-injective domain. $\square$

The theorem of Levy was extended to noncommutative rings by Hajarnavis [105]. Recall that a ring is called *right bounded* if every essential right ideal of R contains an ideal which is essential as a right ideal. A prime Goldie ring R is called an *Asano order* if each nonzero ideal of R is invertible. A ring R is called a *Dedekind prime ring* if it is a hereditary noetherian Asano order. Hajarnavis considered a noetherian, bounded, prime, restricted self-injective ring and proved the following.

Theorem 2.27 *Let R be a noetherian, bounded, prime, restricted self-injective ring. Then R is a Dedekind prime ring.*

Proof Since every proper homomorphic image of R is noetherian and self-injective, every proper homomorphic image of R is quasi-Frobenius. Thus every proper homomorphic image of R is artinian. Let R/A be any proper homomorphic image of R. Since R/A is an artinian ring and R/A itself and each of its factor rings are self-injective, we have that each factor ring of R/A is a quasi-Frobenius ring. It is well known due to Nakayama [181] that a ring is uniserial if and only if each of its factor rings in quasi-Frobenius. Thus we conclude that R/A is a uniserial ring. Therefore R/A is artinian principal right and left ideal ring and consequently R/A has a primary decomposition. Hence R has a primary decomposition. Let M_1, M_2 be two distinct maximal ideals of R. By considering the primary decomposition for $M_1 M_2$ it is easily seen that $M_1 M_2 = M_1 \cap M_2$. By symmetry, we also have $M_2 M_1 = M_1 \cap M_2$. Thus $M_1 M_2 = M_2 M_1$. This shows that the maximal ideals of R commute. Let M be a maximal ideal of R. Since R is a bounded noetherian prime ring in which the nonzero prime ideals are maximal and maximal ideals commute, the localization of R at M is classical. Let R_M be the localization of R at M. The ideal $M^2 \neq 0$ since R is a prime ring. It may be checked that $M^2 R_M$ is an ideal of R_M and $M^2 R_M \cap R = M^2$. Define $\varphi : R/M^2 \to R_M/M^2 R_M$ by $\varphi(r + M^2) = r + M^2 R_M$ for each $r \in R$. Clearly φ is a one-to-one ring homomorphism. Let $rc^{-1} + M^2 R_M$ be an arbitrary element of $R_M/M^2 R_M$. Now c is a regular element of R mod M and R/M is an artinian ring. Therefore $c + M$ is a unit element of R/M. Since M/M^2 is the nilpotent radical of the ring R/M^2 and units can be lifted over the nilpotent radical, it follows that $c + M^2$ is a unit of R/M^2. Hence there exists $d \in R$ such that $cd + M^2 = dc + M^2 = 1 + M^2$. This implies $cd + M^2 R_M = dc + M^2 R_M = 1 + M^2 R_M$. Hence $c^{-1} + M^2 R_M = d + M^2 R_M$. Thus we have $\varphi(rd + M^2) = rc^{-1} + M^2 R_M$. This shows φ is onto and hence an isomorphism. Thus $R/M^2 \cong R_M/M^2 R_M$. As explained above $R_M/M^2 R_M$ is artinian principal right and left ideal ring. So $MR_M/M^2 R_M$ is a principal right (and also left) ideal of the ring $R_M/M^2 R_M$. But MR_M is the Jacobson radical of the ring R_M. Hence by Nakayama's lemma, MR_M is a principal right and left ideal of R_M. This implies that MR_M is an invertible ideal. Therefore by ([106], Proposition 1.3), R_M is a hereditary ring. Hence by ([106], Theorem 2.6), R is an Asano order. But R is bounded. So R is hereditary and thus a Dedekind prime ring by a theorem of Asano and Michler (see [163]). $\qquad\square$

Klatt and Levy [154] later described all commutative rings, not necessarily noetherian, all of whose proper homomorphic images are self-injective. We will first give some definitions. A ring R is said to have *rank* 0 if every prime ideal of R is a maximal ideal. A ring R is said to have *rank* 1 if every proper prime ideal of R is a maximal ideal. We call a uniserial ring R *maximal* if each family of pairwise solvable congruences of the form $x \equiv x_i \pmod{A_i}$ has a simultaneous solution x, where each $x_i \in R$ and each A_i is an ideal of R. A uniserial ring R is said to be *almost maximal* if each proper factor ring of R is maximal.

Theorem 2.28 *A commutative ring R is restricted self-injective if and only if R is one of the following:*

1. *A Prufer domain such that the localization R_M for each maximal ideal M is an almost maximal rank 1 uniserial domain; and every proper ideal is contained in only finitely many maximal ideals; or*
2. *the finite direct sum of self-injective maximal uniserial rings of rank 0; or*
3. *an almost maximal rank 0 uniserial ring; or*
4. *a local ring whose maximal ideal M has composition length 2 and satisfies $M^2 = 0$.*

Proof Let R be a commutative restricted self-injective ring. Note that every nonzero ideal (or element) of R is contained in only finitely many maximal ideals. Assume first that R has an infinite number of maximal ideals M. Let $x\,(\neq 0), y \in R$ with $xy = 0$. Then x is contained in only finitely many maximal ideals, and $xy = 0$ is in every maximal ideal. Since maximal ideals are prime, y is in infinitely many maximal ideals. So $y = 0$. Thus R is a domain. We now show that R is of type (1). For each maximal ideal M, R_M is a local restricted self-injective domain and hence an almost maximal uniserial domain. Since each R_M is a uniserial domain, all of its finitely generated ideals are invertible (in fact, principal) and hence the same is true of R, that is, R is a Prufer domain.

Now consider the case that R is not a domain. Then R has only a finite number of maximal ideals. Since all proper homomorphic images of R are self-injective, every nonzero prime ideal of R is maximal, and since R is not a domain, 0 is not a prime ideal of R. Hence R is the direct sum of a finite number of local rings. If the number of summands is 1, that is R is local, then it may be shown that R must be of type (3) or (4). Observe that rings of type (4) are not self-injective. Now if the number of summands is at least 2, then each of them is a self-injective ring all of whose proper homomorphic images are self-injective, and so it may be shown that each summand must be a maximal rank 0 uniserial ring.

Now we proceed to show the converse part. Note that the rings of type (2) and (3) are clearly restricted self-injective. Rings of type (4) are restricted self-injective since their proper homomorphic images have at most one proper ideal, hence are maximal rank 0 uniserial rings. Finally, let R be a domain of type (1). Then every nonzero prime ideal of R is maximal (since this is true of each R_M). Consequently, for each nonzero ideal A of R, the ring R/A has only a finite number of maximal ideals and all of its primes are maximal. It may be shown that R/A is a direct sum of a finite number of rings of the form R_M/A_M, where M a maximal ideal of R. Note that the ring R_M/A_M is a maximal rank 0 uniserial ring and hence self-injective. Therefore, R/A is self-injective and the proof of the theorem is complete. $\square$

Corollary 2.29 *Every nonnoetherian restricted self-injective ring has cardinality at least* $2^{\aleph_0}$.

Proof First let R be a maximal rank 0 uniserial ring which is not noetherian. In view of the Hopkins–Levitzki Theorem, we know that R is not artinian. Thus R has an infinite decreasing sequence of ideals $A_1 \supset A_2 \supset \cdots$. For each i, let $a_i \in A_i - A_{i+1}$. Then the system of congruences $x \equiv a_1 + a_2 + \cdots + a_{n-1}$ $\pmod{A_n}$ for $n \geq 2$ is pairwise solvable and hence has a solution in R. We may think of the solution as an infinite sum of the form $a_1 + a_2 + \ldots$. Note that if one or more of the a_i are replaced by 0 then the above system of congruences remains solvable. Thus R contains $2^{\aleph_0}$ infinite sums of the form $c_1 + c_2 + \cdots$ where each c_i is either a_i or 0. To complete the proof, it will be sufficient to show that every nonnoetherian restricted self-injective ring R can be mapped onto a nonnoetherian rank 0 maximal uniserial ring. Let R be a nonnoetherian restricted self-injective domain. Then for some proper ideal A of R, R/A is nonnoetherian because if R/A is noetherian for each proper ideal A of R then R will become noetherian. Now, if R/A is a domain, then R/A (being a self-injective domain) will be a field, a contradiction. Therefore R/A is not a domain. Thus R/A is a nondomain nonnoetherian restricted self-injective ring. This shows that the problem for domains is reduced to that for nondomains. By the above theorem, we clearly see that when R is not a domain then a nonnoetherian restricted self-injective ring R can be mapped onto a nonnoetherian rank 0 maximal uniserial ring. This completes the proof. $\square$

As a consequence, we have the following.

Corollary 2.30 *A domain of algebraic numbers is restricted self-injective if and only if it is a Dedekind domain.*

It is known that finitely generated modules over pre-self-injective domains are direct sums of cyclic modules and ideals ([154], Theorem 5.1). We will discuss rings with each module a direct sum of cyclic modules in a later chapter.

2.31 Questions

1. Describe nonsimple prime right noetherian rings R such that each proper ring homomorphic image of R is von Neumann regular.
2. Let R be a restricted unit-regular ring, that is, every proper factor ring of R is unit-regular. When is R a unit-regular ring?

3 Rings each of whose proper cyclic modules has a chain condition

3.1 Rings each of whose proper cyclic modules is artinian

The Hopkins–Levitzki Theorem motivated Camillo and Krause to ask the following problem (see Problem 12 [38]): Is a ring R right noetherian if for each nonzero right ideal A of R, R/A is an artinian right R-module? We will call a ring R a right *Camillo–Krause ring* if for each nonzero right ideal A of R, R/A is a right artinian module. A cyclic right R-module R/A is called *proper cyclic* if R/A is not isomorphic to R_R.

We start with the following useful observation.

Lemma 3.1 *If a right Camillo–Krause ring is not right artinian then it must be a right Ore domain.*

Proof Let R be a right Camillo–Krause ring that is not right artinian. Let $a \neq 0$ be an element of R. We have $R/ann_r(a) \cong aR$. If $ann_r(a) \neq 0$, then aR is right artinian and therefore by assumption $aR \neq R$. Since $aR \neq 0$, R/aR is also right artinian, and consequently so is R, a contradiction. Therefore $ann_r(a) = 0$, and hence R is a domain. Now suppose that there exist two nonzero proper right ideals A and B such that $A \cap B = 0$. Then $(A + B)/B \cong A$ is right artinian. Now as R/A and A are right artinian, this yields that R is right artinian, which is a contradiction. Therefore R is a right Ore domain. $\qquad\square$

Lemma 3.2 *If each proper cyclic right R-module is artinian and R is not right artinian, then R is a domain.*

Proof We first show that R has finite right Goldie dimension. Assume to the contrary that R has infinite right Goldie dimension. Then R contains a direct sum $A \oplus B$ where both A and B are infinite direct sums of nonzero right ideals. If $R/A \cong R$, then R/A is projective, and hence there exists a right ideal C of R such that $R = C \oplus A$. But then the cyclic module R/C is isomorphic to an infinite direct sum of nonzero modules, a contradiction. Thus R/A is proper and hence artinian. Similarly R/B is proper and hence artinian. But then R is a right artinian ring, a contradiction. Thus R has finite right Goldie dimension.

Assume now that $\mathrm{Gdim}(R) = n > 1$. Then there exist closed uniform right ideals U_i such that $\bigoplus_{i=1}^{n} U_i \subset_e R$. Now $\mathrm{Gdim}(R/U_1) = n - 1 = \mathrm{Gdim}(R/U_2)$, and so R/U_1 and R/U_2 are both proper and hence artinian. Thus R is an artinian ring, a contradiction. So must have $\mathrm{Gdim}(R) = 1$, that is R is right uniform.

Now assume R is not a domain. Then there exists a nonzero element $a \in R$ such that $ann_r(a) \neq 0$. Now $R/ann_r(a)$ cannot be isomorphic to R because R is uniform. Thus $R/ann_r(a)$ is proper cyclic and hence artinian. As $R/ann_r(a) \cong aR$, we have that aR is artinian. Note that by assumption, R/aR is artinian. Hence R is artinian, a contradiction. Thus R must be a domain. $\qquad\square$

Theorem 3.3 *The following are equivalent statements for a ring R:*

1. *Every proper cyclic right R-module is artinian.*
2. *Every cyclic right R-module R/I with $I \neq 0$ is artinian.*

Proof Obviously $(2) \Rightarrow (1)$. Under the hypothesis (1), we have shown that R is either artinian or a domain. If R is artinian, there is nothing to prove. So assume R is not artinian. Let I be a nonzero right ideal of R. Then R/I is not isomorphic to R because R is a domain. By (1), R/I is artinian. This proves (2). $\qquad\square$

Shamsuddin [201] gave a partial answer to the Camillo–Krause conjecture in the following theorem. First, recall that a right ideal I of R is called a *prime right ideal* of R, if for all x and y in R, $xRy \not\subseteq I$ whenever $x \notin I$ and $y \notin I$. Equivalently, a right ideal I of R is called a prime right ideal of R, provided that if X, Y are right ideals of R with $XY \subseteq I$ then either $X \subseteq I$ or $Y \subseteq I$.

Theorem 3.4 *Let R be an affine algebra over an uncountable algebraically closed field $\mathbb{F}$. Suppose that R is a right Camillo–Krause domain which is not right primitive. Then R is right noetherian.*

Proof Let I be a nonzero prime right ideal of R. To prove that R is right noetherian, it suffices to show that I is finitely generated (see [174]). By assumption R/I is right artinian. Therefore there exists a right ideal K of R such that K/I is a simple module. Since R is not right primitive, the annihilator P of K/I is nonzero. Now $KP \subseteq I$. Because I is right prime, we have $P \subset I$. Since R/P is right artinian, it is right noetherian. Hence I/P is finitely generated. Since R/P is right primitive, R/P is simple artinian and hence $R/P \cong \mathrm{M}_n(D)$ for some division ring D. Since R is affine and $\mathbb{F}$ is uncountable, D is algebraic over $\mathbb{F}$. But as $\mathbb{F}$ is algebraically closed, we have $D = \mathbb{F}$. Thus $R/P \cong \mathrm{M}_n(\mathbb{F})$. Hence $\dim_{\mathbb{F}}(R/P)$ is finite. This yields that P is finitely generated as a right ideal (see [53]). Hence it follows that I is finitely generated and therefore R must be right noetherian. $\qquad\square$

A module M is called *hollow* if every proper submodule is a small submodule of M. A module M is called *special-hollow* if M is a nonzero, nonlocal, and hollow module with simple essential socle. Noyan Er [68] has proved the following.

Theorem 3.5 *Let R be a right Camillo–Krause ring. Then R is right noetherian if and only if R_R has no countably generated special-hollow subfactor with all nonzero factors faithful.*

Proof Assume R is not right noetherian. Then R_R contains a nonfinitely generated submodule M. Let $a\,(\neq 0) \in M$. Then $N = M/aR$ is a nonfinitely generated artinian module. So N has a minimal non-finitely generated submodule C. Then any proper submodule of C is finitely generated. So the module C is hollow as well as nonlocal. Also we can find a strictly ascending chain Ω of finitely generated submodules of C. By the minimality of C, we conclude that $C = \cup\Omega$. Thus C is countably generated. Let L be a simple submodule of C. Let K be a complement of L in C. As K is finitely generated, $(L+K)/K$ is essential in C/K. Thus it follows that C has a special-hollow factor $D := C/K$. Let E be a nonzero factor of D and I be the right annihilator of E in R. Then E is hollow and nonfinitely generated as well. If $I \neq 0$, R/I is right artinian, so $J(E) \neq E$, a contradiction. Hence $I = 0$. Thus D is a countably generated subfactor of R_R with all nonzero factors faithful. This completes the proof. $\square$

However the question whether a right Camillo–Krause ring is right noetherian is still open.

3.2 Rings with restricted minimum condition

A right R-module M is said to satisfy the *restricted minimum condition* (*RMC* for short) if for each essential submodule E of M, M/E is an artinian right R-module. A ring R is said to satisfy the *right restricted minimum condition* if R_R satisfies *RMC*.

We first prove the following lemma which is quite useful for the study of modules and rings with the right restricted minimum condition.

Lemma 3.6 *If M is a module such that for each essential submodule E of M, M/E has finite Goldie dimension, then $M/Soc(M)$ has finite Goldie dimension.*

Proof Let $A \subseteq B$ be submodules of M such that A is essential in B. Let C be a complement of A in M. Then $A \oplus C$ is essential in M and therefore, by assumption, $M/(A \oplus C)$ has finite Goldie dimension. Since $B/A \cong (B \oplus C)/(A \oplus C)$, B/A has finite Goldie dimension. Now let D be a complement of $Soc(M)$ in M. Then $D \oplus Soc(M)$ is essential in M and so, by assumption, $M/(D \oplus Soc(M))$ has finite Goldie dimension. Now we claim that D has finite Goldie dimension. Assume to the contrary that D contains an infinite direct sum $X = X_1 \oplus X_2 \oplus \cdots$ of nonzero submodules X_i. Since $X_i \cap Soc(M) = 0$, each X_i contains an essential submodule Y_i with $Y_i \neq X_i$. Then $Y = Y_1 \oplus Y_2 \oplus \cdots$ is clearly an essential submodule of X. Therefore X/Y has finite Goldie dimension as observed in the beginning of this proof. But this is impossible, because $X/Y \cong X_1/Y_1 \oplus X_2/Y_2 \oplus \cdots$ where each X_i/Y_i is nonzero. This shows that D has finite Goldie dimension. Thus it follows that $M/Soc(M)$ has finite Goldie dimension. $\square$

Lemma 3.7 *Let M be a finitely generated CS module. If M contains an infinite direct sum of nonzero submodules $N = \oplus_{i\in\mathbb{N}}N_i$, then the factor module M/N does not have finite Goldie dimension.*

Proof Assume to the contrary that M/N has finite Goldie dimension, say n. Partition $\mathbb{N}$ as a disjoint union of infinite sets $\mathcal{I}_1, \mathcal{I}_2, \ldots, \mathcal{I}_{n+1}$ and set $U_j = \oplus_{i \in \mathcal{I}_j} N_i$. Then $N = \oplus_{j=1}^{n+1} U_j$. Let E_j be a maximal essential extension of U_j for each $j \leq n+1$. Since M is a CS module, each E_j is a direct summand of M and hence finitely generated. This implies $E_j/U_j \neq 0$. Now, $M/N = M/(\oplus_{j=1}^{n+1} U_j)$ contains a submodule isomorphic to $E_1/U_1 \oplus E_2/U_2 \oplus \cdots \oplus E_{n+1}/U_{n+1}$. This yields a contradiction to our assumption that $\mathrm{Gdim}(M/N) = n$. Hence the factor module M/N must have infinite Goldie dimension. $\square$

Theorem 3.8 *If R is a right self-injective ring such that R satisfies the right restricted minimum condition, then R is quasi-Frobenius.*

Proof Let R be a right self-injective ring such that R satisfies the right restricted minimum condition. Then $J(R)$ consists precisely of those $x \in R$ with $ann_r(x)$ essential in R and $\bar{R} = R/J(R)$ is von Neumann regular right self-injective and satisfies the right restricted minimum condition. Let $\bar{S} = \mathrm{Soc}(\bar{R}_{\bar{R}})$. By Lemma 3.6, $\bar{R}/\bar{S}$ has finite Goldie dimension. Now, by Lemma 3.7, $\bar{S}_R$ is finitely generated. So $R/J(R)$ has finite Goldie dimension. Since $R/J(R)$ is a von Neumann regular ring with finite Goldie dimension, $R/J(R)$ is semisimple artinian. Let I be any nonzero right ideal of R. If $I \cap J(R) = 0$, then I is embedded in $R/J(R)$, so I_R is semisimple. If $I \cap J(R) \neq 0$, consider $x \, (\neq 0) \in I \cap J(R)$. We have $R/ann_r(x) \cong xR$. But, since $x \in J(R)$, $ann_r(x)$ is an essential right ideal of R, therefore xR must be right artinian. Thus, in any case I has nonzero socle. Hence $\mathrm{Soc}(R_R)$ is essential in R_R. Therefore $R/\mathrm{Soc}(R_R)$ is right artinian. Then, by Lemma 3.7, $\mathrm{Soc}(R_R)$ is finitely generated so artinian. This implies R is right artinian and hence R is quasi-Frobenius. $\square$

Huynh [113] proved the following:

Theorem 3.9 *Let R be a left self-injective ring such that R satisfies the right restricted minimum condition. If R is semiprime or $J(R)$ is nil, then R is quasi-Frobenius.*

Proof Case 1: R is semiprime.
Let $\overline{R} = R/J(R)$, $S = \mathrm{Soc}(R_R), \overline{S}_1 = \mathrm{Soc}(\overline{R}_{\overline{R}})$. Since $\overline{R}$ also has right RMC, it follows from Lemma 3.6 that $\overline{R}/\overline{S}_1$ has finite right Goldie dimension and so $\overline{R}/\overline{S}_1$ is semisimple. But $\overline{S}_1$ is also the left socle of $\overline{R}$, and since $\overline{R}$ is left self-injective, $\overline{R}$ must be semisimple. Since $SJ(R) = 0$, we have $(S \cap J(R))^2 = 0$. Now, since R (being semiprime) has no nonzero nilpotent ideal, we have $S \cap J(R) = 0$. Hence S is embedded in $R/J(R)$ as a right R-module. Therefore S is an artinian right R-module. Thus R/S has finite right Goldie dimension, and hence R has finite right Goldie dimension. So R contains independent uniform right ideals $U_1, \ldots, U_k$ such that $U = U_1 \oplus \cdots \oplus U_k \subset_e R_R$. Thus R/U is an artinian right R-module. Moreover, since R has right RMC, it is easy to check that for each nonzero submodule V_i of U_i, U_i/V_i is artinian. Hence U has Krull dimension at most 1. Thus R has right Krull dimension. Since R is semiprime with right Krull

dimension, R is right Goldie (see Lemma 6.2.8, [180]). Therefore R has ACC on right annihilators. Hence R is quasi-Frobenius (see [70]).

Case 2: $J(R)$ is nil.

By Lemma 3.6, R/S has finite right Goldie dimension. Since R/S also has right RMC, we can use a similar argument to that given above to show that the ring R/S has right Krull dimension at most 1. Since $(J(R) + S)/S$ is a nil ideal in R/S, $(J(R) + S)/S$ must be nilpotent (see [100]). Therefore $J(R)^n \subset S$ for some $n \geq 1$. But $SJ(R) = 0$, so $J(R)^{n+1} = 0$, thus $J(R)$ is nilpotent. Since $R/J(R)$ is von Neumann regular left self-injective and $R/J(R)$ has right RMC, $R/J(R)$ is semisimple, as shown in the first case. Thus R is semiprimary. It follows that R/S is a semiprimary ring with right Krull dimension, so R/S is right artinian. Therefore, R has DCC on essential right ideals, and hence R is quasi-Frobenius (see [14]). $\qquad\square$

Chatters [44] studied a hereditary noetherian ring R which satisfies the restricted minimum condition and proved the following.

Theorem 3.10 *Let R be a left noetherian left hereditary ring, and let I be a finitely generated essential right ideal of R. Then R satisfies the descending chain condition for finitely generated right ideals which contain I.*

Proof Let Q be the maximal right quotient ring of R, and let I be an arbitrary finitely generated essential right ideal of R. Set $I^* = \{x \in Q : xI \subseteq R\}$. Clearly I^* is a left R-submodule of Q. Since a left noetherian left hereditary ring is right semihereditary (see [207]), I is projective. So by the Dual Basis Lemma, there exists a family of elements $\{a_i : i \in \mathcal{J}\} \subseteq I$ and $\{f_i : i \in \mathcal{J}\} \subseteq \mathrm{Hom}(I, R)$ such that for every $y \in I$, we have $f_i(y) = 0$ for all but finitely many $i \in \mathcal{J}$, and $y = \Sigma_{i \in \mathcal{J}} a_i f_i(y)$. Since R is right semihereditary, it follows that R is right nonsingular and hence Q is right self-injective. So there exist elements $x_i \in Q$ such that $f_i(x) = x_i x$ for all $x \in I$, $i \in \mathcal{J}$. Because I is finitely generated, we may take the index set $\mathcal{J}$ to be finite. Set $h = \Sigma_{i \in \mathcal{J}} a_i x_i$. For any $x \in I$ we have $hx = x$. Thus left multiplication by h coincides with left multiplication by 1 on the essential right ideal I. Therefore we have $h = 1$. Thus $1 \in II^*$, because $a_i \in I$ and $x_i \in I^*$ for all i. It follows that I^* is a finitely generated left R-module and that $I = \{x \in Q : I^*x \subseteq R\}$. Now let $I_1 \supseteq I_2 \supseteq \cdots$ be a descending chain of finitely generated essential right ideals all containing the fixed finitely generated essential right ideal A. This gives rise to an ascending chain of left R-modules $I_1^* \subseteq I_2^* \subseteq \cdots$, all contained in the finitely generated left R-module A^*. But A^* satisfies the ascending chain condition, because R is left noetherian. Therefore there exists a positive integer n such that $I_n^* = I_{n+k}^*$ for all $k \in \mathbb{N}$. Since $I_k = \{x \in Q : I_k^*x \subseteq R\}$ for each k, we have $I_n = I_{n+k}$ for each $k \in \mathbb{N}$. This completes the proof. $\qquad\square$

As a consequence, we have the following:

Corollary 3.11 *Let R be a hereditary noetherian ring. Then R satisfies both the right and left restricted minimum condition.*

The following example, due to Small [206], shows that a left hereditary left noetherian ring does not necessarily satisfy the left restricted minimum condition.

Example Let R be the ring of all matrices of the form $\begin{pmatrix} a & 0 \\ b & c \end{pmatrix}$ where a is an integer and b and c are rationals. This ring R is left noetherian and left hereditary. Let I be the ideal of R consisting of all matrices of the form $\begin{pmatrix} 0 & 0 \\ b & c \end{pmatrix}$. Clearly, I is an essential left ideal of R. Now R/I being isomorphic to the ring of integers is not left artinian, and so it follows that R does not satisfy the left restricted minimum condition. $\qquad\square$

Let I be a right ideal of a semiprime right Goldie ring R and let Q be the classical right ring of quotients of R. Define $I^* = \{q \in Q : qI \subseteq R\}$ and $I^{**} = \{q \in Q : I^*q \subseteq R\}$. Then for a finitely generated projective right ideal I of R we have $I = I^{**}$. Somsup et al.[210] considered serial rings with restricted minimum condition and proved the following.

Lemma 3.12 *Let R be a semiprime serial ring with the right restricted minimum condition. Then R is noetherian.*

Proof First note that R has right Krull dimension at most 1 (see Lemma 6.2.8 [170]). Since R is a semiprime ring with right Krull dimension (at most 1), R is a right Goldie ring (see Lemma 6.3.5, [170]). Let $R = e_1R \oplus \cdots \oplus e_nR$, where $\{e_i : 1 \leq i \leq n\}$ is a set of mutually orthogonal primitive idempotents of R. Suppose that $\mathrm{Soc}(e_iR) \neq 0$, $i = 1, \ldots, k$ $(k \leq n)$ and $\mathrm{Soc}(e_jR) = 0$, $j = k+1, \ldots, n$. Since R has the right restricted minimum condition, we have e_iR is artinian for each $i = 1, \ldots, k$. Set $A = e_1R \oplus \cdots \oplus e_kR$, $B = e_{k+1}R \oplus \cdots \oplus e_nR$. Then A is clearly a two-sided ideal of R with A_R artinian and $R_R = A \oplus B$. We will show that B is also a two-sided ideal of R. It is clear that $BA = 0$. Therefore, $(AB)^2 = A(BA)B = 0$. Since R is semiprime, it follows that $AB = 0$. This shows that B is also a left ideal of R, that is B is a two-sided ideal. Thus R has a ring decomposition $R = A \oplus B$ where A is semiprime artinian and $\mathrm{Soc}(B_B) = 0$. Therefore, splitting off the socle, we may assume $\mathrm{Soc}(R) = 0$. Let I be a finitely generated uniserial submodule of R_R. Then there exists an indecomposable idempotent $e \in R$ and an epimorphism $\sigma : eR \to I$. Suppose $\ker(\sigma) \neq 0$. As R satisfies right restricted minimum condition, there exists a right ideal $K \subseteq eR$ such that $\ker(\sigma)$ is maximal in K. Then $\sigma(K)$ is minimal in eR, which is a contradiction. Hence I is projective. Now, any finitely generated right ideal of R is serial and therefore it is projective. Hence R is semihereditary. Since any right semihereditary right serial ring is right Goldie, R is right Goldie. Let $E_1 \subseteq E_2 \subseteq \cdots \subseteq R$ be an ascending chain of finitely generated essential right ideals. Choose a regular element a in E_1. Then we have a chain of left R-modules

$Ra^{-1} \supseteq (E_1)^* \supseteq (E_2)^* \supseteq \cdots \supseteq R$ so that $R \supseteq (E_1)^* a \supseteq (E_2)^* a \supseteq \cdots \supseteq Ra$. This chain of left ideals terminates by the restricted minimum condition. Therefore, $E_n^* a = E_{n+1}^* a$. Thus $E_n^* = E_{n+1}^*$ and hence $E_n = E_{n+1}$ by taking stars again and using the fact that each E_i is projective. Since R is a semiprime right Goldie ring with ACC on finitely generated essential right ideals, R must be right noetherian (see [47]) and hence two-sided noetherian [215]. $\qquad\square$

Theorem 3.13 *Let R be a serial ring which satisfies both the left and right restricted minimum condition. Then R must be a noetherian ring.*

Proof Note that R has right and left Krull dimension at most 1. Let N be the prime radical of R. Then N is nilpotent, by Lenagan [164]. Consider the ring $\overline{R} = R/N$. Then $\overline{R}$ is semiprime and satisfies both the left and right restricted minimum condition. Hence by the above result $\overline{R}$ is noetherian. Therefore, $\overline{R}$ has a ring decomposition $\overline{R} = \overline{S} \oplus \overline{B}$ where $\overline{S}$ is an artinian ring, $\overline{B}$ is a direct sum of hereditary prime rings (see [234]). In particular, $\mathrm{Soc}(\overline{B}_{\overline{B}}) = 0$. Assume that $\overline{S} = S/N$, where S is a two-sided ideal of R. By the above arguments, $\mathrm{Soc}(R/S)_R = 0$. Since R has the right restricted minimum condition, it follows that there is an idempotent e of R such that $S = eR$ (see [46], Corollary 6.6). Using the same arguments, we can find an idempotent f of R such that $S = Rf$. Thus we have $R = S \oplus R'$, with $N \subset S$. Since R' is a semiprime serial ring with the left and right restricted minimum condition, R' is noetherian. For the ring S, since S/N is semiprime artinian, we have $N = J(S)$. Thus $J(S)$ is nilpotent and $S/J(S)$ is artinian. From this, we conclude that S is a semiprimary ring. It follows that S is an artinian ring since S has left and right restricted minimum condition. Therefore R is a noetherian ring. $\qquad\square$

A ring R is said to satisfy the *restricted right socle condition* (*RRS* for short) if for each essential right ideal $E\,(\neq R)$, R/E has nonzero socle. A ring R is called a right *QI-ring* if each quasi-injective right R-module is injective. Boyle proved that a right QI-ring is right noetherian, and conjectured that every right QI-ring is hereditary [30]. This is still an open problem. However, the following is known due to Faith [76].

Theorem 3.14 *Any right QI-ring with restricted right socle condition is right hereditary.*

3.3   Rings each of whose proper cyclic modules is perfect

Definition A right R-module P is *semiperfect* if:

1. $P/J(P)$ is semisimple.
2. $J(P)$ is a small submodule of P.
3. Any decomposition of $P/J(P)$ into simple modules can be lifted to a decomposition of P into local modules.

A right R-module M is called *perfect* if M has a projective cover P such that $P^{(A)}$ is a semiperfect module for any set A. Consequently, if R_R is a perfect module, then R is a right perfect ring.

Abuhlail, Jain, and Laradji [5] have studied rings over which each cyclic module is a perfect module and proved the following.

Proposition 3.15 *Let R be a ring such that each proper cyclic right R-module is a perfect module. Then*

1. *Every nonzero prime ideal is maximal and hence primitive.*
2. *R is either prime or semiperfect with nil Jacobson radical.*

Proof The proof of (1) is obvious. For the proof of (2) we assume that R is not prime. Then there exist nonzero ideals A and B such that $AB = 0$. Now by hypothesis R/A is a perfect ring and each prime ideal is maximal. Thus there are only finitely many prime ideals containing A. Similarly, there are only finitely many prime ideals containing B. Let P be any prime ideal. Then $AB = 0 \subset P$. So either $A \subset P$ or $B \subset P$. This yields that there are only finitely many prime ideals, equivalently, primitive ideals. Hence the Jacobson radical $J(R)$ is nil and $R/J(R)$ is semisimple artinian and therefore R is semiperfect. $\square$

Theorem 3.16 *If R is any ring such that each proper cyclic right R-module is a perfect module, then if R is not right perfect, it is either prime or a local domain.*

Proof By the above proposition, R is either prime or semiperfect. Suppose R is semiperfect with n nonisomorphic simple modules where $n > 1$, that is, R is a direct sum of n local modules. Then, by hypothesis, each local summand is perfect and hence R is perfect. Assume now R is local. Recall R has nil Jacobson radical. Let $a\,(\neq 0) \in R$.

Case 1. aR is isomorphic to R. Then $ann_r(a)$ is either zero or R, because R is projective. Since a is nonzero, we obtain $ann_r(a) = 0$.

Case 2. Assume aR is not isomorphic to R. Thus by hypothesis, aR is perfect. Because there is a canonical homomorphism of R onto aR, with kernel in the radical, the projective cover of aR is a direct summand of R. But since R is local, the projective cover of aR must be R. Then by the definition of perfect module, an arbitrary direct sum of copies of R is semiperfect. Now by Kasch ([153], p. 294), R is perfect. Thus if for any nonzero a, aR is not isomorphic to R, then R is perfect. Thus if R is not perfect, then aR must be isomorphic to R, which implies $ann_r(a) = 0$.

Thus we conclude that if R is not perfect then R is prime and if, in addition, it is local then it must be a local domain. $\square$

3.17 Questions

1. Is a ring R right noetherian if for any nonzero right ideal A of R, R/A is an artinian right R-module?

2. If R is a left self-injective ring such that R/E is an artinian right R-module for any essential right ideal E of R, then must R be a quasi-Frobenius ring?

3. Let R be a serial ring such that R/E is an artinian right R-module for any essential right ideal E of R. Must R be a noetherian ring?

4 Rings each of whose cyclic modules is injective (or CS)

4.1 Rings where each cyclic module is injective

The study of noncommutative rings characterized by the properties of their cyclic modules has a long history. One of the first important contributions in this direction is due to Osofsky [185] who considered rings over which all cyclic modules are injective. It is clear that if each R-module is injective, then R is semisimple artinian. Osofsky showed R is semisimple artinian by simply assuming that all cyclic R-modules are injective.

The initial proof of Osofsky in her dissertation was quite elaborate. Later Osofsky gave a shorter proof of this theorem [186]. Skornyakov [204] also attempted to give a proof of this theorem but unfortunately his proof had an error (see [187]).

We start with the following useful lemma.

Lemma 4.1 *Let R be a ring and let $\{e_i : i \in \mathcal{I}\}$ be an infinite set of nonzero orthogonal idempotents in R. Assume that, for every nonempty subset $\mathcal{A} \subseteq \mathcal{I}$, there exists an element $f_{\mathcal{A}} \in R$ such that $e_i = f_{\mathcal{A}}e_i$ for $i \in \mathcal{A}$ and $e_i f_{\mathcal{A}} = 0$ for $i \in \mathcal{I} \setminus \mathcal{A}$. We set $S = \{r \in R : e_i r = 0, i \in \mathcal{I}\}$. Then the module $R/(\sum_{i \in \mathcal{I}} e_i R + S)$ is not injective.*

Proof We write the set $\mathcal{I}$ as an infinite disjoint union of infinite subsets $\mathcal{I} = \cup_{\mathcal{A} \in \Sigma} \mathcal{A}$, where $\Sigma \subseteq 2^{\mathcal{I}}$. We consider the set T of all elements Ω in $2^{\mathcal{I}}$ that satisfy the following conditions.

(a) If $\mathcal{A} \in \Omega$, then $\mathcal{A}$ is an infinite set.
(b) If $\mathcal{A}, \mathcal{B} \in \Omega$ and $\mathcal{A} \neq \mathcal{B}$, then $\mathcal{A} \cap \mathcal{B}$ is finite.

It is easy to verify that T is an inductively ordered set. By Zorn's Lemma, the set T contains a maximal element Δ with $\Sigma \subset \Delta$. If for an element $r \in R$ we have $\{e_i r : i \in \mathcal{I} \setminus \{i_1, \dots, i_n\}\} = \{0\}$ for some $n \in \mathbb{N}$, then $r - e_{i_1} r - \cdots - e_{i_n} r \in S$. Therefore the set $L = \sum_{i \in \mathcal{I}} e_i R + S$ coincides with the set of all elements in R that annihilate from the right almost all elements of the set $\{e_i : i \in \mathcal{I}\}$. We prove that the sum $\sum_{\mathcal{A} \in \Delta} (f_{\mathcal{A}} R + L)/L$ is a direct sum. Let $\mathcal{A} \in \Delta$, $r \in R$, $\mathcal{A}_j \in \Delta \setminus \{\mathcal{A}\}$, where $1 \leq j \leq n$. We assume that $f_{\mathcal{A}} r \notin L$. There exists an infinite number of elements $i \in \mathcal{A}$ with $e_i f_{\mathcal{A}} r \neq 0$. By construction, the set $\mathcal{A} \cap (\cup_{j=1}^n \mathcal{A}_j)$ is finite. Therefore if $r_0 \in \sum_{j=1}^n f_{\mathcal{A}_j} R + L$, then for almost all elements $i \in \mathcal{A}$, we have $e_i r_0 = 0$. Therefore $f_{\mathcal{A}} r \notin \sum_{j=1}^n f_{\mathcal{A}_j} R + L$. There exists a mapping

$$\varphi \colon \sum_{\mathcal{A} \in \Delta} (f_{\mathcal{A}} R + L)/L \to R/L,$$

such that $\varphi(f_{\mathcal{A}} + L) = f_{\mathcal{A}} + L$ for $\mathcal{A} \in \Sigma$ and $\varphi(f_{\mathcal{A}} + L) = 0$ for $\mathcal{A} \in \Delta \setminus \Sigma$.

We wish to show that R/L is not injective. Assume to the contrary that R/L is an injective module. Therefore the homomorphism φ can be extended to a homomorphism $\psi \colon R/L \to R/L$. Therefore there exists an element $t \in R$ such that $f_{\mathcal{A}} = tf_{\mathcal{A}} + \ell$ for all $\mathcal{A} \in \Sigma$, where $\ell \in L$. Therefore $e_i f_{\mathcal{A}} e_i = e_i t f_{\mathcal{A}} e_i + e_i \ell e_i$ for every element $i \in \mathcal{A}$. For almost all elements $i \in \mathcal{A}$, we have $e_i \ell = 0$ and $f_{\mathcal{A}} e_i = e_i$. Therefore the set $\mathcal{A}_0 = \{i \in \mathcal{A} : e_i = e_i t f_{\mathcal{A}} e_i = e_i t e_i\}$ is infinite. Note that $\mathcal{A} \setminus \mathcal{A}_0$ is finite. Let $\mathcal{C}$ be a set such that the intersection of $\mathcal{C}$ with every element of the set $\{\mathcal{A}_0 : \mathcal{A} \in \Sigma\}$ consists of exactly one element. We assume that $\mathcal{C} \in \Delta$. It is clear that $\mathcal{C} \in \Delta \setminus \Sigma$. Therefore $tf_{\mathcal{C}} \in L$ and $e_c t f_{\mathcal{C}} e_c = 0$ for almost all $c \in \mathcal{C}$. On the other hand, $e_c = e_c t e_c$ for every $c \in \mathcal{C}$. This contradiction shows that $\mathcal{C} \notin \Delta$. Since Δ is maximal, there exists a $\mathcal{D} \in \Delta$ such that the intersection $\mathcal{C} \cap \mathcal{D}$ is infinite and $\mathcal{D} \notin \Sigma$. Therefore $tf_{\mathcal{D}} \in L$ and for almost all elements $i \in \mathcal{I}$, we have $e_i t f_{\mathcal{D}} = 0$. Therefore $e_d t f_{\mathcal{D}} = 0$ for almost all elements $d \in \mathcal{C} \cap \mathcal{D}$. On the other hand, for every element $d \in \mathcal{C} \cap \mathcal{D}$, we have $e_d = e_d t e_d = e_d t f_{\mathcal{D}} e_d = 0$. Therefore the idempotents e_d are equal to zero for an infinite set of elements $d \in \mathcal{C} \cap \mathcal{D}$, a contradiction. Hence R/L is not injective. $\qquad\square$

Lemma 4.2 *Let R be a right self-injective von Neumann regular ring and let C be a countably generated right ideal which is not finitely generated. Then the right R-module R/C is not injective.*

Proof Note that $C = \sum_{i \in \mathcal{I}} e_i R$, where $\{e_i : i \in \mathcal{I}\}$ is a countably infinite set of orthogonal nonzero idempotents of the ring R.

Since R is right self-injective, for any nonempty subset $\mathcal{A}$ of $\mathcal{I}$, there exists an idempotent $f \in R$ such that $\oplus_{i \in \mathcal{A}} e_i R$ is essential in fR. It is easy to check that $fe_i = e_i$ for every $i \in \mathcal{A}$. Moreover, since $\oplus_{i \in \mathcal{A}} e_i R \subseteq_e fR$, the right ideal $J = (\oplus_{i \in \mathcal{A}} e_i R : f)$ is essential in R. Then, for all $j \notin \mathcal{A}$, $e_j f J \subseteq e_j (\oplus_{i \in \mathcal{A}} e_i R) = 0$ and so $e_j f \in Z(R)$, hence $e_j f = 0$. By Lemma 4.1, the module $R/(\sum_{i \in \mathcal{I}} e_i R + S)$, where $S = \{r \in R : e_i r = 0, i \in \mathcal{I}\}$, is not injective.

There exists a direct summand L of R_R such that S is essential in L. Therefore $L = gR$, where g is an idempotent in R. The right ideal $N = \{r \in R : gr \in S\}$ is essential in R_R. Since R is right nonsingular and $e_i g N = 0$ for every $i \in \mathcal{I}$, we have that $e_i g = 0$ for all $i \in \mathcal{I}$. Therefore $g \in S$ and $S = gR$. The module $\oplus_{i \in \mathcal{I}} e_i R$ is essential in some direct summand T of R_R. Since $S \cap T = 0$ and R_R is an injective module, $R_R = S \oplus T \oplus U$, where U is a submodule of R_R.

We assume that $R/ \oplus_{i \in \mathcal{I}} e_i R$ is an injective module. Therefore the module $(R/ \oplus_{i \in \mathcal{I}} e_i R) \oplus U$ is injective. We have the isomorphism

$$R/\left(\sum_{i \in \mathcal{I}} e_i R + S\right) \cong (T/ \oplus_{i \in \mathcal{I}} e_i R) \oplus U.$$

As $(R/\oplus_{i\in\mathcal{I}} e_i R + S)$ is not injective, $T/\oplus_{i\in\mathcal{I}} e_i R$ is not injective. Thus it follows that $R/\oplus_{i\in\mathcal{I}} e_i R$, being isomorphic to $(T/\oplus_{i\in\mathcal{I}} e_i R) \oplus S \oplus U$, is not injective. $\qquad\square$

Lemma 4.3 *Let R be a right self-injective von Neumann regular ring, A be a countably generated right ideal which is not finitely generated, and, for some idempotent $e \in R$, let the module eR be an essential extension of A_R. Then the module $R/(A \oplus (1-e)R)$ does not contain a nonzero injective submodule.*

Proof Let us write $B = A \oplus (1-e)R$. Suppose that the module R/B contains a nonzero injective submodule C/B. Therefore $R/B = C/B \oplus D/B$, say. Since $R/D \cong C/B$, the module R/D is injective. Since B is countably generated and D/B is cyclic, D is countably generated. By Lemma 4.2, D is finitely generated. Therefore D is a cyclic direct summand of R_R. Since B is essential in R_R, we have $D = R$. Therefore $C/B = 0$, a contradiction to our assumption. This shows that the module $R/(A \oplus (1-e)R)$ does not contain a nonzero injective submodule. $\qquad\square$

Theorem 4.4 (*Osofsky's Theorem*) *For a ring R, the following conditions are equivalent.*

1. *Each cyclic right R-module is injective.*
2. *R is a semisimple artinian ring.*

Proof $(1)\Rightarrow(2)$. By assumption, every principal right ideal xR of R is injective and hence a direct summand of R. Thus R is von Neumann regular right self-injective. If R is not semisimple artinian then R possesses a countably infinite set of orthogonal idempotents $\{e_i : i \in \mathcal{I}\}$. Let $A = \oplus_{i\in\mathcal{I}} e_i R$. Then A is a countably generated right ideal which is not finitely generated. By Lemma 4.2, the cyclic module R/A is not injective, a contradiction. Hence R is semisimple artinian.

The implication $(2)\Rightarrow(1)$ is obvious. $\qquad\square$

4.2   Rings each of whose cyclic modules is CS

Recall that a module M is called a CS module if every submodule of M is essential in a direct summand of M. Each uniform module is CS but need not be injective, for example, if R is a right Ore domain which is not a division ring then R_R is CS but not injective. By Osofsky's Theorem, if each cyclic R-module is injective, then each finitely generated R-module has finite uniform dimension. This was extended by Osofsky–Smith in [193] for rings each of whose cyclic modules is CS.

We proceed to prove the Osofsky–Smith Theorem. We start with the following simple observation.

Lemma 4.5 *Let M be a right R-module.*

1. *If $A_1 \subset A_2 \subset \cdots$ is an infinite strictly ascending chain of direct summands of M, then there exists an infinite strictly descending chain of direct summands $B_1 \supset B_2 \supset \cdots$ of M such that $M = A_i \oplus B_i$ for every i.*

2. *If $B_1 \supset B_2 \supset \cdots$ is an infinite strictly descending chain of direct summands of M, then there exist nonzero submodules $C_1, C_2, \ldots$ of M such that $M = C_1 \oplus \cdots \oplus C_n \oplus B_n$ and $\oplus_{i=n+1}^{\infty} C_i \subset B_n$ for every n.*

Proof

(1) There exists a submodule B_1 of M with $M = A_1 \oplus B_1$. By the modular law, we have $A_2 = A_1 \oplus (A_2 \cap B_1)$. Since A_2 is a direct summand of M, for some submodule B_2 of M, we have $B_1 = (A_2 \cap B_1) \oplus B_2$. Therefore $M = A_1 \oplus B_1 = A_1 \oplus (A_2 \cap B_1) \oplus B_2 = A_2 \oplus B_2$ and $B_1 \supset B_2$. We repeat this argument and obtain an infinite strictly descending chain $B_1 \supset B_2 \supset \cdots$ of direct summands of M with $M = A_i \oplus B_i$ for every i.

(2) For any $i \geq 1$, we have $B_i = B_{i+1} \oplus C_{i+1}$, where C_{i+1} is a submodule of M and $M = B_1 \oplus C_1$, where C_1 is a submodule of M. It is clear that for every n, we have $M = C_1 \oplus \cdots \oplus C_n \oplus B_n$ and $\oplus_{i=n+1}^{\infty} C_i \subset B_n$. $\qquad\square$

Theorem 4.6 *For a right R-module M, the following conditions are equivalent.*

1. *Every descending chain of direct summands of M terminates.*
2. *Every ascending chain of direct summands of M terminates.*
3. *The ring $\mathrm{End}(M)$ does not contain an infinite number of nonzero orthogonal idempotents.*

Proof The equivalence (1)$\Leftrightarrow$(2) follows from Lemma 4.5.

(2)$\Rightarrow$(3) Let $e_1, e_2, \ldots$ be orthogonal idempotents of the ring $\mathrm{End}(M)$. Then $e_1 M \subseteq (e_1 + e_2)M \subseteq \cdots$ is an ascending chain of direct summands of M. Therefore there exists a positive integer n such that $(e_1 + \cdots + e_k)M = (e_1 + \cdots + e_n)M$ for every $k \geq n$. Hence $e_k = 0$ for every $k \geq n$.

(3)$\Rightarrow$(1) We assume the contrary. Then there exists an infinite strictly descending chain of direct summands $B_1 \supset B_2 \supset \cdots$ of M. Without loss of generality, we can assume that $B_1 \neq M$. By Lemma 4.5, the module M contains submodules $C_1, C_2, \ldots$ such that $M = C_1 \oplus \cdots \oplus C_i \oplus B_i$ and $\oplus_{k=i+1}^{\infty} C_i \subset B_i$ for every positive integer i. Let π_i be the projection of M onto the submodule C_i with respect to the decomposition $M = C_1 \oplus \cdots \oplus C_i \oplus B_i$. Therefore $\pi_1, \pi_2, \ldots$ is an infinite set of nonzero orthogonal idempotents. This contradicts to our assumption. $\qquad\square$

A module is said to be *I-finite* if it satisfies one of the equivalent conditions of Theorem 4.6.

A module X is called a *subfactor* of a module M if X is a submodule of a factor module of M or, equivalently, X is a factor module of a submodule of M.

Theorem 4.7 *([65, 7.12]) Let M be a cyclic right R-module such that every closed submodule of any cyclic subfactor of M is an essential extension of a cyclic submodule. Then M is I-finite.*

Proof We assume to the contrary that M is not I-finite. By Lemma 4.5 and Theorem 4.6, there exist families of nonzero submodules $N_1, N_2, \ldots$ and $L_1, L_2, \ldots$ of M such that $M = N_1 \oplus \cdots \oplus N_n \oplus L_n$ and $\oplus_{i=n+1}^{\infty} N_i \subset L_n$ for every n. For any i, the submodule N_i is cyclic. Therefore N_i has a maximal submodule A_i. Let $\overline{M} = M/(\oplus_{i=1}^{\infty} A_i)$ and let $S = (\oplus_{i=1}^{\infty} N_i)/(\oplus_{i=1}^{\infty} A_i)$.

Therefore S is a semisimple submodule of $\overline{M}$ and for every n, the submodule $(\oplus_{i=1}^{n} N_i + \oplus_{i=1}^{\infty} A_i)/(\oplus_{i=1}^{\infty} A_i)$ is a direct summand of $\overline{M}$.

Let P be a maximal essential extension of the submodule S in the module $\overline{M}$. Since P is a closed submodule of $\overline{M}$, it follows from our assumption that P contains a cyclic essential submodule T. It is clear that $S = \mathrm{Soc}(T)$. We express S as $S = \oplus_{i=1}^{\infty} S_i$, where the submodules S_i are not finitely generated for every i. For every i, let B_i be a maximal essential extension of the submodule S_i in the module T. By assumption, for every i, the module B_i contains an essential cyclic submodule. Therefore $B_i \neq S_i$ and $\overline{B}_i = (B_i + S)/S \neq 0$ for every i.

Let $\overline{T} = T/S$ and let W be a maximal essential extension of the submodule $\oplus_{i=1}^{\infty} \overline{B}_i$ in $\overline{T}$. Therefore W contains a cyclic essential submodule $\overline{E}$. Let E be a cyclic submodule of T with $(E + S)/S = \overline{E}$. We set $\overline{F}_i = \overline{E} \cap \overline{B}_i$. It is clear that $\overline{F}_i \neq 0$ for every i. We denote by F_i the full pre-image of the submodule $\overline{F}_i$ in B_i with respect to the canonical homomorphism. Therefore $S_i \subset F_i \subset B_i$.

If $S_i \cap E = 0$ for some i, then as S_i is essential in F_i, we have $F_i \cap E = 0$. Since $F_i \subset E + S = E \oplus S_0$, where $S_0 \subset S$, we have that F_i is embeddable in S_0. Therefore F_i is semisimple. Therefore $F_i \subset S$ and $\overline{F}_i = 0$: this is impossible. Thus, for every i, there exists a simple submodule V_i in $\overline{M}$ such that $V_i \subset S_i \cap E$.

Set $V = \oplus_{i=1}^{\infty} V_i$. By our assumption, there exists a cyclic submodule K of E such that K is essential in some closure of V in E. It is clear that $V \neq K$ and $\overline{K} = (K + S)/S \neq 0$.

We prove that $K \cap \oplus_{i=1}^{\infty} B_i \subset S$. For any $n \geq 1$, we have

$$(K \cap \oplus_{i=1}^{n} B_i) \cap S = K \cap (\oplus_{i=1}^{n} B_i \cap S) = K \cap \oplus_{i=1}^{n} S_i \subset \mathrm{Soc}(K) \cap \oplus_{i=1}^{n} S_i =$$
$$(\oplus_{i=1}^{\infty} V_i) \cap \oplus_{i=1}^{n} S_i = \oplus_{i=1}^{n} V_i.$$

Since $\oplus_{i=1}^{n} V_i$ is a direct summand of T, for some submodule $T' \subset T$, we have $T = \oplus_{i=1}^{n} V_i \oplus T'$. If $(K \cap \oplus_{i=1}^{n} B_i) \cap T' \neq 0$, then as S is essential in T, there exists a simple submodule N_0 of S such that $N_0 \subset (K \cap \oplus_{i=1}^{n} B_i) \cap T'$. Therefore $N_0 \subset (K \cap \oplus_{i=1}^{n} B_i) \cap S \subset \oplus_{i=1}^{n} V_i$: this is impossible. This contradiction shows that $(K \cap \oplus_{i=1}^{n} B_i) \cap T' = 0$. Therefore the module $K \cap \oplus_{i=1}^{n} B_i$ is embeddable in $\oplus_{i=1}^{n} V_i$ and it is semisimple. Thus $K \cap \oplus_{i=1}^{n} B_i \subset S$. Since the last inclusion holds for any $n \geq 1$, we have $K \cap \oplus_{i=1}^{\infty} B_i \subset S$. Since $S \subset \oplus_{i=1}^{\infty} B_i$, $\overline{K} \cap \oplus_{i=1}^{\infty} \overline{B}_i = 0$.

This contradicts to the fact that $\oplus_{i=1}^{\infty} \overline{B}_i$ is essential in W. Hence M is I-finite. $\qquad\qquad\square$

Now we are ready to prove the Osofsky–Smith Theorem.

Theorem 4.8 (*Osofsky–Smith Theorem*) *Let M be a cyclic right R-module such that every cyclic subfactor of M is a CS module. Then the module M is a direct sum of uniform submodules.*

Proof By Theorem 4.7, M is I-finite. Therefore M is a finite direct sum of indecomposable submodules. Since every indecomposable CS module is uniform, the theorem follows. $\qquad\qquad\square$

In particular, we have:

Theorem 4.9 *Let R be a ring such that every cyclic right R-module is CS. Then every cyclic right R-module is a finite direct sum of uniform modules.*

As a consequence of Theorem 4.8, we have the following corollary which is a module analog of Theorem 4.4.

Corollary 4.10 *For a right R-module M, the following conditions are equivalent.*

1. *M is a semisimple module.*
2. *Every cyclic module in $\sigma[M]$ is M-injective.*
3. *Every cyclic subfactor of M is M-injective.*

Proof The implications $(1) \Rightarrow (2)$ and $(2) \Rightarrow (3)$ are easy to prove.

$(3) \Rightarrow (1)$. Let N be cyclic submodule of M. It follows from Theorem 4.9 that N is a finite direct sum of uniform submodules. Let U be a uniform direct summand of N. Since every cyclic submodule A of U is M-injective, A is a direct summand of U. Thus we have that U is a simple module. Hence every cyclic submodule of M is semisimple. Therefore the module M is semisimple. $\qquad\qquad\square$

The Osofsky–Smith Theorem was generalized by Dung, Huynh, and Wisbauer in the following theorem in [63].

Theorem 4.11 *Let M be a finitely generated right R-module. Assume that every cyclic subfactor of M is a direct sum of a CS module and a module of finite Goldie dimension. Then every factor module of M has finite Goldie dimension.*

Next, we consider rings each of whose proper cyclic modules is CS. We call a ring R a *right PCCS-ring* if each proper cyclic right R-module is CS. In a private communication, S. Singh has informed us that each right $PCCS$-ring has finite right Goldie dimension. We proceed to present his proof. First, we have the following:

Lemma 4.12 *Let R be a right PCCS-ring. Then the following hold.*

 1. *Any singular cyclic right R-module is of finite Goldie dimension.*
 2. *If some right ideal $A \cong R_R$, then any complement of A in R is semisimple.*

Proof

 1. This is immediate from Theorem 4.8.
 2. Let B be a complement of A in R. Suppose $B \neq 0$. Then $A \oplus B \cong R \oplus B \cong R \oplus B \oplus B \cong R \oplus B \oplus B \oplus B \cong \cdots$. Thus we get an infinite maximal direct sum $T = \oplus_{\alpha \in \Lambda} B_\alpha$, in R, where each $B_\alpha \cong B$. Suppose B is not semisimple. There exists an $F \subset_e B$. We get an infinite direct sum $T_1 = \oplus_{\alpha \in \Lambda} F_\alpha$ with $F_\alpha \subset_e B_\alpha$. Then, every cyclic subfactor of R/T_1 is proper, for otherwise, a complement of T in R will contain a copy of B, and that will give a contradiction. Hence R/T_1 is of finite Goldie dimension. This contradicts the fact that T/T_1 is of infinite Goldie dimension (see Lemma 3.7). Hence B is semisimple. $\square$

Theorem 4.13 *Let R be a right PCCS-ring. Then R has finite right Goldie dimension.*

Proof If R_R is CS, then the result follows by Theorem 4.8. Suppose R_R is not CS. Set $S = \mathrm{Soc}(R_R)$.

Case 1. $S = 0$. Suppose there exist two nonzero right ideals A, B such that $A \cap B = 0$. In view of Lemma 4.12(2), any cyclic subfactor of R/A, and R/B is proper. Thus R/A, and R/B are of finite Goldie dimensions. As R_R embeds in $R/A \oplus R/B$, R is of finite right Goldie dimension.

Case 2. $S \neq 0$. Suppose S is essential in R_R. If S is finitely generated, then the result obviously holds. Therefore, we take S to be not finitely generated. Now $S = A \oplus B$ for some right ideals A, B which are not finitely generated. Then $M = R/A$ is proper, for otherwise A is finitely generated. Consider any factor R/W of M. Here $A \subseteq W$. If $R/W \cong R_R$, then $R = W \oplus W'$, where $W' \cong R_R$. Then W is cyclic and by Lemma 4.12(2), W is semisimple. As $A \subseteq W$, A is also finitely generated, which is a contradiction. Thus $N := R/W$ is proper, hence a CS module. Suppose some cyclic subfactor K of N is not proper. There exists a summand K' of N such that $K \subseteq_e K'$. Then K' is a CS module. We identify R with K. Then $S = \mathrm{Soc}(K') \subset_e K'$. As K' is cyclic, by Lemma 4.12(1), K'/S is of finite Goldie dimension. But, by Lemma 3.7, K'/S is of infinite Goldie dimension, which is a contradiction. Hence every cyclic subfactor of M is proper. By Theorem 4.8, M is of finite Goldie dimension, which contradicts the fact that B embeds in M. Thus, if S is essential in R, then S is finitely generated, and we are done. So, we take S to be not essential in R_R. Let X be a complement of S. Then $X \neq 0$. Let $M' = R/X$. Suppose a cyclic subfactor of M' is not proper. We get a right ideal $A \cong R_R$ with $A \cap X = 0$. By Lemma 4.12(2), X is semisimple, which is a contradiction. Hence M' is of finite Goldie dimension. In particular, S is finitely generated. Let S' be a complement of X containing S. Suppose some cyclic subfactor of R/S' is not proper. Then, by Lemma 4.12(2), we get $R = S \oplus T$ for some right ideal $T \cong R_R$. Thus S is not finitely generated, which is a contradiction. Hence R/S' is proper and hence of finite Goldie dimension. This proves that R is of finite right Goldie dimension. $\qquad\square$

4.14 Question

1. Let M be a cyclic module whose factors are CS. Is M a finite direct sum of uniform modules?

5 Rings each of whose proper cyclic modules is injective

A ring R is called a *right PCI-ring* if each proper cyclic right R-module is injective. Note that this property does not hold for all Dedekind domains. For example, the ring of integers $\mathbb{Z}$ is a Dedekind domain which is not a PCI-domain.

We first state a lemma, more generally, for rings R over which each proper cyclic module is quasi-injective. Thus this lemma holds, in particular, for right PCI-rings also.

Lemma 5.1 *Let R be a prime ring such that each proper cyclic right module is quasi-injective. Then R is either artinian or a right Ore domain.*

Proof We first show that R is right nonsingular. Assume to the contrary that $Z(R_R) \neq 0$. By Theorem 4.9, R has finite right Goldie dimension and, therefore, there is a uniform submodule U of R_R. Consider $S = U \cap Z(R_R)$. Since R is prime and $S \neq 0$, we have $S^2 \neq 0$. Let $a \in S$ be such that $aS \neq 0$ and $0 \neq x \in S \cap ann_r(a)$. Since $x \in Z(R_R)$, $xR \ncong R$. Therefore, by hypothesis, xR is quasi-injective. Let $E(S)$ be the injective hull of S. As U is uniform, $xR \subset_e U$ and hence $xR \subset_e S$. Therefore xR is a fully invariant submodule of $E(S)$. In particular, $S(xR) \subseteq xR$. Thus $(aS)(xR) = a(SxR) \subseteq a(xR) = 0$, while $aS \neq 0$ and $xR \neq 0$, a contradiction to the primeness of R. Therefore R is right nonsingular. Hence R is a right Goldie ring. If R_R is uniform, then either R is a division ring or a right Ore domain. Assume that R_R is not uniform. Let $U_1 \oplus U_2 \oplus \cdots \oplus U_m \subset_e R_R$, where $m > 1$ and each U_i is a uniform right ideal of R. Let $0 \neq a_1 \in U_1$. Then $a_1 R \ncong R$. Therefore $a_1 R$ is quasi-injective. Since R is a prime right Goldie ring, all uniform right ideals of R are subisomorphic to each other (recall that two uniform right ideals A and B are called subisomorphic to each other if A contains an isomorphic copy of B and B contains an isomorphic copy of A). Hence each U_i, $(i \geq 2)$ contains an isomorphic copy $a_i R$ of $a_1 R$. It follows that $A = a_1 R \oplus a_2 R \oplus \cdots \oplus a_m R$ is quasi-injective (see, for example, §16 of [12]). Since A is essential in R_R, A contains a regular element b. Thus, as A is bR-injective and $bR \cong R$, A is injective. Then $A = R$ and R is right self-injective. Hence R is simple artinian. $\qquad\square$

Faith [73] proved the following for right PCI-rings without assuming any chain condition.

Theorem 5.2 *A right PCI-ring R is either semisimple artinian or a simple right semihereditary right Ore domain.*

Proof If A is any nonzero ideal of R, then every cyclic module C of R/A is a proper cyclic module over R, hence C is an injective R-module, and therefore an injective R/A-module. Thus by Theorem 4.4, R/A is semisimple artinian. In particular, every nonzero prime ideal is maximal.

Suppose R is not prime. Then there exist nonzero ideals A and B such that $AB = 0 \subset P$, for any prime ideal P. This implies that either A or B is contained in P. Furthermore, R/A and R/B are semisimple artinian. This means there are only finitely many prime ideals above A and B. Since every prime ideal contains either A or B, it follows that there are only finitely many prime ideals in R. This gives that the prime radical $N(R) = P_1 \cap \cdots \cap P_k$ where $P_1, \ldots, P_k$ are prime ideals in R and $R/N(R) \cong R/P_1 \times \cdots \times R/P_k$. By Theorem 4.4, R/P_i is simple artinian for each i. Thus $R/N(R)$ is semisimple artinian. Since $N(R)$ is nil, this implies that $N(R) = J(R)$. Hence R is semiperfect. Suppose R is not local. Then $R = e_1 R + \cdots + e_n R$, where $e_i R$ are indecomposable right ideals and by hypothesis these are injective. Then R_R is injective. Then, by Theorem 4.4, R is semisimple artinian. Assume now R is local. We claim $N(R) = 0$. Else, let $0 \neq a \in N(R)$. If $aR \not\cong R$, then aR is injective and hence a summand of R. Since R is local, this gives $aR = R$, a contradiction because $a \in N(R)$. Thus $aR \cong R$. This is not possible either because a is a nilpotent element. Thus $N(R) = 0$ and we conclude that R is semisimple artinian if it is not prime.

If R is prime then by the above lemma, R is either artinian or a right Ore domain. Now we proceed to show that R is right semihereditary, that is, each finitely generated ideal of R is projective. The proof is by induction. Let $A = aR + bR$. Then $(aR + bR)/bR$ is cyclic. In case $(aR + bR)/bR$ is isomorphic to R, then $aR + bR = bR \oplus K$, where $K \cong R$. So $aR + bR$ is projective. In the other case, $(aR + bR)/bR$ is injective and so a direct summand of R/bR. This gives $(aR + bR)/bR \oplus X/bR = R/bR$ and so $(aR + bR) + X = R$, where $(aR + bR) \cap X = bR$. Thus $(aR + bR) \times X \cong R \times bR$, proving that $aR + bR$ is projective. By induction, we may deduce that each finitely generated right ideal of R is projective. $\qquad\square$

We give below some examples of left and right PCI-domains. The first example is from Cozzens [54]. This is an example of a right and left noetherian domain which is both right and left PCI-ring. We do not know any example of a right PCI-domain which is not a left PCI-domain.

Example Let k be a universal differential field with derivation D and let $R = k[y, D]$ denote the ring of differential polynomials in the indeterminate y with coefficients in k, that is the additive group of $k[y, D]$ is the additive group of the ring of polynomials in the indeterminate y with coefficients in field k, and multiplication in $k[y, D]$ is defined by: $ya = ay + D(a)$ for all $a \in k$. Let $f = \Sigma_{i=1}^{n} a_i y^i \in k[y, D]$, $a_n \neq 0$. We define the degree $\delta(f)$ of f to be n. Clearly we have the following: (i) $\delta(fg) = \delta(f) + \delta(g)$, (ii) for $f, g \in k[y, D]$, there exist $h, r \in k[y, D]$ such that $f = gh + r$ with $r = 0$ or $\delta(r) < \delta(g)$ (a similar algorithm holds on the left). From (ii) it follows that this ring $R = k[y, D]$ is both a left and right principal ideal domain. The simple right R-modules are precisely of the form $V_\alpha = R/(y-\alpha)R$ where $\alpha \in k$. We claim that $V_\alpha = R/(y-\alpha)R$ is a divisible right R-module for all $\alpha \in k$. For this, it suffices to show that $V_\alpha(y+\beta) = V_\alpha$ for all $\alpha, \beta \in k$. Equivalently, given $h \in R$, $\delta(h) = 0$, there exist $f, g \in R$ such that $f(y+\beta) + (y+\alpha)g = h$. We shall determine $a, b \in k$ such that $a((y+\beta) + (y+\alpha)b = h$. This is equivalent to an equation of the form $D(b) + (\alpha - \beta)b = h$. Since k is a universal differential field, there exists a $b \in k$ satisfying this equation. Hence each simple right R-module is divisible. Since R is a principal right ideal domain, this implies that each simple right R-module is injective. Similarly, it can be shown that each simple left R-module is injective. Hence by (Corollary 10, [30]), R is both a left as well as a right PCI-domain. $\qquad\square$

Osofsky [191] provided another example of a right and left PCI-domain, as follows.

Example Let F be a field of characteristic $p > 0$, and σ an endomorphism of F defined by $\sigma(\alpha) = \alpha^p$ for all $\alpha \in F$. We then form the ring of twisted polynomials with coefficients on the left, $R = F[x, \sigma]$ with $R = \{\Sigma_{i=0}^n \alpha_i x^i : n \in \mathbb{Z}, \alpha_i \in F\}$ under usual polynomial addition and multiplication given by the relation $x\alpha = \sigma(\alpha)x$ for all $\alpha \in F$. It may be observed that this ring R is a left and right principal ideal domain. If F is separably closed then every simple right (left) R-module is divisible and hence injective. Therefore every proper cyclic right (left) R-module is injective, that is, R is both a left and right PCI-domain. $\quad\square$

In [55], Cozzens and Faith asked the question if every right PCI-ring is right noetherian. This question was later answered in the affirmative by Damiano [56].

Theorem 5.3 *Let R be a right PCI-ring. Then R is right noetherian and right hereditary.*

Proof First recall that a right PCI-ring is either semisimple artinian or a simple right Ore domain. Assume R is a right Ore domain which is not a division ring. So each right ideal is essential and the socle is zero. Now, for any nonzero right ideal E, R/E is a proper cyclic module. Hence, by the Osofsky–Smith Theorem (see Theorem 4.9), R/E has finite Goldie dimension. Since every cyclic submodule of R/E is injective, R/E is semisimple. This yields that R is right noetherian. Now, in view of Theorem 5.2, R must be right hereditary. $\qquad\square$

Boyle and Goodearl [31] studied the question of left-right symmetry of PCI-domains. They proved the symmetry for two-sided noetherian domains. In order to present their proof, we will first prove a useful lemma.

Lemma 5.4 *A left Ore right PCI-domain is left noetherian.*

Proof Let R be a left Ore right PCI-domain. Let $I_1 \subseteq I_2 \subseteq \cdots \subseteq I_n \subset R$ be an ascending chain of finitely generated (hence projective) left ideals of R, and let $a \in I_1$. Let Q be the maximal quotient ring of R. The set $I_k^{-1} = \{q \in Q :$

$I_k q \subseteq R\}$ is a right R-submodule of Q, canonically isomorphic to $I_k^* = \mathrm{Hom}(I_k, R)$ since $_R I_k$ is essential. Thus we have $(Ra)^{-1} \supseteq I_1^{-1} \supseteq I_2^{-1} \supseteq \cdots \supseteq R$. This yields $R \supseteq aI_1^{-1} \supseteq aI_2^{-1} \supseteq \cdots \supseteq aR$. Clearly R/aR is artinian. Thus for some $n > 0$, $aI_n^{-1} = aI_{n+j}^{-1}$ for all $j \geq 1$ and hence $I_n^{-1} = I_{n+j}^{-1}$. Now, by projectivity of each I_k, we get $I_n = I_{n+j}$ for all $j \geq 1$. Thus the above ascending chain terminates and hence R is left noetherian. $\qquad\square$

Now we are ready to prove the following

Theorem 5.5 *Let R be a right and left noetherian domain. Then R is a right PCI-domain if and only if R is a left PCI-domain.*

Proof Let R be a left and right noetherian right PCI-domain. In view of the above lemma, we may assume that R is a left Ore domain. Note that each proper cyclic right as well as left module is torsion since R is a right and left Ore domain. Furthermore, the functor $\mathrm{Hom}(-, Q(R)/R)$ defines a duality between the category of finitely generated torsion right modules and the category of finitely generated torsion left modules. Because R is a right PCI-domain, each proper cyclic right module is semisimple. This implies that each proper cyclic left module is semisimple. Therefore R is a left PCI-domain. $\qquad\square$

In [56], Damiano had claimed that a right PCI-ring R is a left PCI-ring if and only if R is a left coherent ring. But this is incorrect (see e.g. Remark 6.3, [133]). Jain, Lam, and Leroy [133] considered twisted differential polynomial rings over a division ring which are right PCI-domains and found equivalent conditions as to when these will be left PCI-domains. However, in general, the question of left–right symmetry of PCI-domains is still open.

5.6 Questions

1. Is every right PCI-domain also a left PCI-domain?
2. Let R be a ring such that, for each proper cyclic right R-module C, the injective dimension of C is less than or equal to n $(n \geq 1)$. Must R as a right R-module have finite injective dimension?
3. Let R be a ring such that each factor of every cyclic right R-module is injective. Must R be right noetherian?

6 Rings each of whose simple modules is injective (or Σ-injective)

6.1 V-rings

A ring R is called a *right V-ring* if each simple right R-module is injective. The class of right V-rings was introduced by Villamayor [175]. It is a well-known unpublished result owed to Kaplansky that a commutative ring is von Neumann regular if and only if it is a V-ring. Clearly, every right PCI-ring a right V-ring.

Theorem 6.1 *For any ring R, the following are equivalent:*

1. *R is a right V-ring.*
2. *For any right R-module M, $J(M) = 0$.*
3. *Every right ideal $A\,(\neq R)$ of R is an intersection of maximal right ideals.*

Proof $(1){\Rightarrow}(2)$. Let R be a right V-ring and let M be a right R-module. We claim that the intersection of all maximal submodules of M is zero. Let $x\,(\neq 0) \in M$. Consider the cyclic right R-module xR. Let N be a maximal submodule of xR. Then $S = xR/N$ is a simple right R-module and hence injective, by assumption. Consider the canonical surjective homomorphism $f : xR \to S$. Since S is injective, f can be extended to $f' : M \to S$. Clearly $\mathrm{Ker}(f')$ is a maximal submodule of M such that $x \notin \mathrm{Ker}(f')$. This shows that the intersection of all maximal submodules of M is zero, that is, $J(M) = 0$.

$(2){\Rightarrow}(3)$. Let $A\,(\neq R)$ be a right ideal of R. Since $J(R/A) = 0$, we conclude that A is an intersection of maximal right ideals.

$(3){\Rightarrow}(1)$. Let S be any simple right R-module. Let I be any right ideal of R. Consider a nonzero homomorphism $f : I \to S$. To prove that S is injective, we need to show that f can be extended to R. Fix an element $x \in I \setminus \mathrm{Ker}(f)$. Since $\mathrm{Ker}(f)$ is the intersection of a family of maximal right ideals of R and $x \notin \mathrm{Ker}(f)$, there exists a maximal right ideal $M \supseteq \mathrm{Ker}(f)$ such that $x \notin M$. Since $I/\mathrm{Ker}(f) \cong S$ is simple, we have $M \cap I = \mathrm{Ker}(f)$. Clearly, $I + M = R$. Now we may extend f to $f' : R \to S$ by defining $f'(a + m) = f(a)$ for any $a \in I$ and $m \in M$. This shows that every simple right R-module is injective and hence R is a right V-ring. $\qquad\square$

Theorem 6.2 *For a ring R, the following are equivalent:*

1. *R is a right V-ring.*
2. *Each right ideal I of R is an idempotent, that is, $I^2 = I$ and all right primitive factor rings of R are right V-rings.*

Proof $(1) \Rightarrow (2)$. Let R be a right V-ring. Clearly every factor ring of R is a right V-ring. Suppose there is a right ideal I of R such that $I^2 \neq I$. By the above theorem, the right ideal I^2 is an intersection of maximal right ideals. Therefore there is a maximal right ideal M of R such that $I^2 \subseteq M$ and $I \not\subseteq M$. Then $R = M + I$. Thus there exist elements $m \in M$ and $a \in I$ with $1 = m + a$. Then $I = (m + a)I \subseteq mI + aI \in M + I^2 = M$, a contradiction. Hence $I^2 = I$.

$(2) \Rightarrow (1)$. Let S be a simple right R-module. Let E be an essential extension of S, and let $A = ann_r(S)$. Since the factor ring R/A is right primitive, by assumption, R/A is a right V-ring. Now we claim that $A = ann_r(E)$. Assume to the contrary that $A \neq ann_r(E)$. Then there is an element $x \in E$ such that $xA \neq 0$. Since E is an essential extension of the simple module S, we have $S \subseteq xA$. Therefore there exists a right ideal I of R such that $I \subseteq A$ and $xI = S$. Now $S = xI = xI^2 = SI \subseteq SA = 0$, a contradiction. Hence $A = ann_r(E)$. Now E is a right R/A-module that is an essential extension of the simple R/A-module S. Since R/A is a right V-ring, $S = E$. Therefore S is injective and hence R is a right V-ring. $\qquad\qquad\square$

As a consequence, we have the following.

Corollary 6.3 *Let R be a von Neumann regular ring such that each primitive factor ring of R is artinian. Then R is a right and left V-ring.*

Now we can prove Kaplansky's result that we mentioned at the beginning of this section.

Theorem 6.4 *A commutative ring R is a V-ring if and only if it is a von Neumann regular ring.*

Proof Let R be a commutative von Neumann regular ring. Since a commutative primitive ring is a field and every right ideal of a von Neumann regular ring is idempotent, from the above theorem, it follows that R is a V-ring. Conversely, suppose R is a commutative V-ring. Then, by the above theorem, each right ideal of R is idempotent. Let $a \in R$. Then $aR = (aR)^2 = a^2R$. So there exists an element b in R such that $a = a^2b = aba$. Hence R is a von Neumann regular ring. $\qquad\qquad\square$

However, in the case of noncommutative rings, the two notions are completely different. We give below an example of a ring which is von Neumann regular but not a right V-ring.

Example Let V be a countably infinite-dimensional left vector space over a division ring D. Let $R = \mathrm{End}(_D V)$ be the ring of right linear operators on V.

Then R is a von Neumann regular ring. We proceed to show that the simple right R-module V is not injective. Assume to the contrary that V_R is injective. Consider a basis $\{v_i : i \in \mathbb{N}\}$ of V. For each $i \in \mathbb{N}$, let us define $f_i \in R$ by $(v_i)f_i = v_i$ and $(v_j)f_i = 0$ for $j \neq i$. Set $A = \Sigma_i f_i R$. Then A is a right ideal of R. Consider a right R-homomorphism $g : A \to V_R$ defined by $g(\Sigma_i f_i r_i) = \Sigma_i v_i r_i$, where $r_i \in R$ is zero for all but finitely many i. Since V_R is injective, there exists $v \in V$ such that $g(f_i) = v f_i$ for every $i \in \mathbb{N}$. This gives $v_i = v f_i$ for every $i \in \mathbb{N}$. Now if $v = d_1 v_1 + d_2 v_2 + \cdots + d_n v_n$, then for any $i \in \mathbb{N} \setminus \{1, \ldots, n\}$, we have $v f_i = 0$, a contradiction. This shows V_R is not injective. Thus R is not a right V-ring as the simple right R-module V is not injective. $\qquad\square$

Next, we give an example of a unit-regular ring which is neither a right nor a left V-ring.

Example Let D be a division ring and let X be an infinite set. Consider the column-finite matrix ring $S = \mathbb{CFM}_X(D)$. We may identify D with the subring of S of all scalar matrices. Let K be subset of S of all matrices with only a finite number of nonzero entries. Let $R = K + D$. It is not difficult to observe that every element of R belongs to a subring of R isomorphic to one of the form $\mathbb{M}_n(D) \times D$, for some positive integer n. Now we will show that R is not a right V-ring. Assume to the contrary that R is a right V-ring. If e is a primitive idempotent of R, then eR_R is simple, faithful, and injective. Now eS_R is an injective hull of eR_R and so we have $eR = eS$. This leads to a contradiction, because $eR \subset K$ and eS contains matrices with infinitely many nonzero entries. This shows R is not a right V-ring. By viewing R as a subring of row-finite matrix ring $\mathbb{RFM}_X(D)$, we see that the above argument applies on the left too. Thus R is neither a right nor a left V-ring. $\qquad\square$

The examples of Cozzens and Osofsky (see the examples on p. 47) are examples of V-rings which are not von Neumann regular.

Example Let V be an infinite-dimensional vector space over F. Set $Q = \mathrm{End}_F(V)$, $J = \{x \in Q : \dim_F(xV) < \infty\}$ and $R = F + J$. Then R is a von Neumann regular right V-ring which is not a left V-ring (see [97], Example 6.19). This shows that von Neumann regular V-rings are not left–right symmetric.

This example also shows that a von Neumann regular right V-ring need not have all primitive factors artinian. $\qquad\square$

6.2 WV-rings

A ring R is called a right *weakly-V ring* (WV-ring for short) if each simple right R-module is R/A-injective for any right ideal A such that $R/A \not\cong R$ (i.e., R/A is proper). Such rings need not be right V-rings, as for example the ring $\mathbb{Z}_{p^2}$ for any prime p is a WV-ring which is not a V-ring. These rings have been studied by Holston, Jain, and Leroy in [110] and Holston and Huynh in [111].

We start with the following useful observation.

Lemma 6.5 *Let R be a right WV-ring, and R/A and R/B be proper cyclic right R-modules such that $A \cap B = 0$. Then R is a right V-ring.*

Proof Since R is a right WV-ring, any simple right R-module is $(R/A) \times (R/B)$-injective. Since R_R embeds in $(R/A) \times (R/B)$, any simple right R-module is R_R injective, that is R is a right V-ring. $\qquad\square$

Theorem 6.6 *Let R be a right WV-ring which is not a right V-ring. Then R must be right uniform.*

Proof Suppose R is a right WV-ring. If R is of infinite right Goldie dimension, then R contains a direct sum $A \oplus B$ where both A and B are infinite direct sums of nonzero right ideals. If $R/A \cong R$, then R/A is projective, and hence there exists a right ideal C of R such that $R = C \oplus A$. But then the cyclic module R/C is isomorphic to an infinite direct sum of nonzero modules, a contradiction. Thus R/A is proper. Similarly R/B is proper, and so R is a right V-ring by Lemma 6.5. Assume now that $\mathrm{Gdim}(R) = n > 1$ is finite. Then there exist closed uniform right ideals U_i such that $\bigoplus_{i=1}^{n} U_i \subseteq_e R$. Now $\mathrm{Gdim}(R/U_1) = n - 1 = \mathrm{Gdim}(R/U_2)$, and so R/U_1 and R/U_2 are proper. Hence R is a right V-ring by Lemma 6.5. So if R is a right WV-ring but not a right V-ring, we must have $\mathrm{Gdim}(R) = 1$, that is R is right uniform. $\qquad\square$

The proof of the following proposition is straightforward.

Proposition 6.7 *Let R be a ring such that R/I is proper for any nonzero right ideal I. Then the following are equivalent:*

1. R is a right WV-ring.
2. $J(R/I) = 0$ for any nonzero right ideal I of R.
3. Any nonzero right ideal $I \neq R$ is an intersection of maximal right ideals.

In particular, the above statements are equivalent when R is right uniform.

Corollary 6.8 *If R is a right WV-ring, then $R/J(R)$ is a right V-ring.*

Proof If R is a right V-ring, then clearly $R/J(R)$ is a right V-ring. So, assume that R is not a right V-ring. Then, by Theorem 6.6, R is right uniform. By Proposition 6.7, every nonzero right ideal $(\neq R)$ of R is an intersection of maximal right ideals. If $J(R) = 0$, then the zero ideal is also an intersection of maximal right ideals, and so $R (= R/J(R))$ is a right V-ring. If $J(R) \neq 0$, then in $R/J(R)$ all right ideals $(\neq R/J(R))$ are intersections of maximal right ideals, and so $R/J(R)$ is a right V-ring. $\qquad\square$

Thus, in particular, a von Neumann regular right WV-ring is a right V-ring.

Proposition 6.9 *If R is a right WV-ring, then the following statements hold:*

1. *If I is a right ideal of R, then either $I^2 = 0$ or $I^2 = I$.*
2. *If R is a domain, then R is simple.*

Proof

(1) If R is a right V-ring, then, by Theorem 6.2, $I^2 = I$ for every right ideal of R. Assume that R is not a right V-ring. Then R is right uniform by Theorem 6.6. Let $I \neq R$ be a right ideal and suppose $I^2 \neq 0$. By Proposition 6.7, both I and I^2 are intersections of maximal right ideals. If $I^2 \neq I$, there must exist a maximal right ideal M such that $I^2 \subseteq M$ but $I \nsubseteq M$. We thus have $R = I + M$ and we can write $1 = x + m$ for some $x \in I, m \in M$. This gives $I = (x + m)I \subseteq xI + mI \subseteq I^2 + M = M$, a contradiction. Hence $I^2 = I$.

(2) Let $0 \neq a \in R$. Since R is a domain, $(aR)^2 \neq 0$, so part (1) gives us $(aR)^2 = aR$, that is $aRaR = aR$. Since R is a domain, this gives $RaR = R$ and hence R is a simple ring. $\square$

Corollary 6.10 *If R is a right WV-domain, then R is a right V-domain.*

Recently, Holston and Huynh have given the complete description of right WV-rings which are not right V-rings [111]. Before proceeding further to present their proof, we recall some definitions.

A right R-module M is called a *V-module* if every simple module in $\sigma[M]$ is M-injective. Thus, a ring R is a right V-ring if and only if every right R-module is a V-module; and a ring R is a right WV-ring if and only if every proper cyclic module is a V-module. The following is a generalization of Theorem 6.1 for V-modules. See [235] for its proof.

Theorem 6.11 *For a right R-module M, the following are equivalent:*

1. *M is a V-module.*
2. *Every proper submodule of M is an intersection of maximal submodules of M.*
3. *$J(N) = 0$ for every $N \in \sigma[M]$.*

In view of Corollary 6.8, it follows that if R is a right WV-ring which is not a right V-ring, then $J(R) \neq 0$.

Lemma 6.12 *Let R be a right WV-ring which is not a right V-ring. Then $J(R)$ is a simple right R-module.*

Proof Let N_R be a proper submodule of $J(R)$. If $R/N \cong R$, then R/N is projective and hence N is a direct summand of R, a contradiction because R is right uniform. Thus R/N is proper cyclic. Since R is a right WV-ring, R/N is a V-module and hence $J(R/N) = 0 = J(R)/N$. Thus $J(R) = N$ and hence $J(R)$ is a simple right R-module. $\square$

Corollary 6.13 *Let R be a right WV-ring which is not a right V-ring. Then every nonzero right ideal of R contains $J(R)$.*

Proof Let I be a nonzero right ideal of R. Since R is right uniform, $0 \neq I \cap J(R) \subseteq J(R)$. Because $J(R)_R$ is simple, we have $I \cap J(R) = J(R)$. Thus, $J(R) \subseteq I$.   □

Lemma 6.14 *Let R be a right WV-ring which is not a right V-ring. Then every nonzero cyclic right ideal of $R/J(R)$ is isomorphic to $R/J(R)$.*

Proof Set $\overline{R} = R/J(R)$. Let $0 \neq \bar{x} \in \overline{R}$. Then there exists $x \in R \setminus J(R)$ such that $\bar{x} = x + J(R)$. If $xR \not\cong R_R$, then as R is a right WV-ring and $J(R)$ is a simple right R-module, $J(R)$ is xR-injective. But $J(R) \subseteq xR$ by Corollary 6.13, so $J(R)$ is a direct summand of xR, a contradiction as R is right uniform. Thus $xR \cong R_R$. Let $f : xR \to R$ be an isomorphism. Since $\overline{R}$ is a right V-ring, $J(xR/J(R)) = 0$. Hence $J(xR) \subseteq J(R)$. Now every nonzero submodule of xR contains $J(R)$ by 6.13, and so $J(xR) \neq 0$. Now, as $J(R)_R$ is simple, $J(xR) = J(R)$. Hence $f(J(R)) \subseteq J(R)$. Since f is an injection, $f(J(R)) \neq 0$, and so $f(J(R)) = J(R)$. Thus $\bar{x}\overline{R} \cong \overline{R}$.   □

Theorem 6.15 *Let R be a right WV-ring which is not a right V-ring. Then*

1. *R has exactly three two-sided ideals $0 \subset J(R) \subset R$. Moreover, for every nonzero element $x \in J(R)$, $Rx \cong_R (R/J(R))$.*
2. *If $_RJ(R)$ is finitely generated, then $R/J(R)$ is a division ring and so, R has exactly three right ideals $0 \subset J(R) \subset R$.*

Proof

1. Let $\overline{R} = R/J(R)$. Let $0 \neq \bar{x} \in \overline{R}$. Then $\bar{x}\overline{R} \cong \overline{R}$ by Lemma 6.14. So $\bar{x}\overline{R}$ is projective. Thus $\overline{R}_{\overline{R}} = \overline{P} \oplus \overline{Q}$ for some $\overline{Q}$, where $\overline{P} \cong \bar{x}\overline{R}$. If $\overline{Q} \neq \bar{0}$, then it is generated by a nontrivial idempotent $\bar{e} \in \overline{R}$. Since $J(R)_R$ is simple, $J(R)^2 = 0$. Since $J(R)$ is nil, idempotents modulo $J(R)$ can be lifted. So we may assume that e is an idempotent in R. But R is right uniform, so this is a contradiction. Thus $\bar{0} = \overline{Q} \cong ann_{\overline{R}}(\bar{x})$. This implies that $\overline{R}$ is a domain. Since $\overline{R}$ is a right V-ring, $\overline{R}$ is a simple ring. Thus it follows that R has exactly three two-sided ideals $0 \subset J(R) \subset R$. Finally, let $0 \neq x \in J(R)$. Then $xR = J(R)$. Since $ann_l^R(J(R))$ is a two-sided ideal, so is $ann_r^R(x)$. Since $J(R)^2 = 0$ and R is simple, we must have $ann_l^R(x) = J(R)$. Thus it follows that $Rx \cong_R (R/J(R))$.

2. Suppose $_RJ(R) = Rx_1 + \cdots + Rx_k$ for some positive integer k, where $0 \neq x_i \in J(R)$ for $1 \leq i \leq k$. Define $\varphi : R_R \to J(R)^{(k)}$ by $\varphi(r) = (x_1 r, \ldots, x_k r)$. Set $I = \text{Ker}(f) = \cap_{i=1}^k ann_r^R(x_i)$. Then $I \subseteq ann_r^R(J(R))$. But $J(R)^2 = 0$ and $\overline{R}$ is simple, so $ann_r^R(J(R)) = J(R) = I$. Now $R/J(R)$ is isomorphic to a submodule of $J(R)^{(k)}$, whence $\overline{R}_{\overline{R}}$ is semisimple. Since $\overline{R}$ is a domain, it must be a division ring and consequently R has exactly three right ideals $0 \subset J(R) \subset R$.   □

Consequently, we have the following.

Corollary 6.16 *Let R be a right WV-ring such that R is not a right V-ring. Assume, in addition, that R is a left WV-ring. Then R is a right and left uniserial ring, and $0 \subset J(R) \subset R$ is the only composition series in R.*

Consequently, if R is a right WV-ring such that R is not a right V-ring, then R is a left WV-ring if and only if $_R J(R)$ is a simple module.

It is known that the property of R being a V-ring is not left–right symmetric. In fact, the property of being a WV-ring is not left–right symmetric either, as shown by the following example, lowed to Faith ([74], p. 335).

Example Let $R = \begin{bmatrix} a & b \\ 0 & \sigma(a) \end{bmatrix}$ where $a, b \in \mathbb{Q}(x)$ and σ is the $\mathbb{Q}$-endomorphism of $\mathbb{Q}(x)$ such that $\sigma(x) = x^2$.

In this ring there are only three left ideals and hence it is a left WV-ring. But this ring has more than three right ideals (it is not even right noetherian) and thus it cannot be a right WV-ring. $\qquad\square$

6.3 Σ-V rings

A ring R is called a *right Σ-V ring* if each simple right R-module is Σ-injective. Recall that a module M is called Σ-injective if $M^{(\alpha)}$ is injective for any cardinal α. The following theorem gives characterizations for an injective module to be Σ-injective.

Theorem 6.17 *For an injective module M_R, the following are equivalent:*

(i) M is Σ-injective.

(ii) $M^{(\aleph_0)}$ is injective.

(iii) Each essential extension of $M^{(\aleph_0)}$ is a direct sum of modules that are either injective or projective.

(iv) R satisfies ACC on the set of right ideals I of R that are annihilators of subsets of M.

Proof We will first show the equivalence of (i) and (iii). Suppose each essential extension of $M^{(\aleph_0)}$ is a direct sum of modules that are either injective or projective. Assume to the contrary that M is not Σ-injective. Then $\oplus_{i \in \mathcal{I}} M_i$, $(M_i \cong M)$ is not injective for some infinite index set $\mathcal{I}$. Thus, by Baer's injectivity criterion, there exists a right ideal A of R and a right R-homomorphism $g : A \to \oplus_{i \in \mathcal{I}} M_i$ such that the set $I' = \{j \in \mathcal{I} : \pi_j \circ g \neq 0\}$ is infinite, where $\pi_j : \oplus_{i \in \mathcal{I}} M_i \to M_j$ is the canonical projection. Because otherwise $Im(g)$ would be contained in a finite direct subsum of $\oplus_{i \in \mathcal{I}} M_i$; and since any finite direct sum of injective modules is injective, the map g would extend to R and this would contradict our assumption that $\oplus_{i \in \mathcal{I}} M_i$ is not injective. Let $\mathcal{J}$ be a countably infinite subset of I'. Now, choose an element $a_j \in A$. Let $b_j = g(a_j)$ and $N_j = b_j R$.

Then N_j is a cyclic submodule of M_j. Since $\mathcal{J}$ is countable and each N_j is cyclic, $\oplus_{j \in \mathcal{J}} N_j$ is countably generated. Denote by Q_j, an injective hull of N_j in M_j. Let $Q = E(\oplus_{j \in \mathcal{J}} Q_j)$ be an injective hull of $\oplus_{j \in \mathcal{J}} Q_j$. Let $\pi : \oplus_{i \in \mathcal{I}} M_i \to \oplus_{j \in \mathcal{J}} Q_j$ be the epimorphism that carries M_i to zero if $i \in \mathcal{I} \setminus \mathcal{J}$; whereas for all $i \in \mathcal{J}$, the restriction of π to M_i, $\pi|_{M_i} = \beta \circ \alpha$ where $\alpha : M_i \to Q_i$ is the natural direct summand projection and $\beta : Q_i \to \oplus_{j \in \mathcal{J}} Q_j$ is the canonical monomorphism. We claim that the homomorphism $f = \pi \circ g : A \to \oplus_{j \in \mathcal{J}} Q_j$ cannot be extended to a homomorphism $h : R \to \oplus_{j \in \mathcal{J}} Q_j$ along the monomorphism $\mu : A \to R$. In particular, we claim that $\oplus_{j \in \mathcal{J}} Q_j$ is not injective. Suppose to the contrary that f admits such an extension h. Since $h(1)$ is contained only in a finite direct subsum of $\oplus_{j \in \mathcal{J}} Q_j$, $Im(f)$ is contained in $\oplus_{j \in \mathcal{F}} Q_j$ for some finite subset $\mathcal{F}$ of $\mathcal{J}$. Thus, $\pi_j \circ f = 0$ for each $j \in \mathcal{J} \setminus \mathcal{F}$. But this is not possible as $\pi_j \circ f : A \to Q_j$ and each Q_j is an injective envelope of N_j in M_j.

Consider the set Ω of submodules P of Q satisfying the following three conditions:

1. $\oplus_{j \in \mathcal{J}} Q_j \subseteq P \subseteq Q$;
2. P is a direct sum of injective submodules of Q;
3. $f = \pi \circ g : A \to \oplus_{j \in \mathcal{J}} Q_j \subseteq P$ cannot be extended to a homomorphism $h : R \to P$ along the monomorphism $\mu : A \to R$.

Clearly, Ω is nonempty as $\oplus_{j \in \mathcal{J}} Q_j \in \Omega$. Define partial order $\leq$ on Ω as $P_1 \leq P_2$ if and only if $P_1 \subseteq P_2$. We claim that Ω is an inductive set under this partial order. Let $\{P_k\}_{k \in \mathcal{K}}$ be a chain in Ω. Let $P = \cup_{k \in \mathcal{K}} P_k$. As $\oplus_{j \in \mathcal{J}} Q_j \subseteq_e Q = E(\oplus_{j \in \mathcal{J}} Q_j)$, we have $\oplus_{j \in \mathcal{J}} Q_j \subseteq_e P$. Hence, $\oplus_{j \in \mathcal{J}} M_j \subseteq_e P$. By assumption, $P = (\oplus_{u \in \mathcal{U}} C_u) \oplus (\oplus_{v \in \mathcal{U}'} C'_v)$, where the C_u are injective modules and C'_v are projective modules. By Kaplansky [150], we know that each projective module is a direct sum of countably generated modules. Hence, we have $P = (\oplus_{u \in \mathcal{U}} C_u) \oplus (\oplus_{v \in \mathcal{V}} D_v)$, where each C_u is an injective module and each D_v, a countably generated module. Moreover, $\mathcal{U}$ and $\mathcal{V}$ are countable sets, because P contains a countably generated submodule $\oplus_{j \in \mathcal{J}} N_j$ such that $\oplus_{j \in \mathcal{J}} N_j \subseteq_e P$. Thus, $D = \oplus_{v \in \mathcal{V}} D_v$ is countably generated. We may write $D = \Sigma_{n \in \mathbb{N}} D'_n$ as a countable sum of finitely generated submodules. Since D'_1 is finitely generated, $D'_1 \subset \cup_{k \in \mathcal{F}} P_k$ for some finite subset $\mathcal{F} \subset \mathcal{K}$. Furthermore, since each P_k is a direct sum of injective submodules, P contains an injective hull $E(D'_1)$ of D'_1. Moreover, $E(D'_1) \cap (\oplus_{u \in \mathcal{U}} C_u) = 0$, because $D'_1 \cap (\oplus_{u \in \mathcal{U}} C_u) = 0$. Thus

$$E(D'_1) \cong \frac{(\oplus_{u \in \mathcal{U}} C_u) \oplus E(D'_1)}{\oplus_{u \in \mathcal{U}} C_u} \subseteq \frac{(\oplus_{u \in \mathcal{U}} C_u) \oplus (\oplus_{v \in \mathcal{V}} D_v)}{\oplus_{u \in \mathcal{U}} C_u} \cong \oplus_{v \in \mathcal{V}} D_v = D.$$

Clearly the above isomorphism fixes D'_1. Thus, D contains the injective hull $E(D'_1)$ of D'_1, and therefore we have a decomposition $D = E(D'_1) \oplus D''_1$. We denote by $D'_{1,n}$ the image of D'_n under the natural projection on D''_1 for $n \geq 2$. Set $D'_{1,1} = D'_1$ for simplicity. It is easy to check that $D = E(D'_{1,1}) \oplus \Sigma_{n \geq 2} D'_{1,n}$. This yields us a decomposition $P = (\oplus_{u \in \mathcal{U}} C_u) \oplus E(D'_{1,1}) \oplus \Sigma_{n \geq 2} D'_{1,n}$. By applying the same construction to P and $D'_{1,2}$ we get $P = (\oplus_{u \in \mathcal{U}} C_u) \oplus E(D'_{1,1}) \oplus E(D'_{2,2}) \oplus \Sigma_{n \geq 3} D'_{2,n}$. Repeating this process, we construct an infinite set $\{E(D'_{n,n})\}_{n \in \mathbb{N}}$ of injective submodules of P such that for each $m \in \mathbb{N}$, we have that $(\oplus_{u \in \mathcal{U}} C_u) \oplus (\oplus_{n=1}^{m} E(D'_{n,n})) \subseteq P$. Moreover, by construction, $D'_m \subseteq \oplus_{n=1}^{m} D'_{n,n}$, for each $m \in \mathbb{N}$. As a consequence, $D \subseteq \oplus_{n \in \mathbb{N}} E(D'_{n,n})$, so $P = (\oplus_{u \in \mathcal{U}} C_u) \oplus (\oplus_{n \in \mathbb{N}} E(D'_{n,n}))$. Thus P satisfies (2). Finally, we proceed to show that the homomorphism $f = \pi \circ g : A \to \oplus_{j \in \mathcal{J}} Q_j \subseteq P$ cannot be extended to a homomorphism $h : R \to P$ along the monomorphism $\mu : A \to R$. Suppose, if possible, that g admits such an extension h. Since $Im(h)$ is finitely generated and $\{P_k\}_{k \in \mathcal{K}}$ is a chain, there exists a $k \in \mathcal{K}$ such that $Im(h) \subseteq P_k$. This yields a contradiction because $P_k \in \Omega$ and therefore, by assumption, f cannot be extended to a homomorphism $R \to P_k$. Hence, $P \in \Omega$. This establishes our claim that Ω is an inductive set and hence by Zorn's Lemma, Ω has a maximal element, say P_0. By hypothesis, $P_0 = \oplus_{t \in \mathcal{T}} W_t$, where each W_t is injective. Let $\varphi_t : P_0 \to W_t$ be the canonical projections. Since, by hypothesis, f cannot be extended to a homomorphism $h : R \to P_0$, there exists an infinite subset $\mathcal{T}' \subseteq \mathcal{T}$ such that $\varphi_t \circ f \neq 0$, for each $t \in \mathcal{T}'$. Because otherwise $Im(f)$ would be contained in $\oplus_{\mathcal{F}} W_t$ where F is a finite set. Since $\oplus_{\mathcal{F}} W_t$ is injective, f would extend to a homomorphism $R \to \oplus_{\mathcal{F}} W_t \subseteq P_0$, yielding a contradiction. Let us write $\mathcal{T}$ as a disjoint union of infinite sets $\mathcal{T}_1$ and $\mathcal{T}_2$. Denote $\varphi_{\mathcal{T}_1} : \oplus_{t \in \mathcal{T}} W_t \to \oplus_{t \in \mathcal{T}_1} W_t$ and $\varphi_{\mathcal{T}_2} : \oplus_{t \in \mathcal{T}} W_t \to \oplus_{t \in \mathcal{T}_2} W_t$. Note that $\varphi_{\mathcal{T}_i} \circ f : A \to \oplus_{t \in \mathcal{T}_i} W_t$ cannot be extended to a homomorphism $h : R \to \oplus_{t \in \mathcal{T}_i} W_t$ for each $i \in \{1, 2\}$. Because otherwise $Im(h) \subset \oplus_{t \in \mathcal{F}} W_t$, where $\mathcal{F}$ is a finite set and hence $\varphi_t \circ f = \varphi_t \circ \varphi_{\mathcal{T}_i} \circ f = 0$, for each $t \in \mathcal{T}_i \setminus \mathcal{F}$, a contradiction. This implies that $\oplus_{t \in \mathcal{T}_1} W_t$ is not injective and hence $\oplus_{t \in \mathcal{T}_1} W_t \neq E(\oplus_{t \in \mathcal{T}_1} W_t)$. Thus, $P_0 = \oplus_{t \in \mathcal{T}} W_t \subsetneq E(\oplus_{t \in \mathcal{T}_1} W_t) \oplus (\oplus_{t \in \mathcal{T}_2} W_t)$. Now, it may be observed that f cannot be extended to a homomorphism $R \to E(\oplus_{t \in \mathcal{T}_1} W_t) \oplus (\oplus_{t \in \mathcal{T}_2} W_t)$, because otherwise $\varphi_{\mathcal{T}_2} \circ f$ would extend to a homomorphism $R \to \oplus_{t \in \mathcal{T}_2} W_t$, a contradiction. Therefore, $E(\oplus_{t \in \mathcal{T}_1} W_t) \oplus (\oplus_{t \in \mathcal{T}_2} W_t) \in \Omega$. But this yields a contradiction to the maximality of P_0. Hence, M must be Σ-injective.

The converse is obvious because if M is Σ-injective, then $M^{(\aleph_0)}$ is injective and hence $M^{(\aleph_0)}$ has no proper essential extension. Thus the statement that each essential extension of $M^{(\aleph_0)}$ is a direct sum of modules that are either injective or projective holds trivially. This completes the proof of equivalence of (i) and (iii).

The equivalence of (i) and (ii) follows immediately from the equivalence of (i) and (iii). We refer the reader to Faith [70] for the proof of the equivalence of (i) and (iv). $\qquad\qquad\square$

An internal direct sum $\oplus_{i \in \mathcal{I}} A_i$ of submodules of a module M is called a *local summand* of M, if given any finite subset $\mathcal{F}$ of $\mathcal{I}$, the direct sum $\oplus_{i \in \mathcal{F}} A_i$ is a direct summand of M.

Let $M = \oplus_{i \in \mathcal{I}} M_i$ be a decomposition of the module M into nonzero summands M_i. This decomposition is said to *complement direct summands* if, whenever A is a direct summand of M, there exists a subset $\mathcal{J}$ of $\mathcal{I}$ for which $M = (\oplus_{j \in \mathcal{J}} M_j) \oplus A$.

Now we are ready to prove the following.

Corollary 6.18 *An arbitrary right R-module M is Σ-injective if and only if each essential extension of $M^{(\aleph_0)}$ is a direct sum of injective modules.*

Proof Suppose each essential extension of $M^{(\aleph_0)}$ is a direct sum of injective modules. Let $E = E(M)$. We have $M^{(\aleph_0)} \subseteq_e E^{(\aleph_0)}$. By assumption, $M^{(\aleph_0)}$ itself is a direct sum of injective modules. Therefore, $M^{(\aleph_0)}$ is a local summand of $E^{(\aleph_0)}$. Since by above theorem, E is Σ-injective, so is $E^{(\aleph_0)}$. Hence $E^{(\aleph_0)}$ has an indecomposable decomposition that complements direct summands. Therefore, any local summand of $E^{(\aleph_0)}$ is a direct summand (see [49]). Hence, $M^{(\aleph_0)}$ is a direct summand $E^{(\aleph_0)}$. Therefore, $M^{(\aleph_0)}$ is injective and thus M is Σ-injective. The converse is obvious. $\qquad\square$

Goursaud and Valette [102] initiated the study of Σ-V rings. Srivastava continued the study of Σ-V rings in [212]. Clearly, every right Σ-V ring is a right V-ring. However, there are examples of right V-rings which are not right Σ-V rings. It follows from the Theorem 6.17(3) that a commutative V-ring must be a Σ-V ring.

It is well-known that a ring R is right noetherian if and only if each injective right R-module is Σ-injective [20]. Thus, a right noetherian right V-ring is a right Σ-V ring. In particular, any right PCI ring is a right Σ-V ring. Since in a right Σ-V ring we require each simple module to be Σ-injective, we expect some kind of finiteness in such rings. We begin with some useful definitions and observations.

A ring R is called right *q.f.d.* (*quotient finite-dimensional*) *relative to a module M* if no cyclic right R-module contains an infinite direct sum of modules isomorphic to submodules of M.

Lemma 6.19 *If a right R-module M is Σ-injective, then R is right q.f.d. relative to M.*

Proof Assume to the contrary that R is not right *q.f.d.* relative to M. Then there exists a cyclic right module C with an infinite independent family $\{V_i : i \in \mathcal{I}\}$ of nonzero submodules of C such that each V_i is isomorphic to a submodule of M and $\oplus_{i \in \mathcal{I}} V_i$ is essential in C. Set $M_i = M$, $i \in \mathcal{I}$. Since M is Σ-injective, the monomorphism $\varphi : \oplus_{i \in \mathcal{I}} V_i \longrightarrow \oplus_{i \in \mathcal{I}} M_i$ such that $\varphi(V_i) \subseteq M_i$ for all $i \in \mathcal{I}$ extends to a monomorphism $f : C \longrightarrow \oplus_{i \in \mathcal{I}} M_i$. Now, since C is cyclic, there exists a finite subset $\mathcal{J} \subseteq \mathcal{I}$ such that $f(C) \subseteq \oplus_{j \in \mathcal{J}} M_j$. Therefore $f(V_k) \cap M_k \subseteq f(C) \cap M_k = 0$ for all $k \notin \mathcal{J}$, a contradiction to the fact that $f(V_i) = \varphi(V_i) \subseteq M_i$ for all i.

Thus R is right *q.f.d.* relative to M. $\qquad\square$

Lemma 6.20 *Let R be a right Σ-V ring and let S be any simple right R-module. Then R is right q.f.d. relative to S.*

Proof This follows from the above lemma. $\square$

Recall that a ring R is called *directly finite* if for each $x, y \in R$, $xy = 1$ implies $yx = 1$. It is not difficult to observe that a ring R with finite right Goldie dimension must be directly finite. The standard technique to prove this is to assume to the contrary that R is not directly finite. This gives rise to an infinite set of orthogonal idempotents in R yielding an infinite direct sum of right ideals contained in R, a contradiction to the finite right Goldie dimension of R. In the next result we generalize this argument.

Lemma 6.21 *Let R be a ring which is right q.f.d. relative to every simple R-module. Then R must be directly finite.*

Proof Assume to the contrary that R is not directly finite. Then there exist $x, y \in R$ such that $xy = 1$ and $yx \neq 1$. Set $e_{ij} = y^{i-1}x^{j-1} - y^i x^j$ for all $(i, j) \in \mathbb{N} \times \mathbb{N}$. Then $\{e_{ij} : (i, j) \in \mathbb{N} \times \mathbb{N}\}$ is an infinite set of nonzero matrix units, that is $e_{ij}e_{kl} = \delta_{jk}e_{il}$ for all $i, j, k, l \in \mathbb{N}$. For each $n \in \mathbb{N}$, we will produce cyclic submodules $C_{n,i}$ of R_R, where $C_{n,i} \cong C_{n,j}$ for all $i, j = 2, 3, \ldots, n$. We produce these cyclic submodules by induction.

Let $n > 1$. Let us denote by Q, the maximal right ring of quotients of R. Since $R \subseteq_e Q$, there exists a nonzero cyclic submodule $C_{n,1}$ of R_R such that $C_{n,1} \subseteq e_{n^2,n^2}Q \cap R$. Now we choose $0 \neq x_2 \in e_{n^2+1,n^2}C_{n,1} \cap R$. Then $x_2 = e_{n^2+1,n^2}x_1$, where $x_1 \in C_{n,1}$. Denote $C_{n,2} = x_2R$ and redefine $C_{n,1}$ by setting $C_{n,1} = x_1R$. Define the module homomorphism $\varphi : C_{n,1} \longrightarrow C_{n,2}$ by $\varphi(x) = e_{n^2+1,n^2}x$. Clearly, φ is an isomorphism (with inverse given by left multiplication by e_{n^2,n^2+1}), and so $C_{n,1} \cong C_{n,2}$. Suppose now that we have defined cyclic submodules $C_{n,1} \cong C_{n,2} \cong \ldots \cong C_{n,j-1}$ in R, where $C_{n,i} = x_iR$, $i = 1, 2, \ldots, j - 1$. Next, we choose x_j such that $x_j \in e_{n^2+j-1,n^2+j-2}C_{n,j-1} \cap R$ and write $x_j = e_{n^2+j-1,n^2+j-2}x_{j-1}r_{j-1}$ where $r_{j-1} \in R$. Let $x'_{j-1} = x_{j-1}r_{j-1}$, and set $C_{n,j} = x_jR$. Now redefine $C_{n,j-1} = x'_{j-1}R$ (which is contained in the previously constructed $C_{n,j-1}$). Then $C_{n,j-1} \cong C_{n,j}$ under the isomorphism that sends $x \in C_{n,j-1}$ to $e_{n^2+j-1,n^2+j-2}x$. We redefine preceding $C_{n,1}$, $C_{n,2}, \ldots, C_{n,j-2}$ accordingly so that they all remain isomorphic to each other and to $C_{n,j-1}$. Note that the family $\{C_{n,i} : n = 2, 3, \ldots, \ i = 1, 2, \ldots, n\}$ is independent since $\{e_{ij}Q : i, j \in \mathbb{N} \times \mathbb{N}\}$ is independent. By our construction, $C_{n,i} \cong C_{n,j}$ for all $n = 2, 3, \ldots$ and $1 \leq i, j \leq n$. Therefore there exist maximal submodules $M_{n,i}$ of $C_{n,i}$, $n = 2, 3, \ldots$ and $1 \leq i \leq n$, such that $C_{n,i}/M_{n,i} \cong C_{n,j}/M_{n,j}$ for all n, i, j. Set $M = \oplus_{n,i}M_{n,i}$, and $S_n = C_{n,1}/M_{n,1}$. Clearly, we have $R/M \supset (\oplus C_{n,i})/(\oplus M_{n,i}) \cong C_{2,1}/M_{2,1} \times C_{2,2}/M_{2,2} \times C_{3,1}/M_{3,1} \times \cdots$. Thus R/M is a cyclic right R-module that contains an infinite direct sum of modules each isomorphic to the simple module S_n. This yields a contradiction to our hypothesis. Therefore R must be directly finite. $\square$

Theorem 6.22 *Every right Σ-V ring is directly finite.*

Proof Let R be a right Σ-V ring. Let S be any simple right R-module. Then, by Lemma 6.20, R is right $q.f.d.$ relative to S. Therefore, by Lemma 6.21, R must be directly finite.

We give a direct shorter proof below.

Assume to the contrary that R is not directly finite. Then $R = R_1 \oplus A_1$, $R_1 = R_2 \oplus A_2$, and so on, where for each n, $R_n \cong R_R$, and $A_1 \cong A_2 \cong \cdots$. Then we can find, in each A_n, a maximal submodule M_n such that A_n/M_n are all isomorphic to the same simple module. Then, by assumption, $\oplus_n A_n/M_n$ is injective and hence it splits in $R/(\oplus_n M_n)$, a contradiction, since $R/(\oplus_n M_n)$ is cyclic. Therefore R is directly finite. $\qquad\square$

Example A right V-ring need not be directly finite. Let K be a field and let T be the endomorphism ring of a countably infinite-dimensional K-vector space V. Let M be the maximal ideal of T, that is, M consists of those endomorphisms of V whose images are of finite dimension. Set $A = T/M$ and write $\bar{x}$ for $x + M$. Let $\{e_i\}_{i \geq 0}$ be a K-basis for V and define $a, b \in T$ by $a(e_0) = 0$, $a(e_{i+1}) = e_i$ and $b(e_i) = e_{i+1}$ for all $i \geq 0$. Clearly $ab = 1$ and $(ba - 1)(V) = Ke_0$, consequently $\bar{a}$ is a unit in A. Let $p(x) \in K[x]$ be any nonzero polynomial. We claim that the element $p(\bar{a})$ is a unit in A. Since $\bar{a}$ is a unit, we may assume that $p(x) = \alpha_0 + \cdots + \alpha_n x^n$, $\alpha_i \in K$ and $\alpha_0 \neq 0$. But then $p(x)$ has an inverse $\Sigma_{i \geq 0} \beta_i x^i \in K[[x]]$. Note that $a^m(e_i) = 0$ for all $m > i$ so that $\Sigma_{i \geq 0} \beta_i a^i$ defines an element of T which is a two-sided inverse of $p(a)$. Thus we have shown that F, the field of quotients of $K[\bar{a}]$, is contained in A. Define R to be the subring of T such that $R/M = F$. It may be shown that R is a right V-ring but R is not directly finite.

This is an example of a ring which is a right V-ring but not a right (or left) Σ-V ring. $\qquad\square$

Recall that a von Neumann regular ring is called an *abelian regular ring* if all its idempotents are central. An idempotent e in a von Neumann regular ring R is called an *abelian (a directly finite) idempotent* if the corner ring eRe is abelian (directly finite). An idempotent e in a regular right self-injective ring is called a *faithful idempotent* if 0 is the only central idempotent orthogonal to e, that is, $ef = 0$ implies $f = 0$ when f is a central idempotent.

Let R be a von Neumann regular right self-injective ring. The ring R is said to be of Type I provided it contains a faithful abelian idempotent. The ring R is said to be of Type II provided R contains a faithful directly finite idempotent but R contains no nonzero abelian idempotents. The ring R is said to be of Type III if it contains no nonzero directly finite idempotents. The ring R is said to be of (i) Type I_f if R is of Type I and is directly finite, (ii) Type I_∞ if R is of Type I and is purely infinite, (iii) Type II_f if R is of Type II and is directly finite, (iv) Type II_∞ if R is of Type II and is purely infinite (see [97], pp. 111–115). If R is a von Neumann regular right self-injective ring of Type I_f then $R \simeq \Pi R_n$

where each R_n is an $n \times n$ matrix ring over an abelian regular self-injective ring (see [97], Theorem 10.24).

The *index of nilpotence* of a nilpotent element x in a ring R is the least positive integer n such that $x^n = 0$. The index of nilpotence of a two-sided ideal I in R is the supremum of the indices of nilpotence of all nilpotent elements of I. If this supremum is finite, then I is said to have *bounded index of nilpotence.*

The next lemma provides the structure of maximal right ring of quotients of any right nonsingular ring which is right *q.f.d.* relative to each of its simple modules.

Lemma 6.23 *Let R be a right nonsingular ring which is right q.f.d. relative to every simple R-module. Then $Q^r_{\max}(R)$, the maximal right ring of quotients of R, is a finite direct product of matrix rings over abelian regular self-injective rings.*

Proof We know that $Q = Q^r_{\max}(R)$ is a von Neumann regular right self-injective ring. By Lemma 6.20, the ring R is directly finite and hence so is the ring Q. Hence by the type theory of von Neumann regular right self-injective rings, $Q = Q_1 \times Q_2$ where Q_1 is of Type I_f and Q_2 is of Type II_f (see [97], Theorem 10.22). Now we claim that Q must be of type I_f, that is, $Q_2 = 0$. Assume to the contrary that $Q_2 \neq 0$. Then, by ([97], Proposition 10.28), there exists an idempotent $e_3' \in Q_2$ such that $(Q_2)_{Q_2} \cong 3(e_3'Q_2)$. Therefore $Q_2 = e_1 Q_2 \oplus e_2 Q_2 \oplus e_3 Q_2$ where $e_1, e_2, e_3 \in Q_2 \subseteq Q$ are nonzero orthogonal idempotents such that their sum is the identity of the ring Q_2. Clearly, $e_i Q = e_i Q_2$, and $e_j Q_2 = e_j Q$ for all $1 \leq i, j \leq 3$ and so $e_i Q \cong e_j Q$ for all $1 \leq i, j \leq 3$. Therefore there exist nonzero cyclic submodules $C_{2i} \subseteq e_i Q \cap R$, $i = 1, 2$, such that $C_{21} \cong C_{22}$. By ([97], Corollary 10.9), $e_3 Q_2 e_3 = \operatorname{End}(e_3 Q_2)$ is of Type II_f and so as above there exist nonzero orthogonal idempotents $f_1, f_2, f_3, f_4 \in e_3 Q_2 e_3$ such that $f_i(e_3 Q_2 e_3) \cong f_j(e_3 Q_2 e_3)$ for all $1 \leq i, j \leq 4$. Hence $f_i(e_3 Q e_3) \cong f_j(e_3 Q e_3)$ for all i, j. By ([160], Proposition 21.20), there exist $a \in f_i(e_3 Q e_3)f_j$ and $b \in f_j(e_3 Q e_3)f_i$ such that $f_i = ab$ and $f_j = ba$. Then, for all i, j, we have $f_i Q \cong f_j Q$ under the mapping which sends $f_i x$ to $b f_i x$ for each $x \in Q$. Furthermore, there exist nonzero cyclic submodules $C_{3i} \subseteq f_i Q \cap R$, $i = 1, 2, 3$ such that $C_{3i} \cong C_{3j}$ for all $1 \leq i, j \leq 3$. Continuing in this fashion, we construct an independent family $\{C_{ij} : i = 2, 3, \ldots; 1 \leq j \leq i\}$ of nonzero cyclic submodules of R such that $C_{ij} \cong C_{ik}$ for all $1 \leq j, k \leq i$; $i = 2, 3, \ldots$. Therefore there exist maximal submodules M_{ij} of $C_{ij}, 1 \leq j \leq i; i = 2, 3, \ldots$ such that $C_{ij}/M_{ij} \cong C_{ik}/M_{ik}$ for all i, j, k. Setting $M = \oplus_{i,j} M_{ij}$, and $S_i = C_{i1}/M_{i1}$, we get that the cyclic right module R/M contains an infinite direct sum of modules each isomorphic to S_i, a contradiction to our hypothesis. Therefore $Q_2 = 0$ and Q is of Type I_f. Hence $Q = \Pi_{i=1}^{\infty} Q_i$ where each Q_i is an $i \times i$ matrix ring over an abelian regular self-injective ring (see [97], Theorem 10.24). Now we claim that this product must be a finite product. Assume to the contrary that the product is infinite. Then, for any positive integer n, there exists an index $m \geq n$ such that $Q_m \neq 0$. Now, for any fixed k, we have matrix units $\{e_{ij}^k : 1 \leq i, j \leq k\}$ which are $k \times k$ matrices.

Thus we have an infinite family of nonzero matrix units $\{\{e_{ij}^k\colon 1 \le i,j \le k\}\colon k = 2,3,\ldots\} \subseteq Q$. Since $R \subseteq_e Q$, there exists a nonzero cyclic submodule $C_{k,1}$ of R such that $C_{k,1} \subseteq e_{1,1}^k Q \cap R$ and then, starting with $C_{k,1}$, we construct an independent family $\{C_{k,i}\colon k = 2,3,\ldots,\; 1 \le i \le k\}$ of cyclic submodules of R such that $C_{k,i} \cong C_{k,j}$ for all k,i,j (exactly as shown in the proof of part (a)). Therefore there exist maximal submodules $M_{k,i}$ of $C_{k,i}$, $k = 2,3,\ldots$ and $1 \le i \le k$, such that $C_{k,i}/M_{k,i} \cong C_{k,j}/M_{k,j}$ for all k,i,j. Set $M = \oplus_{k,i} M_{k,i}$, and $S_k = C_{k,1}/M_{k,1}$. Note that the cyclic right module R/M contains an infinite direct sum of modules each isomorphic to S_k, which contradicts our hypothesis. Therefore $Q = Q_{\max}^r(R)$ is the direct product of a finite number of matrix rings over abelian regular self-injective rings. $\qquad\square$

Theorem 6.24 *Let R be a right nonsingular, right Σ-V ring. Then the maximal right ring of quotients of R, $Q_{\max}^r(R)$ is a finite direct product of matrix rings over abelian regular self-injective rings.*

Proof This follows from Lemma 6.20 and Lemma 6.23. $\qquad\square$

As a consequence of the above, we have the following.

Corollary 6.25 *Let R be a right nonsingular right Σ-V ring. Then R must have bounded index of nilpotence.*

Proof By the above theorem, the maximal right ring of quotients $Q_{\max}^r(R)$ has bounded index of nilpotence and hence R must have likewise. $\qquad\square$

A right nonsingular right V-ring need not have bounded index of nilpotence. The following result of Tyukavkin [232] gives a condition for a von Neumann regular right self-injective right V-ring to have bounded index of nilpotence.

Theorem 6.26 *(Tyukavkin, [232]) Let R be a von Neumann regular right self-injective right V-ring such that the dimension of every simple right R-module S over the division ring of endomorphisms of S is less than $2^{2^{\aleph_0}}$. Then R has bounded index of nilpotence.*

Next we discuss von Neumann regular right Σ-V rings.

Lemma 6.27 *Let R be a von Neumann regular right self-injective right Σ-V ring. Then R is a finite direct product of matrix rings over abelian regular self-injective rings.*

Proof Since von Neumann regular rings are right nonsingular, this follows from Lemma 6.21(b). $\qquad\square$

The next theorem characterizes von Neumann regular right Σ-V rings.

Theorem 6.28 *For a von Neumann regular ring R, the following are equivalent:*

1. R is a right Σ-V ring.
2. For any prime ideal P, R/P is artinian.
3. R is a left Σ-V ring.

Proof (1)$\Rightarrow$(2). Let R be a von Neumann regular right Σ-V ring. Then by Corollary 6.25 R has bounded index of nilpotence. Let P be any prime ideal of R. Then R/P is a prime von Neumann regular ring with bounded index of nilpotence. Hence R/P must be simple artinian (see [97], Theorem 7.9).

(2)$\Rightarrow$(1). Let S be a simple right R-module. Let $P = ann_r(S)$. By assumption R/P is artinian. Therefore S is Σ-injective as an R/P-module. Since R is a von Neumann regular ring, R/P is flat as a right R-module. Then by ([97], Lemma 6.17), S is Σ-injective as a right R-module as well. Hence R is a right Σ-V ring.

The proof of the equivalence of (2) and (3) is similar. $\square$

Remark 6.29 *The above theorem shows that the class of von Neumann regular Σ-V rings is left–right symmetric.*

Proposition 6.30 *Let R be a right nonsingular prime right Σ-V ring. Then R is a simple right Goldie ring.*

Proof By Corollary 6.25, $Q^r_{\max}(R)$ has bounded index of nilpotence. Thus $Q^r_{\max}(R)$ is a prime von Neumann regular ring with bounded index of nilpotence, hence simple artinian (see [97], Theorem 7.9). Therefore R is a simple right Goldie ring. $\square$

Next, we have the following characterization of right Σ-V rings.

Theorem 6.31 *A ring R is a right Σ-V ring if and only if for each simple right R-module S, each essential extension of $S^{(\aleph_0)}$ is a direct sum of injective modules.*

Proof This follows from Corollary 6.18. $\square$

6.4 *CSI* rings

A ring R is called a *right CSI ring* if, for each cyclic right R-module C, the injective hull $E(C)$ is Σ-injective. It is still an open question whether every right CSI ring is right noetherian. In the following theorem Faith [80] showed that right CSI rings are right noetherian in some special cases.

First, recall that a ring R is called a right *Kasch ring* if it contains a copy of every simple right R-module. A right ideal I of R is called *colocal* if R/I is subdirectly irreducible.

Theorem 6.32 *A right CSI ring R is right noetherian under any of the following conditions:*

1. *R is commutative.*
2. *R has only finitely many simple modules up to isomorphism, for example R is semilocal.*
3. *R satisfies the ACC on colocal right ideals.*
4. *R or $R/J(R)$ is right Kasch.*

5. $R/J(R)$ *is von Neumann regular, for example R is right continuous.*

6. *The injective hull of any countably generated semisimple right R-module is Σ-injective.*

Proof It follows from Theorem 6.17(iv) that if R is a right CSI ring then R/I is right Goldie for any ideal I of R. A well-known result due to Camillo [39] says that a commutative ring R is noetherian if R/I is a Goldie ring for any ideal I of R. This proves (1).

Note that (2) follows easily from a result of Kurshan [159] that says a ring R is right noetherian if and only if every countable direct sum of injective hulls of simple right R-modules is injective.

Now we will prove (5). By Lemma 6.19, it follows that a right CSI ring is right qfd. Now assume $R/J(R)$ is von Neumann regular. Since a von Neumann regular ring with finite Goldie dimension is semisimple artinian, we have that a right CSI ring R such that $R/J(R)$ is von Neumann regular must be semilocal and hence the result follows from (2).

The other parts are easy to prove. $\square$

A nonzero module M is called *hollow* if every proper submodule of M is small in M. A hollow module M is called *special-hollow* if M is nonlocal with simple essential socle.

Example Any local module is clearly an example of a hollow module. For any prime p, the $\mathbb{Z}$-module $\mathbb{Z}_{p^\infty}$ is a hollow module which is not local. $\square$

Recently N. Er [68] has shown that a right CSI ring is right noetherian if R satisfies any of the following conditions: (i) there are finitely many simple right R-modules that are socles of special-hollow subfactors of R_R, (ii) there are finitely many injective hulls of special-hollow subfactors of R_R.

In [42], Chairat, Huynh, and Somsup have studied a related notion. They call a ring R a right CSE ring if, for each cyclic right R-module C, the injective hull $E(C)$ is Σ-CS. They show that the class of right CSI rings and the class of right CSE rings coincide.

6.33 Questions

1. Is every right CSI ring right noetherian?
2. We have already seen that the von Neumann regular Σ-V rings are left–right symmetric. Which other classes of Σ-V rings are left–right symmetric?
3. Does every exchange right Σ-V ring have bounded index of nilpotence?
4. Characterize a simple right Goldie right Σ-V ring.
5. Let $R = K[t, \sigma, \delta]$ be a twisted differential polynomial ring over a division ring K. Suppose R is a left V-domain. Then R is a right PCI-domain if and only if σ is onto. Does there exist an example when R is a left V-domain with σ not onto?

7 Rings each of whose (proper) cyclic modules is quasi-injective

7.1 Rings each of whose cyclic modules is quasi-injective

Recall that a module M is said to be *quasi-injective* if M is M-injective. A module M is called *π-injective* if the following two conditions hold: (i) every submodule of M is essential in a direct summand of M, and (ii) if N_1 and N_2 are direct summands of M with $N_1 \cap N_2 = 0$ then $N_1 \oplus N_2$ is also a direct summand of M. It is well-known that every quasi-injective module is continuous; and every continuous module is π-injective (see p. 10).

In [73] Faith proposed the study of the class of rings which satisfy the following condition:

(P) *Every proper cyclic module is injective modulo its annihilator ideal.*

Clearly, every right PCI-ring satisfies condition (P). Commutative rings with condition (P) are precisely the restricted self-injective rings. It may be observed that in a right self-injective ring with condition (P), each cyclic module is quasi-injective. A ring R is called a *qc-ring* if each cyclic R-module is quasi-injective. These rings were studied by Ahsan [6] and Koehler ([156], [157]) among others. Koehler provided a complete characterization of qc rings.

First, we will prove a more general result that holds for rings over which cyclic modules are π-injective. The reader will come across the application of the lemma at several places. To prove the lemma, we need the following.

Lemma 7.1 *Let A and B be R-modules such that A is embeddable in B. If $A \oplus B$ is π-injective then the exact sequence $0 \to A \to B$ splits.*

Proof Let $M = A \oplus B$ be a π-injective module and $\sigma : A \to B$ be an embedding. Then $A' = \{a + \sigma(a) : a \in A\}$ is a direct summand of M and $A \cap A' = 0$. Thus $A + A'$ is also a summand. But $A + A' = A + \sigma(A)$. Hence $\sigma(A)$ is a summand of M. But $\sigma(A) \subseteq B$, hence $\sigma(A)$ is a summand of B. $\qquad\square$

Lemma 7.2 *Let R be a ring over which each cyclic module is π-injective. If e and f are indecomposable orthogonal idempotents in R such that $eRf \neq 0$, equivalently $\mathrm{Hom}_R(fR, eR) \neq 0$, then eR and fR are isomorphic minimal right ideals of R.*

Proof Let $0 \neq eaf \in eRf$. Now $eafR \oplus eR \cong (fR/(ann_r(ea) \cap fR)) \oplus eR \cong (e+f)R/(ann_r(ea) \cap fR)$. This gives by hypothesis, $eafR \oplus eR$ is π-injective.

Since $eafR \subset eR$, and eR is indecomposable, by the above lemma, $eafR = eR$. Then $eR \cong fR/(ann_r(ea) \cap fR)$ gives $eR \cong fR$. To prove that eR is minimal, let $0 \neq eb \in eR$. If $eb(1 - e) \neq 0$, then as before $eR = eb(1 - e)R$. This implies $eR = ebR$. On the other hand, if $eb(1 - e) = 0$, then $ebe \neq 0$. Since $ebeR \oplus fR = (ebe + f)R$, $ebeR \oplus fR$ is π-injective. Furthermore, $eR \cong fR$ implies $ebeR \oplus fR \cong ebeR \oplus eR$. Therefore $ebeR \oplus eR$ is π-injective, and so as explained earlier $ebeR = eR$. This proves eR is a minimal right ideal. $\qquad\square$

Recall that a ring R is said to have rank 0 if every prime ideal of R is maximal. Now we are ready to prove the theorem of Koehler.

Theorem 7.3 *For a ring R the following are equivalent:*

1. *Each cyclic right R-module is quasi-injective.*
2. *$R = A \oplus B$ where A is semisimple artinian and B is a finite direct sum of self-injective, rank 0, uniserial, duo rings with nil Jacobson radical.*
3. *Each cyclic left R-module is quasi-injective.*

Proof If R is prime then, by Lemma 5.1, R is either simple artinian or an Ore domain. But because R is also right self-injective, R is simple artinian in both cases. This implies that each prime ideal is maximal. By the same argument as in Theorem 5.2, R is semiperfect. If e_iR and e_jR are indecomposable quasi-injective (hence uniform) right ideals such that $e_iR \oplus e_jR$ is quasi-injective and $\mathrm{Hom}_R(e_iR, e_jR) \neq 0$, then by Lemma 7.2, e_iR and e_jR are isomorphic minimal right ideals. Write $R = e_1R \oplus \cdots \oplus e_nR$ as a direct sum of indecomposable quasi-injective right ideals. We can group all isomorphic indecomposable right ideals as explained above, and write after renumbering, if necessary, $R = [e_1R] \oplus \cdots \oplus [e_kR]$ as a direct sum of ideals, where each bracket represents the sum of indecomposable right ideals isomorphic to the term in the bracket. By taking endomorphism rings of R-modules on both sides, we obtain $R = M_{n_1}(e_1Re_1) \oplus \cdots \oplus M_{n_k}(e_kRe_k)$, where each $M_{n_i}(e_iRe_i)$ represents the $n_i \times n_i$ matrix ring over the division ring e_iRe_i. So for $n_i \geq 2$, the matrix ring $M_{n_i}(e_iRe_i)$ is simple artinian. It then follows that all matrix rings in the decomposition of R are simple artinian except those which are local rings. It is easy to show a right self-injective local ring with nil Jacobson radical is duo. This proves (1)$\Rightarrow$(2). The proof of (2)$\Rightarrow$(1) is straightforward. By the right–left symmetry of (2), we conclude that the proof of the equivalence (2) and (3) is similar. $\qquad\square$

7.2 Rings each of whose proper cyclic modules is quasi-injective

Jain, Singh, and Symonds [147] generalized the notion of right qc-ring and studied rings where each proper cyclic module is quasi-injective. A ring R is called a *right PCQI-ring* if each proper cyclic right R-module is quasi-injective. Clearly rings with property (P) mentioned at the beginning of this chapter are right $PCQI$-rings.

Proposition 7.4 *Let R be a right $PCQI$-ring. If I is a right ideal of R such that $R/I \cong R$, then I is contained in every nonzero two-sided ideal of R.*

Proof Let S be a nonzero two-sided ideal of R. Then R/S is a qc-ring and hence R/S is semiperfect. Let $f : R/I \to R$ be an isomorphism. Since $1 + I$ generates R/I, $R = xR$, where $x = f(1 + I)$. Then $I = ann_r(x) = \{r \in R : xr = 0\}$. So there exists $y \in R$ such that $xy = 1$. Since R/S is semiperfect, $(x + S)(y + S) = 1 + S = (y + S)(x + S)$. Then $1 - yx \in S$. Let $a \in I$. Then $(1 - yx)a = a - yxa = a$, hence $a \in S$. This shows $I \subseteq S$. $\qquad\square$

We have already proved the following.

Theorem 7.5 *If R is a right $PCQI$-domain, then R is a right Ore domain.*

Proof See Lemma 5.1. $\qquad\square$

Proposition 7.6 *Let R be a right $PCQI$-ring. Then either R is prime or R is semiperfect with nil Jacobson radical.*

Proof Suppose R is not prime. Let $P \neq 0$ be a prime ideal of R. Then R/P is a qc-ring, and hence each right ideal of R/P is quasi-injective. Since a prime ring with the property that each right ideal is quasi-injective must be simple artinian (to be shown in Corollary 19.8), it follows that R/P is simple artinian. Thus P is maximal and hence primitive. So $J(R)$ is nil. Since R is not prime, there exist nonzero ideals A and B such that $AB = 0$. Since R is a right $PCQI$-ring, R/A and R/B are semiperfect, hence each of them has finitely many prime ideals. Since every prime ideal of R contains A or B, it follows that R has finitely many prime ideals as well. Thus $R/J(R)$ is semisimple artinian, and since $J(R)$ is already shown above to be nil, R is semiperfect. $\qquad\square$

Lemma 7.7 *Let R be a nonlocal semiperfect right $PCQI$-ring, and let $R = \oplus_{i=1}^{n} e_i R$ where $\{e_i : 1 \leq i \leq n\}$ is a maximal set of primitive orthogonal idempotents in R with $n \geq 2$.*

  1. If $\sigma \in \mathrm{Hom}_R(e_i R, e_j R)$ such that $\sigma \neq 0$, where $i \neq j$, then $\mathrm{Ker}(\sigma) = 0$.
  2. If $n > 2$, then $\mathrm{Hom}_R(e_i R, e_j R) \neq 0$ if and only if $e_i R \cong e_j R$.
  3. If $n > 2$, then R is a qc-ring.

Proof

(1) Suppose $\mathrm{Ker}(\sigma) \neq 0$. Then $R/\mathrm{Ker}(\sigma) \cong \oplus_{k=1, k \neq i}^{n} e_k R \times Im(\sigma)$, and $R/\mathrm{Ker}(\sigma)$ is quasi-injective. Since $Im(\sigma) \subseteq e_j R$, the inclusion map $i : Im(\sigma) \to \oplus_{k=1, k \neq i}^{n} e_k R$ is a monomorphism. Since $R/\mathrm{Ker}(\sigma)$ is quasi-injective, the inclusion map splits. So $Im(\sigma)$ is a direct summand of $e_j R$. Hence $Im(\sigma) = e_j R$. Since $e_j R$ is projective, $\sigma : e_i R \to e_j R$ splits. Thus $\mathrm{Ker}(\sigma) = 0$.

(2) Let $\sigma \in \text{Hom}_R(e_i R, e_j R)$ such that $\sigma \neq 0$. By part (1), $\text{Ker}(\sigma) = 0$. Since $n > 2$, $e_i R \oplus e_j R \cong R/(\oplus_{k=1, k \neq i,j}^{n} e_k R)$ is quasi-injective. Then σ splits, and $0 \neq Im(\sigma)$ is a direct summand of $e_j R$. So $Im(\sigma) = e_j R$, and σ is an isomorphism. The converse is trivial.

(3) For each i, $e_i R \cong R/(\oplus_{k=1, k \neq i}^{n} e_k R)$. So $e_i R$ is quasi-injective for each i. Let A_i be the sum of all those $e_i R$ which are isomorphic to each other. Then $R = \oplus_{i=1}^{m} A_i$. We claim that each A_i is a two-sided ideal of R. Clearly, A_i is a right ideal. Consider $e_j R$ such that $e_j R \not\subseteq A_i$. Define $f : e_i R \to e_j R$, where $e_i R \subseteq A_i$, by $f(e_i r) = e_j x e_i r$, for $x \in R$. Then $f \in \text{Hom}_R(e_i R, e_j R)$. Since $e_i R$ and $e_j R$ are not isomorphic, $f = 0$ by part (2) above. So, for $e_j R \subsetneq A_i$, $e_j R A_i = 0$. Therefore $R A_i \subset A_i$. Since A_i is a finite direct sum of isomorphic quasi-injective right ideals, A_i is quasi-injective. Thus R is a qc-ring. $\qquad\square$

Lemma 7.8 *Let R be a semiperfect right PCQI-ring such that $R = e_1 R \oplus e_2 R$.*

1. If $e_1 R \cong e_2 R$, then R is a qc-ring.
2. If $e_1 R e_2 = 0$ and $e_2 R e_1 = 0$, then R is a qc-ring.
3. If $e_1 R e_2 \neq 0$ and $e_2 R e_1 \neq 0$, then R is a qc-ring.

Proof

(1) Since both $e_1 R$ and $e_2 R$ are quasi-injective and isomorphic to each other, $R = e_1 R \oplus e_2 R$ is quasi-injective. Thus R is right self-injective and hence a qc-ring.

(2) If $e_1 R e_2 = 0$ and $e_2 R e_1 = 0$, then $e_1 R$ and $e_2 R$ are two-sided ideals of R. Thus $e_1 R \cong R/e_2 R$ and $e_2 R \cong R/e_1 R$ are qc-rings. Then R is a qc-ring.

(3) If $e_1 R e_2 \neq 0$ and $e_2 R e_1 \neq 0$, then there exist nonzero homomorphisms, hence monomorphisms by Lemma 7.7, from $e_1 R$ to $e_2 R$ and from $e_2 R$ to $e_1 R$. Thus by Bumby [34] $e_1 R \cong e_2 R$ and, by (1), the result follows. $\qquad\square$

Recall that a uniserial ring R is called a *maximal ring* if every family of pairwise solvable congruences of the form $x \equiv x_\alpha \pmod{A_\alpha}$ has a simultaneous solution where $x_\alpha \in R$ and each A_α is an ideal in R. In the next theorem we provide the structure of a nonlocal semiperfect right $PCQI$-ring.

Theorem 7.9 *Let R be a semiperfect nonlocal ring. Then R is a right PCQI-ring if and only if either*

1. $R = \oplus_{i=1}^{n} R_i$, where R_i is semisimple artinian or a rank zero duo maximal uniserial ring, or

2. $R = \begin{pmatrix} D & D \\ 0 & D \end{pmatrix}$ where D is a division ring.

Moreover, a nonlocal semiperfect right PCQI-ring is also a left PCQI-ring

Proof Let R be a semiperfect nonlocal right $PCQI$-ring. By the two lemmas above, it follows that R is a qc-ring unless $R = e_1 R \oplus e_2 R$, where $e_1 R$ and $e_2 R$

are not isomorphic and exactly one of $e_1 R e_2$ or $e_2 R e_1$ is zero, say $e_1 R e_2 \neq 0$ and $e_2 R e_1 = 0$. If R is a qc-ring then we get (1) by Theorem 7.3. Otherwise, we have $R \cong \begin{pmatrix} e_1 R e_1 & e_1 R e_2 \\ 0 & e_2 R e_2 \end{pmatrix}$. We claim that $e_1 R e_1$ and $e_2 R e_2$ are isomorphic division rings and $M = e_1 R e_2$ is a D-D bimodule such that $\dim_D M = 1 = \dim M_D$, where $D \cong e_1 R e_1 \cong e_2 R e_2$. We already know that R is not prime and $J(R)$ is nil. Thus $e_2 J(R) e_2$ is nil. We claim that $e_2 J(R) e_2 = 0$. Let $e_2 x e_2 \in e_2 J(R) e_2$. Define $\sigma : e_2 R \to e_2 R$ by $\sigma(e_2 y) = e_2 x e_2 y$. Then $\sigma \in \mathrm{End}_R(e_2 R)$ and, since $e_2 x e_2$ is nilpotent, σ is not a monomorphism. So $\mathrm{Ker}(\sigma) \neq 0$. Since $\mathrm{Hom}_R(e_2 R, e_1 R) \neq 0$, there exists an embedding $\eta : e_2 R \to e_1 R$. Now $\eta \sigma : e_2 R \to e_1 R$, and since $\mathrm{Ker}(\sigma) \neq 0$, $\mathrm{Ker}(\eta \sigma) \neq 0$. Then it may concluded that $\eta \sigma = 0$. Since η is a monomorphism, we have $\sigma = 0$. Thus $e_2 x e_2 = 0$ and $e_2 J(R) e_2 = 0$. So $e_2 R e_2$ is a division ring. Similarly, it can be shown that $e_1 R e_1$ is a division ring. Now using the fact that $\mathrm{End}_R(e_1 R)$ is a division ring and that $e_1 R$ is quasi-injective, it follows that every element of $\mathrm{End}_R(e_1 R e_2 R)$ admits a unique extension to an endomorphism of $e_1 R$. Further, every endomorphism of $e_1 R$ maps $e_1 R e_2 R$ into itself since $e_1 R e_2 R$ is the unique minimal subideal of $e_1 R$. Thus $\mathrm{End}_R(e_1 R e_2 R) \cong \mathrm{End}_R(e_1 R)$. Since $e_1 R e_2 R \cong e_2 R$, we obtain $e_1 R e_1 \cong e_2 R e_2$. Now $e_1 J(R) = e_1 J(R) e_2$ because $e_1 J(R) e_1 = 0$. Since $e_1 R e_2 R \subseteq e_1 J(R)$, we get $e_1 J(R) = e_1 R e_2 = e_1 R e_2 R$. Thus $M = e_1 R e_2$ is a one-dimensional right vector space over $D = e_2 R e_2$. We can also show that M is a one-dimensional left vector space over $e_1 R e_1$. Thus, for each $d \in e_1 R e_1$, there exists a unique $d' \in e_2 R e_2$ such that $dx = xd'$. Define $\theta : e_1 R e_1 \to e_2 R e_2$ by $\theta(d) = d'$. Then θ is an isomorphism and we may identify d and d'. Then $\eta : \begin{pmatrix} D & D \\ 0 & D \end{pmatrix} \to \begin{pmatrix} D & M \\ 0 & D \end{pmatrix}$ defined by $\eta \begin{pmatrix} a & b \\ 0 & c \end{pmatrix} = \begin{pmatrix} a & bx \\ 0 & c \end{pmatrix}$ is an isomorphism.

The converse is obvious. The last statement follows easily from the structure of nonlocal semiperfect right $PCQI$-ring as obtained above. $\qquad\square$

The following example of Jain, Singh, and Symonds [147] shows that a local right $PCQI$-ring need not be a left $PCQI$-ring.

Example Let F be a field which has a monomorphism $\sigma : F \to F$ such that $[F : \sigma(F)] > 2$. Take x to be an indeterminate over F. Make $V = xF$ into a right vector space over F in the natural way. Let $R = \{(\alpha, x\beta) : \alpha, \beta \in F\}$. Define

$$(\alpha_1, x\beta_1) + (\alpha_2, x\beta_2) = (\alpha_1 + \alpha_2, x\beta_1 + x\beta_2)$$

and

$$(\alpha_1, x\beta_1)(\alpha_2, x\beta_2) = (\alpha_1 \alpha_2, x(\sigma(\alpha_1)\beta_2 + \beta_1 \alpha_2))$$

Then R is a local ring with maximal ideal $M = \{(0, x\alpha) : \alpha \in F\}$. In fact, M is also a maximal right ideal with $M^2 = 0$ and hence R is a right $PCQI$-ring.

Further, if $\{\alpha\}_{i \in I}$ is a basis of F as a vector space over $\sigma(F)$ then $M = \oplus R(0, x\alpha_i)$ is a direct sum of irreducible left modules $R(0, x\alpha_i)$. Since $|I| > 2$, R is not a left $PCQI$-ring.

Recall the condition (P) mentioned in the beginning of this chapter.

(P): *Every proper cyclic module is injective modulo its annihilator ideal.*

In an attempt to understand rings R satisfying condition (P), Jain and Singh [146] obtained the following.

Theorem 7.10 *Let R be a ring satisfying condition (P). Then R is either a right Ore domain or is semiperfect. Furthermore, a semiperfect ring R satisfies condition (P) if and only if R is a right PCQI-ring.*

7.11 Questions

1. Give a characterization of right $PCQI$-domains. They are known to be right noetherian for simple domains or V-domains.
2. Characterize a right Ore domain where every proper cyclic module C is injective modulo its annihilator ideal.

8 Rings each of whose (proper) cyclic modules is continuous

8.1 Rings each of whose cyclic modules is continuous

Recall that a right R-module M is called a *continuous module* if the following two conditions hold: (i) every submodule of M is essential in a direct summand of M, and (ii) every submodule of M isomorphic to a direct summand of M is itself a direct summand of M. A ring R is called a *right cc-ring* if each cyclic right R-module is continuous. Jain and Mohamed [140] generalized Koehler's Theorem and gave the structure of a right cc-ring, as follows.

Theorem 8.1 *A ring R is a right cc-ring if and only if $R = A \oplus B$ where A is semisimple artinian and B is a finite direct sum of right uniserial right duo rings with nil Jacobson radical.*

Proof Since R is a continuous ring, $J(R) = Z(R_R)$ and idempotents modulo $J(R)$ can be lifted. Also, $R/J(R)$ is a von Neumann regular ring that has finite Goldie dimension (Theorem 4.9). Therefore $R/J(R)$ is semisimple artinian and hence R is semiperfect.

In what follows, we shall use that if $A \oplus B$ is a direct sum of indecomposable projective right ideals such that for some right ideal $C \subset A$, A/C embeds in B and $A/C \oplus B$ is continuous, then using mutual relative injectivity of A/C and B, we get $C = 0$ and $A \cong B$. Since R is semiperfect, we may write $R = e_1 R \oplus \cdots \oplus e_n R$ as a direct sum of indecomposable continuous right ideals. We may group all isomorphic indecomposable right ideals as explained in the proof of Theorem 7.3, and write after renumbering, if necessary, $R = [e_1 R] \oplus \cdots \oplus [e_k R]$ as a direct sum of ideals, where each bracket represents the sum of indecomposable right ideals (possibly only one term in the sum) isomorphic to the term in the bracket. By taking endomorphism rings of R-modules on both sides, we obtain $R = \mathbb{M}_{n_1}(e_1 R e_1) \oplus \cdots \oplus \mathbb{M}_{n_k}(e_k R e_k)$, where each $\mathbb{M}_{n_i}(e_i R e_i)$ represents the $n_i \times n_i$ matrix ring over $e_i R e_i$ which is local. It is known that any matrix ring of size greater than 1 is continuous if and only if it is self-injective and so $\mathbb{M}_{n_i}(e_i R e_i)$ is self-injective if $n_i \geq 2$. Furthermore, it can be shown that if $eR \oplus fR$ is continuous with $\mathrm{Hom}_R(eR, fR) \neq 0$, and every cyclic is continuous, then $eR \cong fR$ is minimal (by Lemma 7.2). All this shows that $\mathbb{M}_{n_i}(e_i R e_i)$ is semisimple artinian for each $n_i \geq 2$.

Now we consider the ring direct summand S of R which is a local cc-ring. We show S is right duo. Let xS be a right ideal and $a \in S$. If $a \notin J(S)$, then a is unit and so $axS \cong xS$. If $xS \subset axS$, then xS is a direct summand of axS which is impossible as axS is indecomposable because S is local. Hence $axS \subseteq xS$. If $a \in J(S)$, then $1 - a$ is a unit therefore $(1 - a)xS \subseteq xS$ and hence $axS \subset xS$. So xS and thus every right ideal of S is two-sided. Finally we show $J(S)$ is nil. We take a non-nilpotent element $a \in J(S)$ and then using a standard Zorn's Lemma argument we find a maximal element P in the family of right ideals that does not contain any power of a. This ideal P is a prime ideal. Thus S/P is a prime local cc-ring and hence right duo. It follows that S/P is a continuous integral domain which is then a division ring. This yields $P = J(S)$, a contradiction because $a \in J(S)$. Thus all elements of $J(S)$ are nilpotent and, therefore, $J(S)$ is nil. Hence S is a local right duo ring with nil Jacobson radical.

The converse is straightforward. $\qquad\qquad\qquad\qquad\qquad\qquad\qquad\qquad\qquad\qquad\square$

8.2 Rings each of whose proper cyclic modules is continuous

A ring R is called a *right PCC-ring* if every proper cyclic right R-module is continuous. Jain and Mueller [142] studied right PCC-rings and obtained the following.

Lemma 8.2 *Let R be a right PCC-ring. Then R is either a prime ring or a semiperfect ring whose Jacobson radical is nil.*

Proof Assume R is not a prime ring. Then there exist nonzero ideals A and B in R such that $AB = 0$. Hence any prime ideal P lies above A or B. But since both R/A and R/B are rings over which cyclic module is continuous, their prime ideals are maximal and finite in number. Thus we obtain that R has only finitely many maximal ideals. Therefore R is a semiperfect ring whose Jacobson radical is nil. $\qquad\qquad\qquad\qquad\qquad\qquad\square$

Theorem 8.3 *Let R be a semiperfect right PCC-ring. Then R is of one of the following types:*

1. *$R = \oplus_{i=1}^{k} A_i$, where each A_i is a simple artinian right uniserial right duo ring with nil Jacobson radical, or a local ring whose maximal ideal M satisfies $M^2 = 0$ and $l(M) = 2$.*

2. *$R = \begin{pmatrix} \Delta & V \\ 0 & D \end{pmatrix}$, where D and Δ are division rings and V is a one-dimensional right vector space over D.*

Proof Assume that R is a semiperfect ring over which each proper cyclic right module is continuous.

Case 1. Assume R is local. Let I be a nonzero right ideal of R. Then R/I is continuous and hence uniform. We claim that if there exist nonzero right ideals A and B of R such that $A \cap B = 0$ then A, B are minimal right ideals and $S = A \oplus B$ where S is the right socle of R. If X is a nonzero right ideal of R and $X \subset A$ then R/X is uniform. But $A/X \cap (B + X/X) = 0$ gives $A = X$. It is immediate now that $S = A \oplus B$. Let M be the unique maximal ideal of R, and let $x \in M$, $x \notin S$. Then xR must be an essential right ideal, for otherwise xR will be minimal. This implies $S \subset xR$, and thus xR cannot be indecomposable. Therefore $xR = X_1 \oplus X_2$ for some nonzero right ideals X_1, X_2. But then $S = X_1 \oplus X_2$, a contradiction since $x \in S$. Hence $S = M$, which gives $M^2 = 0$ and $l(M) = 2$.

Next, if each nonzero right ideal is essential, it follows immediately that R is a right uniserial ring. To prove that R is right duo, consider a principal right ideal aR and $x \in R$. Then either $xaR \subset aR$ or $aR \subset xaR$. If $xaR \subset aR$, for all $x \in R$, then aR is two-sided. So assume $aR \subset xaR$ for some $x \in R$. In case $x \notin M$ then $xaR \cong aR$. Since xaR is continuous, we get $xaR = aR$. In case $x \in M$, consider $1 - x$ and proceed as before. Thus aR is a two-sided ideal. Hence R is a right uniserial right duo ring.

Case 2. Assume $R = \oplus_{i=1}^{n} e_i R$ where $\{e_i : 1 \leq i \leq n\}$ is a maximal family of primitive idempotents with $\Sigma_{i=1}^{n} e_i = 1$ and each $e_i R e_i$ is a local ring with $n \geq 3$. Now $R \not\cong e_i R \oplus e_j R$ and $e_i R e_j \neq 0$. Then $e_i R \times e_j R \cong e_i R \times \sigma(e_j R)$, where $0 \neq \sigma \in \mathrm{Hom}(e_j R, e_i R e_j)$. Since $R \not\cong e_i R \oplus e_j R$, $e_i R \times \sigma(e_j R)$ is continuous. But then $\sigma(e_j R) = e_i R$, and so $e_R \cong e_j R$. Now we group all isomorphic indecomposable right ideals and write after renumbering, if necessary, $R = [e_1 R] \oplus \cdots \oplus [e_k R]$ as a direct sum of ideals, where each bracket represents the sum of indecomposable right ideals (possibly only one term in the sum) isomorphic to the term in the bracket. If we set $A_i = [e_i R]$, then we get $R = \oplus_{i=1}^{n} A_i$ as a finite direct sum of rings A_i. Each A_i is continuous as R-module (and hence A_i-module). Thus all cyclic A_i-modules of A_i are continuous. Then each A_i is simple artinian or right uniserial right duo by Theorem 8.1.

Case 3. Consider the case $R \cong e_i R \oplus e_j R$. Assume $e_i R e_j \neq 0$. Then we have $\mathrm{Hom}_R(e_j R, e_i R) \neq 0$ and so there is a monomorphism from $e_j R$ to $e_i R$. If $e_j R e_i$ is also not zero then similarly there is a monomorphism from $e_i R$ to $e_j R$. Thus $e_i R$, $e_j R$ are subisomorphic to each other and hence it follows that $e_i R \cong e_j R$. So every nonzero homomorphism $e_j R \to e_i R$ is a monomorphism. Next, we claim $e_j J(R) e_j = 0$. Else, choose a nonzero element $e_j a e_j \in e_j J(R) e_j$. This induces an R-homomorphism $f : e_j R \to e_j R$ given by $f(e_j x) = e_j a e_j(e_j x)$. Then f is not a monomorphism because $e_j a e_j$ is a nilpotent element. Let $0 \neq g \in \mathrm{Hom}_R(e_j R, e_i R)$. Then $0 \neq gf \in \mathrm{Hom}_R(e_j R, e_i R)$. But gf is not a monomorphism, a contradiction. Hence $e_j J(R) e_j = 0$ and so $e_j R e_j$ is a division ring. This proves that R is simple artinian.

Consider now the case $e_j R e_i = 0$. Then $e_j J(R) = e_j J(R)(e_i + e_j) = e_j J(R) e_j = 0$, and so $e_j R$ is a minimal right ideal. This implies that $e_j R e_j$ is a division ring. In this case, since $e_i R$ is uniform, for all $0 \neq e_i x e_j \in e_i R e_j$, $e_i x e_j R$ is the unique minimal right ideal in $e_i R$. But then $e_i x e_j R = e_i R e_j R$, for all $0 \neq e_i x e_j \in e_i R e_j$. We proceed to prove that $e_i J(R) e_i = 0$ and thus $e_i R e_i$ is also a division ring. If possible, let $0 \neq e_i x e_i \in e_i J(R) e_i$. Consider the mapping $\sigma : e_i R \to e_i R$ given by left multiplication with $e_i x e_i$. Since $e_i x e_i$ is nilpotent, σ is not one-to-one. Thus $\mathrm{Ker}(\sigma)$ contains the unique minimal right ideal $e_i R e_j R$. Therefore $e_i x e_i R e_j R = 0$. But then $e_i x e_i R e_j = 0$. Now the unique minimal right ideal $e_i R e_j R$ is contained in every nonzero right ideal. So $e_i R e_j R \subset e_i x e_i R$. Therefore $e_i R e_j R e_j \subset e_i x e_i R e_j = 0$. Since $e_j R e_j$ is a division ring, we obtain $e_i R e_j = 0$, a contradiction. Thus $e_i J(R) e_i = 0$, which gives $e_i R e_i$ is a division ring.

We now prove that $e_i R e_j$ is a one-dimensional right vector space over $e_j R e_j$. Note that $e_i J(R) = e_i J(R) e_j$ because $e_i J(R) e_i = 0$. Since $e_i R e_j R$ is a unique minimal right ideal in $e_i R$, $e_i J(R) \supseteq e_i R e_j R$. Because $e_j R e_i = 0$, $e_i R e_j R = e_i R e_j R e_j$. This implies $e_i R e_j R = e_i R e_j$ because $e_j R e_j$ is a division ring. Furthermore, $e_i R e_j$ is a right ideal as shown above and $(e_i R e_j)^2 = 0$. This implies $e_i R e_j \subset e_i J(R)$, and so $e_i R e_j = e_i J(R) = e_i R e_j R = e_i x e_j R e_j$. This proves $e_i R e_j$ is a one-dimensional right vector space over $e_j R e_j$. Hence
$$R = \begin{pmatrix} e_i R e_i & e_i R e_j \\ 0 & e_j R e_j \end{pmatrix} \cong \begin{pmatrix} \Delta & V \\ 0 & D \end{pmatrix}$$
where D and Δ are division rings and V is a one-dimensional right vector space over $e_j R e_j$. This completes the proof. $\qquad\square$

9 Rings each of whose (proper) cyclic modules is π-injective

9.1 Rings each of whose cyclic modules is π-injective

The class of right cc-rings was further generalized by Goel and Jain [89] who studied rings over which each cyclic module is π-injective. A ring R is called a right πc-*ring* if each cyclic right R-module is π-injective.

We start with the following basic observation.

Lemma 9.1 *Let M be a module such that all factor modules of M are π-injective.*

1. *If all factor modules of the module M are indecomposable, then M is a uniserial module.*
2. *If M is a cyclic module over a local ring, then M is a uniserial module.*

Proof

(1) Let X and Y be two submodules in M, $X \nsubseteq Y$, and let $h\colon M \to M/(X \cap Y)$ be the natural epimorphism. The indecomposable π-injective module $h(M)$ is uniform. In addition, $h(X) \cap h(Y) = 0$ and $h(X) \neq 0$. Then $h(Y) = 0$ and so $Y = X \cap Y \subseteq X$.

(2) Since all factor modules of any cyclic module are cyclic, (2) follows from (1) and the property that all cyclic modules over local rings are indecomposable. $\qquad\square$

In the next theorem, we provide the structure of a right πc-ring.

Theorem 9.2 *If R is a right πc-ring, then $R = R_1 \oplus R_2 \oplus \cdots \oplus R_k$, where each R_i is either a simple artinian ring or a right uniform ring.*

Proof By Theorem 4.9, R is a direct sum of uniform right ideals. Write $R = e_1 R \oplus \cdots \oplus e_k R$ as a direct sum of uniform π-injective right ideals. In Lemma 7.2, we have already seen that if $\mathrm{Hom}(e_i R, e_j R) \neq 0$ then $e_i R$ and $e_j R$ are isomorphic minimal right ideals. If $\mathrm{Hom}(e_i R, e_j R) = 0$ and $\mathrm{Hom}(e_j R, e_i R) = 0$ for all $j \neq i$ then $e_i R$ is a two-sided ideal and is a ring direct summand of R. This gives us the decomposition $R = R_1 \oplus R_2 \oplus \cdots \oplus R_k$, where each R_i is either simple artinian or right uniform. $\qquad\square$

As a consequence, we have the following corollary.

Corollary 9.3 *For a ring R, we have the following:*

1. *If R is right self-injective, then R is a right πc-ring if and only if $R = A \oplus B$ where A is semisimple artinian and B is a finite direct sum of right self-injective right uniserial rings.*
2. *If R is semiperfect, then R is a right πc-ring if and only if $R = A \oplus B$ where A is semisimple artinian and B is a finite direct sum of right uniserial rings.*
3. *If R is a right nonsingular right πc-ring, then $R = A \oplus B$ where A is semisimple artinian and B is a finite direct sum of right Ore domains.*

Huynh and Wisbauer [124] extended the above results by studying finitely generated quasi-projective modules each of whose factor modules is π-injective, and gave the complete structure of such modules.

Theorem 9.4 *Let M be a finitely generated quasi-projective module where each factor module is π-injective. Then there is a decomposition $M = M_0 \oplus M_1 \oplus M_2$ where M_0, M_1, M_2 are fully invariant submodules such that M_0 is a semisimple module, $M_1 \cong N_1^{k_1} \oplus \cdots \oplus N_r^{k_r}$ for fully invariant summands $N_i^{k_i}$ with N_i a nonsimple uniserial module, $\mathrm{End}_R(N_i)$ a division ring, and $M_2 = U_1 \oplus \cdots \oplus U_k$ for fully invariant uniform modules U_i with $\mathrm{End}_R(U_i)$ not a division ring.*

In particular, if R is a ring for which each cyclic module is π-injective then $R = R_0 \oplus R_1 \oplus \cdots \oplus R_k$ where R_0 is a semisimple ring and $R_1, \ldots, R_k$ are rings which are uniform as right modules and any R_i is right uniserial if and only if it is local.

The above theorem is a generalization of Tuganbaev ([219], Lemma 8).

Huynh, Jain, and López-Permouth considered simple rings with the property that every cyclic singular module is CS in [119] and proved the following.

Theorem 9.5 *If R is a simple ring such that every cyclic singular right R-module is CS, then R is right noetherian.*

Proof Let R be a simple ring. If $\mathrm{Soc}(R_R) \neq 0$, then $R = \mathrm{Soc}(R_R)$, proving that R is simple artinian. Hence we consider the case $\mathrm{Soc}(R_R) = 0$. Moreover, since R is simple, R is right nonsingular.

Now assume that every cyclic singular right R-module is CS. Let I be an essential right ideal of R. Then every cyclic subfactor of R/I is singular and so CS. Hence, by Theorem 4.8, R/I has finite Goldie dimension. Now we proceed to show that $R/\mathrm{Soc}(R_R)$ has finite Goldie dimension. Let D be a submodule of R_R such that $\mathrm{Soc}(R_R) \oplus D$ is essential in R_R. Hence $R/(\mathrm{Soc}(R_R) \oplus D)$ and therefore $(R/\mathrm{Soc}(R_R))/\overline{D}$, where $\overline{D} = (\mathrm{Soc}(R_R) \oplus D)/\mathrm{Soc}(R_R)$, has finite Goldie dimension. From this fact and since $D \cong \overline{D}$, we only need to show that D has finite Goldie dimension. Assume, on the contrary, that D contains an essential submodule $V = \oplus_{i=1}^{\infty} V_i$, an infinite direct sum of nonzero submodules. Since $\mathrm{Soc}(D) = 0$, every V_i contains an essential proper submodule W_i, and so $W = \oplus_{i=1}^{\infty} W_i$ is an essential submodule of V. Furthermore, since $\mathrm{Soc}(R_R) \oplus W$ is an essential submodule of R_R, $R_R/(\mathrm{Soc}(R_R) \oplus W)$ has finite Goldie dimension. On the other hand, $R/(\mathrm{Soc}(R_R) \oplus W)$ contains

$$(\mathrm{Soc}(R_R) \oplus V)/(\mathrm{Soc}(R_R) \oplus W) \cong V/W \cong \oplus_{i=1}^{\infty}(V_i/W_i),$$

where each V_i/W_i is nonzero, a contradiction. Therefore D has finite Goldie dimension. Thus it follows that $R/\mathrm{Soc}(R_R)$ has finite right Goldie dimension, and hence R has finite right Goldie dimension. Thus R is a right Goldie ring. Let A be an arbitrary right ideal of R. Let B be a complement of A in R. Then $A \cap B = 0$ and $A \oplus B$ is an essential right ideal of R. As A, B and $R/(A \oplus B)$ have finite Goldie dimension, we conclude that R/A has finite Goldie dimension. Thus R is a right *q.f.d.* ring.

Now let E be an essential right ideal of R and let $M = R/E$. We will show that M is noetherian. Let α be an ordinal. Then the socle series of M is defined inductively as $S_1 = \mathrm{Soc}(M)$, $S_\alpha/S_{\alpha-1} = \mathrm{Soc}(M/S_{\alpha-1})$, and $S_\alpha = \cup_{\beta<\alpha}S_\beta$ if α is a limit ordinal. Let $S = \cup_\alpha S_\alpha$. Then the socle of $U := M/S$ is zero. We claim that U is noetherian. Assume that $U \neq 0$. Since U is CS and has finite Goldie dimension, U is a direct sum of finitely many uniform submodules by Theorem 4.8. Hence without loss of generality, we may assume that U is uniform. Let V be an arbitrary singular simple right R-module. Then, since R is a simple right Goldie ring, it may be shown that the external direct sum $T = U \oplus V$ is cyclic (see [47], Theorem 14.1). Hence, by hypothesis, T is CS. For convenience, we may consider U and V as submodules of T. It is clear that $\mathrm{Soc}(T) = V$, a complement of T. Let A be a submodule of U and assume that there is a nonzero homomorphism φ of A to V. Let $W = \{a - \varphi(a) : a \in A\}$. Then W is contained as an essential submodule in a direct summand W^* of T. Write $T = W^* \oplus B$ for some submodule B of T. Since $W^* \cap V = 0$, and $V = \mathrm{Soc}(T)$, it follows that $V \subseteq B$. But, since V is a complement submodule of T and B is uniform, we must have $V = B$. Thus $T = W^* \oplus V$. Let π be the projection of T onto V along this decomposition. It is easy to check that $\pi|_U$ is an extension of φ from U to V. This shows that V is U-injective, and hence every singular simple right R-module is U-injective. If U is not noetherian, then there is an infinite strictly ascending chain of submodules in U

$$x_1 R \subset x_1 R + x_2 R \subset \cdots$$

Let $X = \cup_{i\in\mathbb{N}}(x_1 R + \cdots + x_i R)$. Then, since every singular simple right R-module is X-injective, we can find a submodule Y of X such that X/Y is a direct sum of infinitely many simple modules. This is a contradiction to the fact that U/Y has finite Goldie dimension. Thus U is noetherian.

Now, to show that M is noetherian, it is enough to verify that S has finite composition length. Since every cyclic right R-module has finite Goldie dimension, S_1 and each $S_{\alpha+1}/S_\alpha$ have finite composition length. Assume that $S_2 \neq S_3$. Then there exists $y \in S_3$ such that $yR \not\subseteq S_2$ and $(yR + S_2)/S_2$ is simple. Since yR is CS, $yR = K_1 \oplus \cdots \oplus K_m$, where each K_i is uniform. There is some

K_i, say K_1, with $K_1 \not\subseteq S_2$. Again, since $K_1/\mathrm{Soc}(K_1)$ is CS, there are finitely many nonzero submodules of K_1, say $H_1, \ldots, H_n$, such that

$$K_1/\mathrm{Soc}(K_1) = (H_1/\mathrm{Soc}(K_1)) \oplus \cdots \oplus (H_n/\mathrm{Soc}(K_1)),$$

where each $H_j/\mathrm{Soc}(K_1)$ is simple or uniform of composition length 2. Then there is some H_j, say H_1, such that $l(H_1/\mathrm{Soc}(K_1)) = 2$. Then H_1 is a cyclic uniserial module with the unique composition series $\mathrm{Soc}(K_1) \subset H \subset H_1$. It is easy to check that $H_1 \oplus (H/\mathrm{Soc}(K_1))$ is not CS. However, $H_1 \oplus (H/\mathrm{Soc}(K_1))$ is a cyclic singular right R-module (see [47], Theorem 14.1) and hence CS by hypothesis. This contradiction shows that $S_3 = S_2$ and so $S = S_2$. Thus S has finite composition length. We have shown that for each essential right ideal E of R, R/E is noetherian. Hence by [96, Proposition 3.6], $R/\mathrm{Soc}(R_R)$ is right noetherian. But $\mathrm{Soc}(R_R) = 0$. So R is right noetherian. $\square$

As a consequence, it follows:

Corollary 9.6 *A simple right PCQI-ring is right noetherian.*

Huynh, Jain, and López-Permouth [119] proved that a simple ring R is right PCI if and only if R is right $PCQI$. Barthwal, Jhingan, and Kanwar extended it and showed that a simple ring R is right PCI if and only if each proper cyclic right-R module is right continuous [17]. Huynh, Jain, and López-Permouth [120] later showed that a simple ring R is Morita equivalent to a right PCI-domain if and only if every cyclic singular right R-module is π-injective. Before presenting their proof, we define right SI-rings. A ring R is called a *right SI-ring* if every singular right R-module is injective. The right PCI and right SI conditions are equivalent for domains.

Theorem 9.7 *A simple ring R is Morita equivalent to a right PCI-domain if and only if every cyclic singular right R-module is π-injective.*

Proof Let R be a simple ring whose cyclic singular right R-modules are π-injective. Then by Theorem 9.5, R is right noetherian. First, consider the case $\mathrm{Soc}(R_R) \neq 0$. Then $R = \mathrm{Soc}(R_R)$ and hence R is a simple artinian ring. Clearly then R is Morita equivalent to a right PCI-domain.

Next, we consider the case $\mathrm{Soc}(R_R) = 0$. Let M be a nonzero artinian right R-module. It is clear that M_R is singular. Let N be a cyclic submodule of M. It may be seen that $N \oplus \mathrm{Soc}(N)$ is cyclic. Therefore $N \oplus \mathrm{Soc}(N)$ is π-injective. So $\mathrm{Soc}(N)$ is N-injective and hence $\mathrm{Soc}(N)$ splits in N. Thus $N = \mathrm{Soc}(N) \subset M$. This shows that M is semisimple.

Now we prove that every singular cyclic R-module is semisimple, that is, for each essential right ideal C of R, R/C is semisimple. To prove R/C is semisimple, it suffices to show that R/C is artinian. Assume to the contrary that there is an essential right ideal E of R such that R/E is not artinian. Since R is noetherian, there exists an essential right ideal L of R maximal with respect to the property that $U = R/L$ is not artinian. It follows that U is uniform and $\mathrm{Soc}(U) = 0$.

Moreover, for any nonzero submodule V of U, U/V is semisimple. Let A and B be submodules of U with $0 \neq A \subset B \subset U$, with $A \neq B \neq U$. Then B/A is a direct sum of finitely many simple modules. Consider the module $Q = U \oplus B$. Since U is cyclic and $Q/(0 \times A) \cong U \oplus B/A$, it can be shown by induction on the number of simple direct summands of B/A, that $Q/(0 \times A)$ is cyclic. Let $x \in Q$ such that $[xR + (0 \times A)]/(0 \times A) = Q/(0 \times A)$. Then xR is not uniform. We may choose x so that xR contains $(U \times 0)$. Hence $xR = U \oplus W$ where $(0 \times W) = xR \cap (0 \times B) \neq (0 \times 0)$. Since xR is π-injective, W is U-injective. Therefore W splits in U, a contradiction. Hence R is Morita equivalent to a right SI-domain. Since right SI-domains are the same as right PCI-domains, the proof is complete. $\square$

As a consequence of this, Huynh, Jain, and López-Permouth [120] also showed the following.

Theorem 9.8 *Let R be a simple ring. If every proper cyclic module is π-injective, then R is Morita equivalent to a right PCI domain and has right uniform dimension at most 2.*

Theorem 9.9 *For a right V-ring R with $\mathrm{Soc}(R_R) = 0$, the following conditions are equivalent:*

1. *Every cyclic singular right R-module is π-injective.*
2. *R has a ring-direct decomposition $R = R_1 \oplus R_2 \oplus \cdots \oplus R_n$, where each R_i is Morita equivalent to a right PCI-domain.*

Proof $(1) \Rightarrow (2)$. Let E be an essential right ideal of R and let $M = R/E$. Then M is a singular right R-module and hence every cyclic submodule of any homomorphic image of M is π-injective. By Theorem 4.9, each factor module of M has finite Goldie dimension. Since R is assumed to be a right V-ring, it follows that M is noetherian. Hence, by [96, Proposition 3.6], $R/\mathrm{Soc}(R_R)$ is noetherian. But $\mathrm{Soc}(R_R) = 0$, so R is right noetherian. As R is a right V-ring, it follows from ([76], Theorem 2) that R has a ring direct sum decomposition $R = R_1 \oplus R_2 \oplus \cdots \oplus R_n$, where each R_i is a simple ring. By Theorem 9.7, each R_i is Morita equivalent to a right PCI-domain.

$(2) \Rightarrow (1)$. This is clear because every ring summand in (2) has the property that every singular right R-module is injective. $\square$

9.2 Rings each of whose proper cyclic modules is π-injective

A ring R is called a right *pcπ-ring* if each proper cyclic right R-module is π-injective. The following lemma easily follows from the definition of π-injective modules.

If A, B are continuous R-modules such that A is embeddable in B and B is embeddable in A, then it is well-known that $A \cong B$. However, this is not true for π-injective modules, in general. But, if both A and B are indecomposable projective and $A \oplus B$ is π-injective, then $A \cong B$. The following lemma was proved by Goel and Jain [89].

Lemma 9.10 *If M_1 and M_2 are π-injective such that $M_1 \oplus M_2$ is π-injective and $E(M_1) \cong E(M_2)$ then $M_1 \cong M_2$.*

Proof Let $\varphi : E(M_1) \to E(M_2)$ be an isomorphism. Consider an idempotent

$$\begin{bmatrix} 1 & 0 \\ \varphi & 0 \end{bmatrix} \in \begin{bmatrix} \mathrm{Hom}(E(M_1), E(M_2)) & \mathrm{Hom}(E(M_2), E(M_1)) \\ \mathrm{Hom}(E(M_1), E(M_2)) & \mathrm{Hom}(E(M_2), E(M_2)) \end{bmatrix}$$

$$\cong \mathrm{Hom}(E(M_1) \oplus E(M_2), E(M_1) \oplus E(M_2)).$$

Since $M_1 \oplus M_2$ is π-injective, $\begin{bmatrix} 1 & 0 \\ \varphi & 0 \end{bmatrix} \begin{bmatrix} m_1 \\ m_2 \end{bmatrix} \in M_1 \oplus M_2$ for all $m_1 \in M_1$ and $m_2 \in M_2$. This gives $\varphi(m_1) \in M_2$ and hence $\varphi(M_1) \subseteq M_2$. Similarly, we can prove $\varphi^{-1}(M_2) \subseteq M_1$. Thus $\varphi(M_1) = M_2$ proving $M_1 \cong M_2$. $\quad\square$

Mueller and Rizvi [180] generalized the above result and showed that if M_1 and M_2 are direct summands of a π-injective module with $E(M_1) \cong E(M_2)$ then $M_1 \cong M_2$.

Theorem 9.11 *Let $R = \oplus_{i \in I} e_i R$ be a semiperfect ring with nil Jacobson radical, where the e_i's are orthogonal idempotents, the $e_i R$ are indecomposable right ideals and $|I| \geq 3$. Then R is right a $pc\pi$-ring if and only if $R = \oplus R_i$ where each R_i is either simple artinian or a right uniserial ring.*

Proof Since the number of summands is more than 2, for each pair of indecomposable right ideals $e_i R$ and $e_j R$, we have that $e_i R \oplus e_j R$ is π-injective. Suppose $e_i R e_j \neq 0$. Then by Lemma 7.2, $e_i R$ and $e_j R$ are isomorphic minimal right ideals. If $e_i R e_j = 0$ and $e_j R e_i = 0$ for all $j \neq i$ then $e_i R$ is a two-sided ideal and is a ring direct summand of R which is a right uniserial ring. This proves the theorem. $\quad\square$

Lemma 9.12 *Let $R = e_1 R \oplus e_2 R$ be a semiperfect ring with nil Jacobson radical $J(R)$, where $e_1 R, e_2 R$ are indecomposable right ideals such that $e_1 R e_2 \neq 0$ and $e_2 R e_1 \neq 0$. Then R is a right $pc\pi$-ring if and only if it is simple artinian.*

Proof Suppose R is a right $pc\pi$-ring. By hypothesis, there exists a nonzero homomorphism $f : e_1 R \to e_2 R$. This gives $R/\mathrm{Ker}\, f \cong e_1 R/\mathrm{Ker}\, f \oplus e_2 R$. If $R/\mathrm{Ker}\, f \cong R$ then $R \cong R \oplus \mathrm{Ker}\, f$, and this implies $\mathrm{Ker}\, f = 0$, because R is semiperfect. If $R/\mathrm{Ker}\, f \ncong R$, then $e_1 R/\mathrm{Ker}\, f \oplus e_2 R$ is π-injective and $e_1 R/\mathrm{Ker}\, f$ embeds in $e_2 R$. This implies $e_1 R/\mathrm{Ker}\, f \cong e_2 R$ and so $\mathrm{Ker}\, f = 0$. In any case, $e_1 R$ is embeddable in $e_2 R$ and similarly $e_2 R$ is embeddable in $e_1 R$. This yields either $e_1 R \cong e_2 R$, or $e_1 R$ embeds in $e_2 J(R)$ and $e_2 R$ embeds in $e_1 J(R)$. Therefore either $e_1 R \cong e_2 R$ or R embeds in $J(R)$. The latter is impossible. Hence $e_1 R \cong e_2 R$.

We now show $e_1 J(R)e_1 = 0$. Since $J(R)$ is nil, there exists $(e_1 x e_1)$ such that $(e_1 x e_1)^{k-1} \neq 0$ but $(e_1 x e_1)^k = 0$. Consider the mapping $g : e_1 R \to e_1 R$ given by left multiplication with $e_1 x$. Then the image of $(e_1 x e_1)^{k-1}$ is zero, a contradiction because this mapping naturally induces a mapping from $e_1 R$ to $e_2 R$ which as proved above must be one-to-one. Hence $e_1 J(R)e_1 = 0$. This yields R as the 2×2 matrix ring over the division ring $e_1 R e_1$. $\square$

Theorem 9.13 *Let $R = e_1 R \oplus e_2 R$ be a semiperfect right $pc\pi$-ring with nil Jacobson radical $J(R)$, where $e_1 R, e_2 R$ are indecomposable right ideals such that $e_1 R e_2 \neq 0$ and $e_2 R e_1 = 0$. Then $e_2 R$ is a minimal right ideal and $ann_{e_1 R e_1}(e_1 R e_2) = e_1 J(R)e_1$.*

Proof As in the previous theorem any nonzero map from $e_2 R$ to $e_1 R$ is a monomorphism. We show $e_2 J(R) = 0$. Choose $e_1 x e_2 \neq 0$ and suppose, by way of contradiction, that $e_2 y \neq 0$ for some $y \in J(R)$. Since $J(R)$ is nil, we can assume $(e_2 y)^2 = 0$. Then $(e_1 x e_2 y e_2)e_2 y = 0$. This yields $e_1 x e_2 y e_2 = 0$. Again $e_2 y e_2 = 0$. This gives $e_2 y = 0$ since $e_2 R e_1 = 0$, a contradiction. Hence $e_2 J(R) = 0$, and so $e_2 R$ is unique minimal right ideal in $e_2 R$ because $e_2 R$ is uniform. The last part is straightforward. $\square$

Lemma 9.14 *Under the hypothesis of the above lemma, $A = \begin{bmatrix} e_1 J(R)e_1 & 0 \\ 0 & 0 \end{bmatrix}$ is an ideal in $S = \begin{bmatrix} e_1 R e_1 & e_1 R e_2 \\ 0 & e_2 R e_2 \end{bmatrix} \cong R$ and $S/A \cong \begin{bmatrix} e_1 R e_1 / e_1 J(R)e_1 & e_1 R e_2 \\ 0 & e_2 R e_2 \end{bmatrix}$ is not π-injective as an S- or S/A-module and therefore R is not a right $pc\pi$-ring if $A \neq 0$.*

Proof It follows from the above lemma that $e_1 R e_2$ is a one-dimensional right vector space over $e_2 R e_2$. Thus we may write $\begin{bmatrix} e_1 R e_1 / e_1 J(R)e_1 & e_1 R e_2 \\ 0 & e_2 R e_2 \end{bmatrix} = \begin{bmatrix} K & D \\ 0 & D \end{bmatrix} \cong \begin{bmatrix} K & D \\ 0 & 0 \end{bmatrix} \oplus \begin{bmatrix} 0 & 0 \\ 0 & D \end{bmatrix}$ as S/A-modules, where K and D are division rings. Since the second summand can be embedded in the first but the embedding is not onto, S/A cannot be π-injective. This completes the proof. $\square$

Theorem 9.15 *Let $R = e_1 R \oplus e_2 R$ be a semiperfect ring with nil Jacobson radical such that e_1, e_2 are primitive orthogonal idempotents, $e_1 R e_2 \neq 0, e_2 R e_1 = 0$. Then R is a right $pc\pi$-ring if and only if $R \cong \begin{bmatrix} K & V \\ 0 & D \end{bmatrix}$ where K and D are division rings and V is a one-dimensional right vector space over D.*

Proof By Theorem 9.13, $e_2 R$ is a minimal right ideal and hence $e_2 R e_2$ is a division ring. We have embedding $\sigma : e_2 R \to e_1 R$. Now $\sigma(e_2) = e_1 x e_2$. Suppose there exists $0 \neq e_1 y e_1 \in e_1 J(R)e_1$ and λ be the endomorphism of $e_1 R$ given by $e_1 y e_1$, this λ cannot be one-to-one as $e_1 y e_1$ is nilpotent. Thus the minimal submodule $e_1 x e_2 R \subseteq \ker(\lambda)$. This gives $e_1 y e_1 R e_2 = 0$. At the same time $e_1 x e_2 R \subseteq e_1 y e_1 R$, as $e_1 R$ is uniform. Thus $e_1 x e_2 = e_1 y e_1 z e_2$, $e_1 y e_1 R e_2 \neq 0$, which is a contradiction. Thus it follows that $e_1 J(R) e_1 = 0$. Thus $R \cong \begin{bmatrix} K & V \\ 0 & D \end{bmatrix}$ where K and D are division rings and V is a one-dimensional right vector space over D. Thus the "only if" part of the theorem is completed. The converse is straightforward. $\square$

To complete our discussion we need to give a characterization of local $pc\pi$-rings, This is contained in the following theorem whose proof is similar to the first part of the proof of Theorem 8.3. However, note that a local right $pc\pi$-ring need not be right duo.

Theorem 9.16 *Let R be a local ring with unique maximal ideal M. Then R is a right $pc\pi$-ring if and only if R is either a right uniserial ring or $M^2=0$, and the composition length of M is 2.*

In summary, we have obtained the following description of nonlocal right $pc\pi$-rings.

Theorem 9.17 *Let R be a nonlocal semiperfect right $pc\pi$-ring. Then R is a direct sum of rings of the following types (not necessarily all)*

1. *Semisimple artinian ring.*
2. *Right uniserial ring with maximal ideal M satisfying $M^2 = 0$, and $l(M) = 2$.*
3. $\begin{pmatrix} K & V \\ 0 & D \end{pmatrix}$ *where K and D are division rings and V is one-dimensional right vector space over D.*

Gómez Pardo and Guil Asensio studied families of modules that are π-injective with respect to pure-essential sequences [95]. A right module M is called a *pqc*-module when (i) every pure submodule of M is purely essential in a direct summand of M, and (ii) if U, V are direct summands of M such that $U \cap V = 0$ and $U \oplus V$ is pure in M, then $U \oplus V$ is a direct summand of M. A module M is called a *completely pqc-module* if every pure factor of M is a *pqc*-module.

Fuchs [84] defined a module M to be ADS (short for *Absolute Direct Summand*) if for every decomposition $M = S \oplus T$ of M and every complement T' of S we have $M = S \oplus T'$. Clearly any commutative ring R has the property that R_R is ADS. Moreover, if a ring R is commutative then every cyclic R-module is ADS. We note that every π-injective module is ADS, but not conversely. However, an ADS module which is also CS is π-injective. We refer the reader to the paper by Alahmadi, Jain, and Leroy [9] that gives a list of equivalent conditions for a module to be ADS. A module M is said to be *completely ADS* if each of its subfactors is ADS. It is shown by Alahmadi, Jain, and Leroy [9] that a completely ADS semiperfect module is a direct sum of semisimple and local modules which, in particular, provides an alternative proof of the characterizations of semiperfect πc-rings.

9.18 Questions

1. Study rings over which every proper quotient of any module M is a finitely presented (i) completely *pqc*-module, or (ii) completely pure injective module.

   Gómez Pardo and Guil Asensio showed that if M is a finitely presented completely *pqc*-module, then M has ACC (and DCC) on direct summands and so it is a (finite) direct sum of indecomposable modules (see [95], Theorem 2.9).
2. Let R be a ring such that R_R and $_RR$ are both ADS. Is R directly-finite?
3. Does an ADS module which is also directly-finite satisfy an internal cancellation property?

10 Rings with cyclics $\aleph_0$-injective, weakly injective, or quasi-projective

10.1 Rings each of whose cyclic modules is $\aleph_0$-injective

Let $\aleph$ be a cardinal number. A right R-module M is said to be $\aleph$-*injective* if for any $\aleph$-generated right ideal B of the ring R, each homomorphism $B_R \to M$ can be extended to a homomorphism $R_R \to M$.

In Theorem 4.4, we proved that the ring R is a semisimple artinian ring provided all cyclic right R-modules are injective. However, there exist nonartinian rings R such that all cyclic R-modules are $\aleph_0$-injective. For example, we can take $R = \prod_{i=1}^{\infty} A_i$, where all A_i are fields (see Lemma 10.1).

A module M is said to be *regular* (in the sense of von Neumann) if all finitely generated submodules of M are direct summands in M. A ring R is von Neumann regular if and only if the module R_R (or equivalently, $_RR$) is regular.

Lemma 10.1 *Let R be a ring and let $\aleph$ be a cardinal number.*

1. *If $\{M_i\}_{i \in I}$ is a set of $\aleph$-injective right R-modules, then $\prod_{i \in I} M_i$ is an $\aleph$-injective module.*
2. *If all $\aleph$-generated right ideals of the ring R are projective, then all homomorphic images of any direct product of $\aleph$-injective right R-modules are $\aleph$-injective.*
3. *If the ring R is right $\aleph$-injective and all $\aleph$-generated right ideals of the ring R are projective, then all finitely generated right R-modules are $\aleph$-injective.*

Proof

(1) Let $\pi_i \colon \prod_{i \in I} M_i \to M_i$ be the natural projection for each $i \in I$, B be an $\aleph$-generated right ideal of the ring R, and let $f \colon B \to M$ be a homomorphism of right R-modules. Since each of the modules M_i is $\aleph$-injective, the homomorphisms $\pi_i f \colon B \to M_i$ can be extended to homomorphisms $g_i \colon R_R \to M_i$. Then the homomorphism $\prod_{i \in I} g_i \colon R_R \to \prod_{i \in I} M_i$ is an extension of the homomorphism f. Therefore $\prod_{i \in I} M_i$ is an $\aleph$-injective module.

(2) Let $\{M_i\}_{i \in I}$ be some set of $\aleph$-injective right R-modules, $M = \prod_{i \in I} M_i$, $h \colon M \to \overline{M}$ be an epimorphism, B be an $\aleph$-generated right ideal in R, and let $\overline{f} \colon B \to \overline{M}$ be a homomorphism of right R-modules. Since B is

a projective R-module, we have that $\overline{f} = hf$ for some homomorphism $f\colon B_R \to M$. By (1), M is an $\aleph$-injective right R-module. Therefore the homomorphism f can be extended to some homomorphism $g\colon R_R \to M$. Then $hg\colon R_R \to \overline{M}$ is an extension of the homomorphism $\overline{f}$. Therefore the module $\overline{M}$ is $\aleph$-injective.

(3) Let M be a finitely generated right R-module. There exists a finite set $\{R_i\}_{i=1}^{n}$ of isomorphic copies of the free cyclic module R_R such that M is a homomorphic image of the module $\oplus_{i=1}^{n} R_i$. Since $\oplus_{i=1}^{n} R_i = \prod_{i=1}^{n} R_i$, it follows from (2) that M is an $\aleph$-injective module. $\square$

Lemma 10.2 *([97], 2.14, 2.15) Let R be a von Neumann regular ring and let B be an $\aleph_0$-generated right ideal in R. Then the right R-module B is projective and the ring R contains a countable set $\{e_i\}_{i=1}^{\infty}$ of orthogonal idempotents such that $B = \oplus_{i=1}^{\infty} e_i R$.*

Lemma 10.3 *For a ring R, the following conditions are equivalent.*

1. *All finitely generated right R-modules are $\aleph_0$-injective.*
2. *All cyclic right R-modules are $\aleph_0$-injective.*
3. *R is a right $\aleph_0$-injective von Neumann regular ring.*
4. *The ring R is isomorphic to a direct product of right $\aleph_0$-injective von Neumann regular rings.*

Proof The implications $(1) \Rightarrow (2)$ and $(3) \Rightarrow (4)$ are obvious.

$(2) \Rightarrow (3)$. Since the module R_R is cyclic, it follows from (2) that the ring R is right $\aleph_0$-injective. Let $a \in R$ and let $f\colon aR \to aR \subseteq R$ be the identity mapping. Since the cyclic module aR is $\aleph_0$-injective, f can be extended to a homomorphism $g\colon R_R \to aR$. Then $aR = g(R) \subseteq R$, $g^2 = g$, so aR is a direct summand of the module R_R, and the ring R is von Neumann regular.

$(4) \Rightarrow (3)$. Since the ring R is isomorphic to a direct product of von Neumann regular rings, R is von Neumann regular. With the use of Lemma 10.1(1), we can verify that R is a right $\aleph_0$-injective ring.

$(3) \Rightarrow (1)$. By Lemma 10.2, all $\aleph_0$-generated right ideals of the ring R are projective. By Lemma 10.1(3), all finitely generated right R-modules are $\aleph_0$-injective. $\square$

Lemma 10.4 *For a von Neumann regular ring R, the following conditions are equivalent.*

1. *R is a right $\aleph_0$-injective ring.*
2. *For any countable set $\{e_i\}_{i=1}^{\infty}$ of orthogonal idempotents in R, and each countable subset $\{a_i\}_{i=1}^{\infty}$ of R, there exists an element $a \in R$ such that $ae_i = a_i e_i$ for all i.*

Proof $(1) \Rightarrow (2)$. There exists a homomorphism $f\colon \oplus_{i=1}^{\infty} e_i R \to R_R$ such that $f(\sum e_i x_i) = \sum a_i e_i x_i$. By assumption, f can be extended to a homomorphism $g\colon R_R \to R_R$. We set $a = g(1)$. Then $ae_i = a_i e_i$ for all i.

$(2) \Rightarrow (1)$. Let B be an $\aleph_0$-generated right ideal of the ring R and let $f \colon B \to R$ be a homomorphism of right R-modules. By Lemma 10.2, the ring R contains a countable set $\{e_i\}_{i=1}^{\infty}$ of orthogonal idempotents such that $B = \oplus_{i=1}^{\infty} e_i R$. We set $a_i = f(e_i)$. Then $f(e_i) = f(e_i)e_i = a_i e_i$, $i = 1, 2, \ldots$. By (2), there exists an element $a \in R$ such that $ae_i = a_i e_i = f(e_i)$ for all i. There exists a homomorphism $g \colon R_R \to R_R$ such that $g(x) = ax$, $x \in R$. Let $b = \sum_{i=1}^{n} e_i x_i \in B$, $x_i \in R$. Then $g(b) = ab = \sum_{i=1}^{n} ae_i x_i = \sum_{i=1}^{n} f(e_i)x_i = f(b)$. Therefore g is an extension of f and the ring R is right $\aleph_0$-injective. $\qquad\square$

The next theorem is due to Tuganbaev [228].

Theorem 10.5 *For a ring R, the following conditions are equivalent.*

1. *All cyclic right R-modules are $\aleph_0$-injective.*
2. *All finitely generated right R-modules are $\aleph_0$-injective.*
3. *R is a von Neumann regular ring and for any countable set $\{e_i\}_{i=1}^{\infty}$ of orthogonal idempotents in R, and each countable set $\{a_i\}_{i=1}^{\infty}$ of elements in R, there exists an element $a \in R$ such that $ae_i = a_i e_i$ for all i.*

Proof This follows easily from Lemmas 10.3 and 10.4. $\qquad\square$

If we replace finitely generated modules by $\aleph_0$-generated modules in the above theorem, then we obtain semisimple rings.

Theorem 10.6 *For a ring R, the following conditions are equivalent.*

1. *All $\aleph_0$-generated right R-modules are $\aleph_0$-injective.*
2. *In the ring R, all $\aleph_0$-generated right ideals are $\aleph_0$-injective.*
3. *R is a von Neumann regular ring without infinite sets of orthogonal idempotents.*
4. *R is a semisimple ring.*
5. *All right R-modules and all left R-modules are injective and projective.*

Proof The equivalence of conditions (3), (4), and (5) is well-known (See [75, 19.26A], [153, 8.2.2]). The implications $(4) \Rightarrow (1)$ and $(1) \Rightarrow (2)$ are obvious.

$(2) \Rightarrow (3)$. By Theorem 10.5, the ring R is von Neumann regular. We assume that R contains some infinite set $\{e_i\}_{i=1}^{\infty}$ of nonzero orthogonal idempotents. We denote by B the right ideal $\oplus_{i=1}^{\infty} e_i R$ of the ring R. Let $f \colon B \to B \subseteq R$ be the identity mapping. Since the $\aleph_0$-generated right ideal B is $\aleph_0$-injective, f can be extended to a homomorphism $g \colon R_R \to B$. Then $B = g(R)$, $g^2 = g$, B is a direct summand of the module R_R, and so $R_R = (\oplus_{i=1}^{\infty} e_i R) \oplus fR$ for some idempotent f, which is impossible since R has an identity. $\qquad\square$

Our next purpose is to show that right $\aleph_0$-injective von Neumann regular rings coincide with right $\aleph_0$-algebraically compact von Neumann regular rings (see Proposition 10.8 below).

Let R be a ring, M be a right R-module, and let $L(a_{ij}, m_i, M_R)$ be a system consisting of a countable number of linear equations $\{\sum_{j=0}^{t(i)} x_j a_{ij} = m_i\}_{i=0}^{\infty}$ with coefficients $a_{ij} \in R$. We assume that $L(a_{ij}, m_i, M_R)$ contains a countable number of variables $\{x_j\}_{j=0}^{\infty}$, which take values in M, and any finite subsystem of the system is solvable. Such a system $L(a_{ij}, m_i, M_R)$ is called a *countable finitely solvable* system. A right R-module M is said to be $\aleph_0$-*algebraically compact* if any countable finitely solvable system $L(a_{ij}, m_i, M_R)$ has a solution in M.

Lemma 10.7 *For a ring R and a right R-module M, the following conditions are equivalent.*

1. *M is an $\aleph_0$-algebraically compact module.*
2. *For any countable finitely solvable system $L(a_{ij}, m_i, M_R) = L$, there exists an element $y_0 \in M$ such that each finite subsystem L^* in L has a partial solution $\overline{y^*} = \{y_0, y_1^*, \ldots\}$, where y_0 depends on the system L only.*

Proof The implication $(1) \Rightarrow (2)$ is easily verified.

$(2) \Rightarrow (1)$. We have to prove that there exists a solution $\overline{y} = \{y_0, y_1, \ldots\}$ of the system L. We inductively prove that the elements $y_n \in M$ exist. By assumption, the element y_0 exists. We assume that $y_0, \ldots, y_n$ have been obtained. We consider the system $N = \{\sum_{j=n+1}^{t(i)} x_j a_{ij} = m_i - \sum_{j=0}^{n} y_j a_{ij}\}_{i=0}^{\infty}$. By the assumption applied to N and the variable x_{n+1}, there exists $y_{n+1} \in M$ such that any finite subsystem N^* in N has a solution $\overline{z^*} = \{y_{n+1}, y_{n+2}^*, \ldots\}$, where y_{n+1} depends on the system N, but y_{n+1} is not changed if we pass to another finite subsystem of N. We prove that $\overline{y} = \{y_0, y_1, \ldots\}$ is the required solution of the system L. It is sufficient to prove that $\overline{y}$ is a solution of any finite subsystem L^* in L. The subsystem L^* contains only a finite number of variables $x_0, \ldots, x_n$. It follows from the method of the choice of the sequence $\{y_m\}_{m=0}^{\infty}$ that there exists a solution $\overline{y}^*$ of the system L^* such that $y_0, \ldots, y_n \in M$ are the first $n + 1$ components of this solution. Therefore $\overline{y}$ is a solution of the system L^*. $\square$

Proposition 10.8 *([220]) For a von Neumann regular ring R, the following conditions are equivalent.*

1. *R is a right $\aleph_0$-injective ring.*
2. *R is a right $\aleph_0$-algebraically compact ring.*

Proof $(1) \Rightarrow (2)$. Let L be a countable, finitely solvable system $\{\sum_{j=0}^{t(i)} x_j a_{ij} = m_i\}_{i=0}^{\infty}$ such that $a_{ij} \in R$, $m_i \in R$, and for any $n \geq 0$, the finite subsystem $L_n = \{\sum_{j=0}^{t(i)} x_j a_{ij} = m_i\}_{i=0}^{n}$ in L has a partial solution $\overline{x(n)} = \{x_j(n)\}_{j=0}^{m(n)}$, where $m(n) = \max_{0 \leq i \leq n}\{t(i)\}$. By Lemma 10.7, it is sufficient to prove that there exists a set $\{\overline{y(n)}_{n=0}^{\infty}\}$ of partial solutions of the system L_n such that $\overline{y(n)} = \{b, y_1(n), y_2(n), \ldots\}$, where b depends on L only and b is not changed with change of n. We fix the integer n. Let $H_n = \{\sum_{j=0}^{t(i)} x_j a_{ij} = 0\}_{i=0}^{n}$ be the nonhomogeneous finite system corresponding to the nonhomogeneous system L_n, G_n be the set of all solutions of the system H_n, and let T_n be the left ideal in R generated by all first components from G_n. Then G_n is the kernel of some endomorphism h of the free left R-module F of rank $m(n)$. The module F is regular

[97, 2.11]. Therefore $h(F)$ is a direct summand in F and the module G_n is finitely generated, since G_n is isomorphic to a direct summand of the finitely generated module F. The left ideal T_n of the von Neumann regular ring R is a homomorphic image of the module G_n, and then T_n is finitely generated. Then $R = T_n \oplus R f_n$, where f_n is an idempotent. We begin to change n. Since $T_n \supseteq T_{n+1}$, there exists an idempotent e_{n+1} such that $T_n = T_{n+1} \oplus R e_{n+1}$ and $f_n = f_{n+1} + e_{n+1}$. We set $e_0 = f_0$. Then $\{e_i\}_{i=0}^{\infty}$ is a countable set of orthogonal idempotents and $f_n = \sum_{i=0}^{n} e_i$ for $n \geq 0$. We set $a_i = x_0(i) \in R$. Then $a_i - a_n \in T_i$ for $i \leq n$, since $x(i) - x(n) \in H_i$. Therefore $a_i f_i = a_n f_i$ for $i \leq n$. By Theorem 10.5, there exists $b \in R$ such that $b e_i = a_i e_i$ for all i. Then

$$(b - a_n) f_n = \sum_{i=0}^{n} (b - a_n) e_i = \sum_{i=0}^{n} (b - a_i) e_i = 0,$$

$$(b - a_n)(1 - f_n) = b - a_n, \quad b - a_n \in T_n.$$

Therefore, for all i, there exists a solution $\overline{z(n)} = \{z_j(n)\}_{j=0}^{m(n)}$ of the nonhomogeneous system H_n such that $b = a_n + z_0(n)$. If $\overline{y(n)}$ is the sum of the solution $\overline{x(n)}$ of the nonhomogeneous system L_n and the solution $\overline{z(n)}$ of the nonhomogeneous system H_n, then $\overline{y(n)}$ is a solution of the nonhomogeneous system L_n and $\overline{y(n)} = \{b, y_1(n), y_2(n), \ldots\}$.

$(2) \Rightarrow (1)$. By Lemma 10.4, it is sufficient to prove that the countable system $L = \{x e_i = a_i e_i\}_{i=1}^{\infty}$ is finitely solvable. This is equivalent to the property that the finite system $L_n = \{x e_i = a_i e_i\}_{i=1}^{n}$ is solvable for any n. Since all e_i are orthogonal to each other, $\sum_{i=0}^{n} a_i e_i$ is a partial solution of the system L_n. $\square$

We give below an example of an $\aleph_0$-injective ring which is not self-injective.

Example Let $\{\mathbb{F}_i : i \in \mathbb{N}\}$ be an infinite family of fields and $R = \frac{\prod_{i \in \mathbb{N}} \mathbb{F}_i}{\oplus_{i \in \mathbb{N}} \mathbb{F}_i}$. Then R is a commutative von Neumann regular $\aleph_0$-injective ring but R is not a self-injective ring. $\square$

Next, we give an example of a commutative von Neumann regular ring which is not $\aleph_0$-injective.

Example Let $\mathbb{F}$ be a field and R be the ring consisting of all eventually constant sequences $f = (f_n)_{n=0}^{\infty}$ of elements of $\mathbb{F}$ (for each $(f_n) \in R$, there exists an integer $N = N(f)$ such that $f_n = f_N$ for all $n \geq N$. Then the ring R is a commutative von Neumann regular ring which is not $\aleph_0$-injective. $\square$

10.2   Rings each of whose cyclic modules is weakly injective

A right R-module M is called *weakly N-injective* if for each homomorphism $f : N \to E(M)$, there exists a submodule X of $E(M)$ such that $f(N) \subset X \cong M$. A right R-module M is called *weakly injective* if M is weakly N-injective for each finitely generated right R-module N.

We give below examples of weakly-injective modules which are not injective.

Example

   (i) $\mathbb{Z}$ is weakly-injective but not injective as a $\mathbb{Z}$-module.

   (ii) Let F be a field and K be a proper subfield. Let $S = \prod_{i=1}^{\infty} F_i$, where each $F_i = F$. Let $R = \{(x_i) \in S : \text{all but finite many } x_i \in K\}$. Then R_R is weakly R-injective but not injective.

We will start with some basic observations on weak injectivity.

Lemma 10.9 *A cyclic right R-module M is weakly injective if M is weakly R^2-injective.*

Proof Suppose the cyclic right R-module M is weakly R^2-injective. To show that M is weakly injective, we need to show that M is weakly R^n-injective for each $n \in \mathbb{N}$. We will show this by induction. Suppose M is weakly R^{n-1}-injective and let $x_1, \ldots, x_{n-1} \in E(M)$. By the induction hypothesis, there exists $xR \subseteq E(M)$ such that $x_1, \ldots, x_{n-1} \in xR$. By the weak R^2-injectivity of M, there exists $X \cong M$ such that $x, x_n \in X$. Hence $x_1, \ldots, x_n \in X$. This completes the proof. $\qquad\square$

We now record some properties of weakly-injective modules established in [10]. Any finite direct sum of weakly injective modules is weakly injective but a direct summand of a weakly injective module need not be weakly injective. If a right R-module M is weakly injective and N is an essential extension of M then N is also weakly injective. In [11] Al-Huzali, Jain, and López-Permouth showed that every direct sum of weakly injective right R-modules is again weakly injective if and only if R is a right *q.f.d.* ring.

The next lemma is a simple observation.

Lemma 10.10 *If a right R-module M is weakly injective and quasi-injective, then M is injective.*

Proof Let $a \in E(M)$. As M is weakly-injective, there exists $X \subseteq E(M)$ such that $a \in X \cong M$. As M is quasi-injective, $X = M$. Therefore $a \in M$. Thus $M = E(M)$, hence M is injective. $\qquad\square$

In [137] Jain, López-Permouth, and Singh studied rings where each cyclic module is weakly injective and proved the following.

Theorem 10.11 *The following conditions on a ring R are equivalent:*

   1. R is right weakly semisimple, that is, each right R-module is weakly injective.

   2. Each finitely generated right R-module is weakly injective and R is right noetherian.

   3. Each cyclic right R-module is weakly R^2-injective and R is right noetherian.

    4. Each cyclic uniform right R-module is weakly R^2-injective and R is right noetherian.

Proof (1)$\Rightarrow$(2). Let R be a right weakly semisimple ring. We will show that R must be right noetherian. Let M be a quasi-injective right R-module. By assumption, M is weakly injective. Then, by the above lemma, M must be injective. Thus R is a right QI ring and hence by [30], R must be right noetherian.

    Clearly (2)$\Rightarrow$(3) and (3)$\Rightarrow$(4).

    (4)$\Rightarrow$(1). Suppose each cyclic right R-module is weakly R^2-injective and R is right noetherian. Let M be any right R-module and $x_1, \ldots, x_n \in E(M)$. Let $K = \Sigma_{i=1}^n x_i R$. It follows that $E(M) = E(K) \oplus L$ for some submodule L of $E(M)$ and $E(K)$ has finite Goldie dimension. Therefore $M \cap E(K)$, being essential in $E(K)$, has finite Goldie dimension. Let N be a finite direct sum of uniform cyclic modules which is essential in $M \cap E(K)$. By Lemma 10.9, we know that if a cyclic right R-module is weakly R^2-injective then it is weakly injective. Since every finite direct sum of weakly injective modules is again weakly injective, it follows that N is weakly injective. Thus there exists an automorphism $\sigma : E(K) \to E(K)$ with $K \subseteq \sigma(N) \subseteq \sigma(M \cap E(K))$. Clearly σ extends to an automorphism φ of $E(M)$ which is the identity on L. This automorphism satisfies $K \subseteq \varphi(M)$. Hence M is weakly injective. Thus every right R-module is weakly injective. Therefore R is right weakly semisimple. $\qquad\square$

    Jain, López-Permouth, and Singh [137] also showed the following.

Theorem 10.12 *If a ring R satisfies the restricted right minimum condition and is right weakly semisimple then R is also left weakly semisimple.*

10.3   Rings each of whose cyclic modules is quasi-projective

A right R-module M is called *quasi-projective* if for every submodule K of M, the induced sequence $\text{Hom}(M, M) \to \text{Hom}(M, M/K) \to 0$ is exact. The study of rings over which each cyclic module is quasi-projective was initiated by Koehler [156]. A ring R is called a *right q^*-ring* if each cyclic right R-module is quasi-projective.

    Koehler [156] proved the following.

Theorem 10.13 *For a ring R, we have the following:*

    1. If R is semiperfect, then R is a left q^*-ring if and only if every left ideal in the Jacobson radical of R is an ideal.

    2. Let R be a left self-injective semiperfect left q^*-ring. Then R has a ring direct sum decomposition given as $R = A \oplus B$, where A is semisimple artinian and B is a finite direct sum of pairwise nonisomorphic uniform left ideals.

    3. If R is a prime semiperfect left q^*-ring, then R is either semisimple artinian or R is local.

4. *Let R be a quasi-Frobenius ring. Then R is a left q^*-ring if and only if R is a right q^*-ring.*

5. *A proper matrix ring $\mathbb{M}_n(R)$ $(n > 1)$ is a left q^*-ring if and only if R is semisimple artinian.*

Proof

(1) To prove this we first note the following useful facts proved by Miyashita [176] and Wu and Jans [236]. Let P be a projective module, $\varphi : P \to M$ be an epimorphism and $S = \mathrm{End}(P)$. Then

 (a) M is quasi-projective if $\mathrm{Ker}(\varphi)$ is invariant under S, and

 (b) $\mathrm{Ker}(\varphi)$ is invariant under S if $\mathrm{Ker}(\varphi)$ is small in P and M is quasi-projective.

 Let R be a semiperfect left q^*-ring. Let I be a left ideal of R contained in $J(R)$. Consider $\varphi : R \to R/I$. Since I is small in R and R/I is quasi-projective, I is invariant under $\mathrm{End}(R) \cong R$ by (b) above. Hence I is an ideal. Conversely, assume that every left ideal in $J(R)$ is an ideal. Let K be a left ideal in R. Then R/K has a projective cover $\varphi : P \to R/K$, and P can be considered to be a direct summand of R. Since $\mathrm{Ker}(\varphi)$ is small in P, it is small in R and contained in $J(R)$. If $f \in \mathrm{End}(P)$, then there is an $r \in R$ such that $f(x) = xr$ for every $x \in P$. Therefore $\mathrm{Ker}(\varphi)$ is invariant under $\mathrm{End}(P)$. Hence by (b) above, R/I is quasi-projective. Therefore R is a left q^*-ring.

(2) Since R is semiperfect, $R = Re_1 + \ldots + Re_k + Re_{k+1} + \ldots + Re_n$, where $e_1, \ldots, e_n$ are orthogonal indecomposable idempotents. We may assume $Re_1, \ldots, Re_k$ are all the simple components of the decomposition. By (1), $J(R)e_i e_i Re_j = 0$ if $i \neq j$. Since $\mathrm{Hom}(Re_i, Re_j) = e_i Re_j$, $Re_i \not\cong Re_j$, for $i, j > k$ and $i \neq j$. Let $A = Re_1 + \ldots + Re_k$, and $B = Re_{k+1} + \ldots + Re_n$. To complete the proof we show that A and B are ideals. Let $i \leq k$ and $j > k$. Then $Re_i e_i Re_j$ is 0 or simple and is contained in Re_j. Since Re_i is a simple injective module, and Re_j is not simple and is indecomposable, $Re_i e_i Re_j = 0$. Similarly $Re_j e_j Re_i = 0$ because Re_i is a simple projective module.

(3) Since R is semiperfect, $R = Re_1 + \ldots + Re_n$ where $e_1, \ldots, e_n$ are nonzero orthogonal indecomposable idempotents. If $n = 1$, then R is local because $J(R)e_1$ is a unique maximal left ideal in Re_1. If $n > 1$, then from (1), $J(R)e_i J(R)e_j \subseteq J(R)e_i \cap J(R)e_j = 0$ for $i \neq j$. Hence $J(R)e_i = 0$ for all but at most one i, say $i = k$, because R is prime. Thus $J(R)e_k \neq 0$. For an $i \neq k$, we get an epimorphism $\sigma : Re_k \to Re_i$. As Re_i is simple, and σ is an isomorphism, we get $J(R)e_k = 0$, which is a contradiction. Hence R is simple artinian.

(4) Let R be a quasi-Frobenius left q^*-ring. We first claim that each essential left ideal of R is an ideal. Let I be an essential left ideal of R. Then R/I has a projective cover $\varphi : P \to R/I$. Let $f : R \to R/I$ be the canonical

homomorphism. Then there is an epimorphism $f' : R \to P$ such that $\varphi \circ f' = f$ because P is projective and $\mathrm{Ker}(\varphi)$ is small. Since P is projective and f' is onto, $R = Re_1 \oplus Re_2$ with $Re_1 \cong P$. Also, it can be seen that $I = K \oplus Re_2$ where $K \cong \mathrm{Ker}(\varphi)$ and $K \subseteq Re_1$. The left ideal K is small in R, and R is a left q^*-ring. Thus $K \subseteq J(R)$, and K must be an ideal by (1). Now as I is an essential left ideal of R, the left socle $\mathrm{Soc}(_RR)$ of R is contained in I. The left ideal $J(R)e_2$ is an ideal in R. So $Re_2e_2Re_1 \subseteq \mathrm{Soc}(_RR)$ because $J(R)e_2e_2Re_1 = 0$ and R is semiperfect. Therefore I is an ideal. Next, we claim that each left ideal of R is quasi-injective. Let A be any left ideal of R. If B is a complement of A in R then $A \oplus B$ is an essential left ideal of R. We have just now shown that $A \oplus B$ must be an ideal of R. Since R is left self-injective, $A \oplus B$ is quasi-injective and hence A is quasi-injective. Thus each left ideal of R is quasi-injective. Now we proceed to show that R must be right q^*. Let L be a right ideal contained in $J(R)$ and $a \in L$. Since R is left self-injective, $J(R) = \{a : ann_l(a)$ is an essential left ideal of $R\}$ and $ann_r(ann_l(aR)) = aR$. In addition, $ann_l(aR) = ann_l(a)$ which is an essential left ideal of R. Therefore aR is an ideal, and so L is an ideal. Hence, by (1), R is a right q^*-ring. The converse is similar.

(5) Let $\mathbb{M}_n(R)$ be a left q^*-ring. To show that R is semisimple artinian, it suffices to show that every simple left R-module is projective. Let M be a maximal left ideal of R. The set of all matrices in $\mathbb{M}_n(R)$ with entries in M will be denoted by $\mathbb{M}_n(M)$. Let I be the left ideal $\mathbb{M}_n(M)e_{11}$ of $\mathbb{M}_n(R)$. Set $A = \mathbb{M}_n(R)e_{11}/I$, and $P = (\Sigma_{i=2}^n \mathbb{M}_n(R)e_{ii}) + I/I$. Note that P is projective as $P \cong \Sigma_{i=2}^n \mathbb{M}_n(R)e_{ii}$. Also, $\mathbb{M}_n(R)/I = P \oplus A$ as $\mathbb{M}_n(R)$-modules. Define an $\mathbb{M}_n(R)$-epimorphism $\varphi : P \to A$ by $\varphi([a_{ij}] + I) = [b_{ij}] + I$, where $b_{ij} = 0$ for $j > 1$ and $b_{i1} = a_{i2}$ for $i = 1, 2, \ldots, n$. Since $P \oplus A$ is quasi-projective, there exists a monomorphism $f : A \to P$ such that φf is identity on A. Let $\pi : R \to R/M$ be the natural map. The simple module R/M will be projective if there is a homomorphism $g : R/M \to R$ such that πg is the identity on R/M. Define $g : R/M \to R$ by $g(r + M) = a_{12}$ where $f(re_{11} + I) = [a_{ij}] + I$. Then clearly g is an R-homomorphism and πg is the identity on R/M. Thus the simple module R/M is projective and hence R is semisimple artinian. The converse is obvious. $\qquad\square$

Koehler [156] gave the following example of a ring which is both left and right artinian, a right q^*-ring but not a left q^*-ring.

Example Let R be the ring of matrices of the form $\begin{bmatrix} \bar{a} & \bar{b} \\ 0 & \bar{c} \end{bmatrix}$ such that $\bar{a} \in \mathbb{Z}_4$ and $\bar{b}, \bar{c} \in \mathbb{Z}_2$. Clearly, $J(R)$ consists precisely of the matrices of the form $\begin{bmatrix} \bar{a} & \bar{b} \\ 0 & \bar{0} \end{bmatrix}$ such that $\bar{a} = \bar{0}$ or $\bar{2} \in \mathbb{Z}_4$ and $\bar{b} \in \mathbb{Z}_2$. Every right ideal in $J(R)$ is an ideal. However the left ideal I consisting of just the two elements $\begin{bmatrix} \bar{0} & \bar{0} \\ 0 & \bar{0} \end{bmatrix}$ and $\begin{bmatrix} \bar{2} & \bar{1} \\ 0 & \bar{0} \end{bmatrix}$ is not an ideal. Therefore by (1) of the above theorem, R is a right q^*-ring but not a left q^*-ring. $\qquad\square$

10.14 Questions

1. By Baer's criterion, it is clear that $\aleph$-injective R-modules are injective provided $\aleph$ is the cardinality of R. Let R be an infinite ring of cardinality $\tau > \aleph_0$. Let us assume that $\aleph_0 < \aleph < \tau$ and all cyclic right R-modules are $\aleph$-injective. What can we say about the ring R? When is R semisimple?
2. Which group rings (monoid rings) are right $\aleph_0$-injective?

11 Hypercyclic, q-hypercyclic, and π-hypercyclic rings

11.1 Hypercyclic rings

A ring R is called a *right hypercyclic ring* if each cyclic right R-module has a cyclic injective hull. Caldwell [37] initiated the study of such rings. A ring R is called a *restricted right hypercyclic ring* if R is right hypercyclic and $R/J(R)$ is artinian.

We start with the following basic lemma.

Lemma 11.1 *A commutative hypercyclic ring is self-injective.*

Proof Let R be a commutative hypercyclic ring. Then by hypothesis, $E(R) \cong R/I$ where I is an ideal of R. Thus it follows that R embeds in R/I. So $I = 0$ and hence $R \cong E(R)$. Thus R is self-injective. $\qquad\square$

Lemma 11.2 *If $\{e_i : i \in \mathcal{I}\}$ is any set of idempotents in a commutative hypercyclic ring R, and if $A = \Sigma_{i \in \mathcal{I}} e_i R$, then R/A is an injective R-module.*

Proof Let $E(R/A) = R/K$. Then R/A embeds essentially in R/K, so $K \subseteq A$. Now if e is any idempotent in R, then

$$\frac{R}{K} = \frac{(1-e)R + K}{K} \oplus \frac{eR + K}{K}.$$

The sum is direct since R is commutative. Let $\varphi : R/A \to R/K$ denote the embedding. Now for any $x \in R$, if $\varphi(x + A) \in (eR + K)/K$, write $\varphi(x + A) = y + K$. Then $ye + K = ey + K = y + K$, so $\varphi(x + A) = \varphi(x + A)e$. Since $\varphi((x + A)e) = \varphi(xe + A) = \varphi(ex + A) = \varphi(0 + A) = 0$ whenever $e \in A$, $(eR + K)/K \cap \varphi(R/A) = 0$ whenever $e \in A$. Since R/A is essential in R/K, $eR \subseteq K$. Hence $A \subseteq K$. So $A = K$. Hence R/A is injective. $\qquad\square$

Theorem 11.3 *A commutative hypercyclic ring is a restricted hypercyclic ring.*

Proof Let R be a commutative hypercyclic ring. Then by Lemma 11.1, R is self-injective. Therefore $R/J(R)$ is a self-injective von Neumann regular ring. Clearly, $R/J(R)$ is either semisimple artinian or it has an infinite set of orthogonal idempotents. Suppose $R/J(R)$ has an infinite set of orthogonal idempotents. Since R is self-injective, orthogonal idempotents can be lifted orthogonally modulo $J(R)$ and so R has an infinite set of orthogonal idempotents, say $\{e_i : i \in \mathcal{I}\}$. Let $A = \Sigma_{i \in \mathcal{I}} e_i R$. Then R/A is noninjective by Lemma 4.2. By the above lemma, R/A is injective. Thus we arrive at a contradiction. Therefore $R/J(R)$ is semisimple artinian and hence R is a restricted hypercyclic ring. $\square$

Next, we consider commutative local hypercyclic ring and show that it must be uniserial. But, first we need the following basic lemma which is true for any local ring not necessarily commutative.

Lemma 11.4 *If R is a local ring, then cyclic right R-modules are indecomposable.*

Proof Let R be a commutative local ring. Consider any cyclic right R-module R/A. If $R/A = C/A \oplus D/A$, then $C + D = R$, and since every right ideal different from R is contained in $J(R)$, either C or D is superfluous, so that $R = C$, say, hence $D \subseteq A$ and R/A is indecomposable. $\square$

Proposition 11.5 *A commutative local hypercyclic ring is uniserial.*

Proof Let R be a commutative local hypercyclic ring. Let A and B be ideals in R. If $A/A \cap B$ and $B/A \cap B$ were both nonzero, then $A/A \cap B$ would have nonzero complement, say C in $R/A \cap B$. But then $A/A \cap B \oplus C$ would lead to a decomposition of $E(R/A \cap B)$, and since the injective hull of $R/A \cap B$ is cyclic, it is indecomposable by above lemma. Hence $A/A \cap B = 0$ or $B/A \cap B = 0$. Thus $A \subseteq A \cap B$ or $B \subseteq A \cap B$. Hence, $A \subseteq B$ or $B \subseteq A$. This shows that R is uniserial. $\square$

Osofsky [190] extended this result to noncommutative rings. We start with the following easy observation.

Lemma 11.6 *Let $R/J(R)$ be artinian, and I be a right ideal of R. Suppose $R/I = \oplus_{i=1}^{n} M_i$, where the modules M_i are nonzero. Then n is less than or equal to the composition length of $R/J(R)$. Moreover, if $M_i/M_i J(R)$ is simple for each i, the number of $M_i/M_i J(R)$ isomorphic to a given simple module M is less than or equal to the number of factor modules isomorphic to M in a composition series for $R/J(R)$.*

Proof Note that

$$\oplus_{i=1}^{n} M_i/M_i J(R) \cong (R/I)/(R/I)J(R) \cong R/(I + J(R))$$

$$\cong (R/J(R))/(I + J(R))/J(R)$$

and so the lemma follows easily from Jordan–Holder–Schreier Theorem. $\square$

Corollary 11.7 *Let R be a restricted right hypercyclic ring, $e = e^2 \in R$, and $l(eR/eJ(R)) = m$. Then any independent set of submodules of a factor of eR has at most m elements.*

Proof Let $\{M_i : 1 \leq i \leq k\}$ be an independent family of submodules of eR/eI. Then $R/eI \supseteq (1 - e)R \oplus (\oplus_{i=1}^{k} M_i$. Therefore $E(R/eI)$ contains a direct sum of $l(R/J(R)) - m + k$ terms. Hence by the above lemma, $k \leq m$. $\square$

Corollary 11.8 *Let R be a restricted hypercyclic ring, $e = e^2 \in R$ such that $eR/eJ(R)$ is a simple module. Then eR is uniserial.*

Proof Let M and N be submodules of eR. By the above corollary, $M/(M \cap N) = 0$ or $N/(M \cap N) = 0$. So $M \subseteq M \cap N$ or $N \subseteq M \cap N$. Thus, $M \subseteq N$ or $N \subseteq M$. $\square$

In particular, we have the following.

Corollary 11.9 *Let R be a local right hypercyclic ring. Then R is a right uniserial ring.*

Next, we prove a useful lemma owed to Osofsky [190].

Lemma 11.10 *Let R be a right hypercyclic ring such that every homomorphic image of R has the ascending chain condition on direct summands. Then R is a right self-injective restricted right hypercyclic ring.*

Proof Since R is right hypercyclic, $E(R) \cong R/I$ for some right ideal I. Let f embed R in R/I, $f(1) = x + I$. Since f is one-to-one, $xR \cap I = 0$, so $E(R) = E(xR) \oplus E(I) \oplus M$ for some submodule M of $E(R)$. Since $xR \cong R$, we have $E(xR) \cong E(R) \oplus E(I) \oplus M$. So $E(R) \cong E(R) \oplus E(I) \oplus M \oplus E(I) \oplus M \cdots \cong E(R) \oplus (\oplus_{i=1}^n E(I) \oplus M)$. Since R/I has ascending chain condition on direct summands, $E(I) \oplus M = 0$, so $I = 0$. Thus $R \cong E(R)$ and hence R is right self-injective. Since then $R/J(R)$ is von Neumann regular, the chain condition on $R/J(R)$ implies that $R/J(R)$ is artinian. This completes the proof. $\square$

As a consequence, we have:

Corollary 11.11 *Let R be a restricted right hypercyclic ring. Then R is right self-injective.*

Proof Let R be a restricted right hypercyclic ring. Then $R/J(R)$ is artinian. Now, by Lemma 11.6, every homomorphic image of R satisfies the ascending chain condition on direct summands. Therefore, by Lemma 11.10, R is right self-injective. $\square$

The above corollary yields the following result of Faith [71].

Lemma 11.12 *A left perfect right hypercyclic ring is right self-injective.*

Proof This is obvious from the above corollary. $\square$

Osofsky [190] continued the study of hypercyclic and restricted hypercyclic rings and showed the following.

Lemma 11.13 *Let R be a restricted right hypercyclic ring. Then $\mathrm{Soc}(R_R) \subseteq_e R_R$.*

Proof Set $\overline{R} = R/J(R)$. We have $\overline{R} = \oplus_{i=1}^n S_i$, where S_i is simple and each simple R-module is isomorphic to some S_i. Hence $E(\overline{R}) = \oplus_{i=1}^n E(S_i)$ is faithful and

$E(S_i) = x_i R$ for some x_i. Let $A_i = \{r \in R : x_i r \in S_i\}$. Then $A_i \subseteq_e R_R$ and so $\cap_{i=1}^n A_i \subseteq_e R_R$. Then $E(\overline{R}).\cap_{i=1}^n A_i \subseteq \overline{R}$ and so $E(\overline{R}).(\cap_{i=1}^n A_i)J(R) = 0$. Thus we have $(\cap_{i=1}^n A_i)J(R) = 0$. So $\cap_{i=1}^n A_i \subseteq_e \mathrm{Soc}(R_R) \subseteq_e R_R$. $\square$

Lemma 11.14 *Let R be right hypercyclic, $R/J(R) \cong \mathbb{M}_n(D)$ where D is a division ring. Then $R \cong \mathbb{M}_n(S)$, where $S/J(S) \cong D$ and so S is a local ring.*

Proof Let $R/J(R) = \oplus_{i=1}^k \bar{e}_i R/J(R)$, where $\{\bar{e}_i\}$ are primitive orthogonal idempotents in $R/J(R)$. These idempotents lift modulo $J(R)$ to yield an orthogonal family $\{e_i\}$ of idempotents of R and we have $R = \oplus_{i=1}^k e_i R$. Since all simple $M_n(D)$-modules are isomorphic, $e_i R \cong e_j R$ for all $1 \leq i, j \leq n$. Then $R \cong M_n(S)$, where $S = \mathrm{Hom}_R(e_i R, e_i R) \cong e_i R e_i$ and $S/J(S) \cong e_i R e_i / e_i J(R) e_i \cong \mathrm{Hom}_{R/J(R)}(\bar{e}_i M_n(D), \bar{e}_i M_n(D)) \cong D$. $\square$

Lemma 11.15 *Let $R = \oplus_{i=1}^n R_i$ be a ring direct sum. Then R is right hypercyclic if and only if each R_i is right hypercyclic.*

Proof If e_i is the identity of R_i then $\{e_i : 1 \leq i \leq n\}$ are orthogonal central idempotents of R. Let I be an ideal of R. Then $I = \oplus_{i=1}^n e_i I$. Now it can be easily checked that $E(R/I) \cong R/K \Leftrightarrow E(e_i R/e_i I) \cong e_i R/e_i K$. $\square$

Lemma 11.16 *Let R be a restricted hypercyclic ring. Then R is a ring direct sum of matrix rings over local rings.*

Proof Since $R/J(R)$ is semisimple artinian, $R/J(R) = \oplus_{i=1}^k M_{n_i}(D_i)$, where each D_i is a division ring, $n_1 \geq n_2 \geq \cdots \geq n_k$, $M_{n_i}(D_i) = \bar{e}_i R/J(R)$ and each $\bar{e}_i$ is a central idempotent of $R/J(R)$. These idempotents lift modulo $J(R)$ to yield an orthogonal family $\{e_i : 1 \leq i \leq k\}$ of idempotents of R such that $\Sigma_{i=1}^k e_i = 1$. In view of Lemmas 11.14 and 11.15, we only need to show that each e_i is central. Assume to the contrary that there exist $i \neq j$ such that $e_i R e_j \neq 0$. Let $e_i r e_j \neq 0$. Consider

$$M = R/(\Sigma_{l \neq i} e_l R + e_j J(R) + e_i r e_j J(R)) \cong e_j R/e_j J(R) \oplus e_i R/e_i r e_j J(R).$$

This contains a direct sum of at least $n_j + 1$ copies of the unique simple $M_{n_j}(D_j)$-module S. Then $E(M) \supseteq \Sigma_{i=1}^{n_j+1} E(S) \cong R/I$ for some I. Since $E(S)/E(S)J(R)$ is the simple $M_{n_l}(D_l)$-module for some l with $n_l > n_j$. Now $E(R/J(R)) \cong R/K$ for some K. Observe that injective hull of any simple right module over a restricted right hypercyclic ring is projective. Since $R/J(R)$ is a finite direct sum of simple modules, $E(R/J(R))$ is projective. Hence $R \cong E(R/J(R)) \oplus K$, where length of $E(R/J(R))/E(R/J(R))J(R)$ equals the length of $R/J(R)$. By Lemma 11.6, $K = 0$, so $\mathrm{Soc}(R) \cong R/J(R)$. But then the number of composition factors of $\mathrm{Soc}(R) \cong S =$ the number of times $E(S)/E(S)J(R)$ appears as a composition factor of $R/J(R) =$ the number of composition factors of $R/J(R) \cong S$. Thus $n_i = n_j$, a contradiction. This shows that each e_i is a central idempotent and hence the proof is complete. $\square$

Next, we state two useful lemmas owed to Osofsky. See [190] for their proof.

Lemma 11.17 *Let $F = \Sigma_{i=1}^{n} x_i R$ be a free R-module with free basis $\{x_i : 1 \leq i \leq n\}$. Then $M_R \mapsto \mathrm{Hom}(F, M)$ provides a category isomorphism between the category of right R-modules and the category of right $\mathbb{M}_n(R)$-modules with the inverse given by $N_{\mathbb{M}_n(R)} \mapsto N \otimes_{\mathbb{M}_n(R)} F$.*

Lemma 11.18 *Let R be right hypercyclic, $R/J(R)$ a simple R-module. Let $F = \oplus_{i=1}^{n} R_i$, $K \subseteq F$. Let $\{N_i : 1 \leq i \leq k\}$ be a family of nonzero independent submodules of F/K. Then $k \leq n$.*

In the next lemma we consider local right hypercyclic rings.

Lemma 11.19 *Let R be a local ring. Then R is right hypercyclic if and only if $\mathbb{M}_n(R)$ is right hypercyclic for some n.*

Proof Let R be a right hypercyclic ring. Let us write $S = M_n(R)$. Let e be a primitive idempotent of S. Identify R with eSe. Then $Se_R \cong \oplus_{i=1}^{n} R_i$ where each $R_i = R$. Since the category isomorphism of Lemma 11.17 takes $R \to \mathrm{Hom}_R(Se, R) \cong eS$, every quotient of eS has an injective hull a quotient of eS. Let I be a right ideal of S. Then $S \to S/I \to 0$ is exact, so $S \otimes_S Se \to S/I \otimes_S Se \to 0$ is exact. Since $S \otimes_S Se_R \cong Se_R$, $S/I \otimes_S Se$ is an R-quotient of Se. By Lemma 11.18, $S/I \otimes_S Se$ is an essential extension of a sum of at most n cyclic right R-modules. By the category isomorphism, S/I is an essential extension of a sum of at most n quotients of eS. Hence its injective hull is a direct sum of at most n quotients of eS, and thus a quotient of S. This shows that $S = M_n(R)$ is a right hypercyclic ring.

Conversely, suppose $S = M_n(R)$ is right hypercyclic for some n, e is a primitive idempotent of S, and $A \subseteq eS$. Then we have $E(S/A) = (1 - e)S \oplus E(eS/A)$, so $E(eS/A)/E(eS/A)J(S)$ is simple, and $E(eS/A)$ is a quotient of eS. By the category isomorphism of Lemma 11.17, every quotient of R has an injective hull also a quotient of R. $\qquad\square$

Consequently, we have the following structure of restricted right hypercyclic rings.

Theorem 11.20 *(Osofsky, [190]) A ring R is restricted right hypercyclic if and only if R is a ring direct sum of matrix rings over right hypercyclic local rings.*

Proof It follows easily from Lemmas 11.15, 11.16, and 11.19. $\qquad\square$

Now as a consequence of Theorems 11.20 and 11.3, we get the following result (see [37]).

Theorem 11.21 *A commutative ring R is hypercyclic if and only if it is a direct sum of finitely many local hypercyclic rings.*

We summarize the structure of a local right hypercyclic ring as follows.

Theorem 11.22 *Let R be a local right hypercyclic ring. Then:*

1. *R is right uniserial.*
2. *R is right self-injective.*
3. *R is left uniserial.*
4. *Every left ideal and every right ideal of R is an annihilator ideal.*
5. *If $J(R)$ is nil, then R is a duo ring.*

Proof

1. See Corollary 11.9.
2. See Corollary 11.11.
3. Let $x_1, x_2 \in R$. Then by (1), $ann_r(x_1) \subseteq ann_r(x_2)$ or $ann_r(x_2) \subseteq ann_r(x_1)$. Now since R is right self-injective, any finitely generated left ideal is the annihilator of some right ideal. So, $ann_l(ann_r(x_1)) = Rx_1 \subseteq ann_l(ann_r(x_2)) = Rx_2$ or $Rx_1 \supseteq Rx_2$. Let I_1, I_2 be two left ideals of R such that $I_1 \not\subseteq I_2$. Let $x \in I_1 - I_2$. For $y \in I_2$, $Ry \not\supseteq Rx$ since $x \notin I_2$. Hence $Ry \subseteq Rx$ or $I_2 \subseteq I_1$.
4. Since R_R is injective and contains a copy of the unique simple right R-module, every right ideal of R is a right annihilator. Let I be any left ideal of R, and let $K = \cap_{Rx \supseteq I} Rx$. Then $K = ann_l(\Sigma_{Rx \supseteq I} ann_r(x))$ since $ann_l(ann_r(x)) = Rx$. If $I \neq K$, let $y \in K - I$. Then $Ry \supseteq I$, and $Rk \supseteq I \implies Rk \supseteq Ry$ for any $k \in R$. Hence Ry/I is a simple left R-module. By Nakayama's Lemma, $J(R)y \neq Ry$. Hence $J(R)y \subseteq I$. Since $Ry/J(R)y$ is also simple, $J(R)y = I$. Take $z \in R$ such that $J(R)yz = 0$ and $Ryz \neq 0$. By the linear ordering of left ideals, $J(R)y \subseteq ann_l(Rz) \subsetneq Ry$. Since $Ry/J(R)y$ is simple, we have $J(R)y = ann_l(Rz)$. Thus $I = ann_l(Rz)$. This shows that every left ideal and every right ideal of R is an annihilator ideal.
5. Assume $J(R)$ is nil, and let $0 \neq x, y \in R$. Suppose $xy \notin Rx$. Then $Rxy \supsetneq Rx$, so $x = zxy$ for some $z \in J(R)$. Since $J(R)$ is nil, $z^n = 0$ for some $n > 1$. Then $x = zxy = z^2xy^2 = \cdots = z^n xy^n = 0$, a contradiction. Hence $xR \subseteq Rx$. By symmetry, $Rx \subseteq xR$. Thus, $xR = Rx$ for all $x \in R$. $\qquad\square$

Remark 11.23 *Caldwell [37, Theorem 2.20] proved that if R is a commutative local hypercyclic ring, then $J(R)$ is nil. However, the corresponding problem for noncommutative rings is still open.*

Finally, we are ready to prove that a ring R is left perfect right hypercyclic if and only if R is an artinian principal left and right ideal ring. This was first proved by Caldwell. We would like to caution the reader here that the original proof of Caldwell given in [37, Theorem 1.2] seems to have errors.

Theorem 11.24 *A ring R is left perfect right hypercyclic if and only if R is an artinian principal left and right ideal ring.*

Proof Let R be a left perfect right hypercyclic ring. Then R is a restricted right hypercyclic ring and therefore, by Theorem 11.20, R is a ring direct sum of matrix rings over local right hypercyclic rings, say S_i. By Corollary 11.9, each S_i is a right uniserial ring. It is easy to verify that a right uniserial left perfect ring is a right artinian ring. Thus each S_i is a right artinian ring and consequently, R is a right artinian ring. By Lemma 11.12, R is right self-injective. Thus R is a quasi-Frobenius ring and therefore each right (resp. left) ideal of R is a right (resp. left) annihilator ideal. Let I be any left ideal of R, and let $A = (I : 0)$. If $C = R/A$, then by hypothesis $E(C)$ is cyclic, so there is an epimorphism $R \to E(C)$. Since over a quasi-Frobenius ring, each injective module is projective, $R \to E(C)$ splits and so $E(C)$ is isomorphic to a direct summand of R_R. Consequently, there is a monomorphism $\sigma : C \to R$. Let $x = \sigma(1 + A)$. Then $(x : 0) = A$. Thus $I' = Rx$ is a left ideal whose right annihilator is A. Since the mapping $I \mapsto (I : 0)$ is one-to-one between the left and right ideals of R, it follows that $I = I' = Rx$ is a principal left ideal. Thus, R is a principal left ideal quasi-Frobenius ring. Hence R is an artinian principal left and right ideal ring. $\qquad\square$

Rosenberg and Zelinsky [199] considered rings over which each cyclic right module has an injective hull of finite length. This led Jain and Saleh [143] to consider rings over which each cyclic module has a finitely generated injective hull.

Lemma 11.25 *For a ring R, we have following:*

1. *If R is a semiprimary ring and $J(R) \neq J(R)^2$, then any simple right R-module S that is not injective can be embedded in $J(R)/J(R)^2$.*
2. *Let M be an injective R-module and I, K be ideals in R, where $K \subseteq I$. Then $\mathrm{Hom}(I/K, M) \cong \mathrm{ann}_M(K)/\mathrm{ann}_M(I)$ as R-modules.*
3. *If every cyclic R-module has finitely generated quasi-injective hull, then every ring homomorphic image of R has this property.*

Proof

(1) If S does not embed in $J(R)/J(R)^2$ then it cannot embed in $J(R)^n/J(R)^{n+1}$ for any n. Since $J(R)$ is nilpotent, it follows that 0 is the only right R-homomorphism from a submodule of $J(R)$ to S. Thus any right R-homomorphism from a right ideal I of R to S is zero on $J(R) \cap I$ and therefore by Baer's criterion S is injective. Thus it follows that if S is not injective, then S can be embedded in $J(R)/J(R)^2$.

(2) The proof follows from the canonical embedding of $\mathrm{ann}_M(K)/\mathrm{ann}_M(I)$ into $\mathrm{Hom}(I/K, M)$ and the Baer criterion for injective modules.

(3) Let A be a two-sided ideal of R. Let $\overline{R} = R/A$ and $\overline{R}/\overline{I}$ be a cyclic $\overline{R}$-module, where $\overline{I} = I/A$. We have $\overline{R}/\overline{I} \cong R/I$ as R-modules. Denote by P the quasi-injective hull of R/I as an R-module. Then $P_1 = \mathrm{ann}_P(A)$ is quasi-injective and it contains R/I. Thus $P = P_1$ is the quasi-injective hull, which is finitely generated as R/A-module. $\qquad\square$

Remark 11.26 *In connection to Lemma 11.25(1) above, we would like to mention here that Faith [75, Problem 17, p. 57] asked as an exercise to prove* *that if R is a ring such that $R/J(R)$ is semisimple and $J(R) \neq J(R)^2$, then any simple right R-module S that is not injective can be embedded in $J(R)/J(R)^2$. It turns out that this exercise is not correct. Osofsky and Singh have both provided us with counter-examples.*

Example (Osofsky) Let R_1 be any artinian ring with $J(R_1)/J(R_1)^2 \neq 0$ and let R_2 be a nondiscrete rank 1 uniserial domain. Then $R = R_1 \times R_2$ still has $J(R)/J(R)^2 \neq 0$ and $R/J(R)$ is semisimple, but the simple module corresponding to R_2 does not embed in $J(R)/J(R)^2$.

Example (Singh) Let R_1 be non-noetherian uniserial ring given by Levy (see the example on p. 24) whose proper homomorphic images are self-injective. In that example the maximal ideal M is such that $M^2 = M$ and $S = R_1/M$ is not injective. Take a noetherian uniserial domain, say R_2. The ring $R = R_1 \times R_2$ is such that $J(R)^2 \neq J(R)$, S is as an R-module which is not injective and it also does not embed in $J(R)/J(R)^2$.

We will denote the quasi-injective hull of a module M by $Q(M)$.

Theorem 11.27 *Let R be a right artinian ring. The following statements are equivalent.*

1. *Every cyclic right R-module can be embedded in a finitely generated injective right R-module.*
2. $\mathrm{Hom}(J(R)/J(R)^2, A)$ *is finitely generated for every simple right R-module A.*
3. *Every cyclic right R-module has a finitely generated quasi-injective hull.*
4. *The injective hull of $R/J(R)^2$ is a finitely generated right R-module.*

Proof $(1) \Rightarrow (3)$. Let M be a cyclic right R-module. Since $Q(M) \subset E(M)$, and R is right artinian, it follows that $Q(M)$ is also finitely generated.

$(3) \Rightarrow (4)$. Let E_1 and E_2 denote respectively the injective hull of $R/J(R)^2$ as $R/J(R)$-module and as R-module. By Lemma 11.25, the quasi-injective hull of every cyclic $R/J(R)^2$-module is finitely generated. In particular, $Q(R/J(R)^2)$ $(= E_1)$ is a finitely generated $R/J(R)^2$-module. This implies $\mathrm{Hom}(J(R)/J(R)^2, I/J(R)^2)$ is also finitely $R/J(R)^2$-module generated for any minimal right ideal $I/J(R)^2$ of $R/J(R)^2$ by Rosenberg and Zelinsky [199]. But then $\mathrm{Hom}(J(R)/J(R)^2, I/J(R)^2)$ is a finitely generated K-module, where $K = (R/J(R)^2)/(J(R)/J(R)^2) \cong R/J(R)$. Therefore $\mathrm{Hom}_{R/J(R)^2}(J(R)/J(R)^2, I/J(R)^2)$ is a finitely generated $R/J(R)$-module, and so $\mathrm{Hom}_R(J(R)/J(R)^2, I/J(R)^2)$ is a finitely generated R-module for each simple submodule $I/J(R)^2$ of $R/J(R)^2$. This shows that E_2, the injective hull of $R/J(R)^2$, is finitely generated as an R-module (see [199], Theorem 1).

$(4) \Rightarrow (2)$. Let A be a simple right R-module. If A is injective, then by the above lemma, $\mathrm{Hom}(J(R)/J(R)^2, A) = 0$, or A which is finitely generated. If A is not injective, then by above lemma, A can be embedded in $J(R)/J(R)^2$. So there exists a right ideal I of R such that $I \subset J(R)$ and $A = I/J(R)^2$. Since $I/J(R)^2$ is a simple submodule of $R/J(R)^2$, and the injective hull of $R/J(R)^2$ as an R-module is finitely generated, $\mathrm{Hom}_R(J(R)/J(R)^2, I/J(R)^2)$ is finitely generated. Therefore $\mathrm{Hom}_R(J(R)/J(R)^2, A)$ is finitely generated.

It remains to see that $(2) \Rightarrow (1)$, and this holds by ([199], Theorem 1) and the above lemma. $\qquad\square$

Next, we consider rings with Krull dimension over which each cyclic right module has a cyclic injective hull. Indeed such rings turn out to be right artinian rings. We shall need the following.

Lemma 11.28 *([100]) If R is a ring with Krull dimension $\mathcal{K}(R)$, then $\mathcal{K}(R) = \mathcal{K}(R/P)$, for some prime ideal P. In fact, P is a minimal prime ideal.*

An ideal P of a ring R is called *completely prime* if $a, b \in R$ such that $ab \in P$, then either $a \in P$ or $b \in P$. The following is straightforward.

Lemma 11.29 *If R is a uniserial ring, and $P = aR$ is a nonzero prime ideal such that $P \neq J(R)$, then P is not completely prime.*

Proof Let $x \in J(R), x \notin P$. Then $Ra \subseteq Rx$. So, there exists $y \in R$ such that $a = yx$. Observe that $y \in J(R)$ because if $y \notin J(R)$, then $y^{-1}a = x$; that is, $x \in Ra \subseteq aR$, which is a contradiction. Suppose, on the contrary, that P is completely prime. Then $yx \in P$, $x \notin P$ implies $y \in P$. Thus there exists $z \in R$ such that $y = az$. Hence $a = azx$. Thus, we have $a(1 - zx) = 0$. Now $zx \in J(R)$ and so $1 - zx$ is invertible. Thus $a = 0$, which is a contradiction. Hence P is not completely prime. $\qquad\square$

Proposition 11.30 *Let R be a local right hypercyclic ring. If R has Krull dimension, then R is right artinian.*

Proof First suppose $J(R)$ is nil. Then, since R has Krull dimension, $J(R)$ is nilpotent. This implies $J(R)$ is the only prime ideal of R. But then by Lemma 11.28, $\mathcal{K}(R) = \mathcal{K}(R/J(R)) = 0$, which proves that R is right artinian. Suppose $J(R)$ is not nil. Then by Osofsky ([190], Theorem 2.12), there exists a nonzero nilpotent ideal $aR \subset J(R)$, $a \in R$ such that aR is the maximal proper two-sided ideal below $J(R)$. Since R has Krull dimension, R satisfies ACC on prime ideals. Thus, if $J(R)$ is not the only prime ideal, there exists a prime ideal Q such that Q is maximal among all the prime ideals different from $J(R)$. Then $Q \subset aR$. Since aR is nilpotent, $Q = aR$. Thus $Q \neq 0$, since aR is not zero. So by Lemma 11.29 and R being uniserial, Q is not completely prime. Consider now the prime ring R/Q. Since R/Q has Krull dimension, it is a prime Goldie ring. Therefore $Z(R/Q)$, the right singular ideal of R/Q, is zero. Thus $ann(\bar{x}) = \bar{0}$, for every $\bar{0} \neq \bar{x} \in R/Q$, since R is a uniserial ring. Hence Q is completely prime, which is a contradiction. This proves that $J(R)$ is the only prime ideal. Therefore $\mathcal{K}(R) = \mathcal{K}(R/J(R)) = 0$, and hence R is right artinian. $\qquad\square$

Theorem 11.31 *Let R be a right hypercyclic ring. Then R has Krull dimension if and only if R is right artinian.*

Proof If R has Krull dimension, then each homomorphic image of R has ACC on direct summands. Thus by Osofsky ([190], Lemma 1.7), R has ACC on direct summands. Thus, by Theorem 11.20, R is a ring direct sum of matrix rings over local right hypercyclic rings and hence R is right artinian. $\square$

Corollary 11.32 *A right hypercyclic ring with Krull dimension is quasi-Frobenius.*

Dauns in [57] and Gómez-Pardo, Dung, and Wisbauer in [90] asked the question whether a right hereditary ring R, whose injective hull $E(R_R)$ is finitely generated, is necessarily right artinian. Dinh, Guil Asensio, and López-Permouth [59] answered this question in the affirmative. To present their proof, we first give some definitions and technical lemmas.

Let $\aleph$ be a cardinal number. A right R-module M is called $\aleph$-*generated* if there exists an epimorphism $f : R^{(\aleph)} \to M$. The module M is said to be $\aleph$-*presented* if the kernel of this epimorphism is also $\aleph$-generated. Given a module M, we denote by $\mathrm{Add}[M]$ the full subcategory of Mod-R consisting of direct summands of direct sums of copies of M.

A cardinal number $\aleph$ is said to be *attained* in a module M if M contains a direct sum of $\aleph$ nonzero submodules. The *cofinality* of a cardinal number $\aleph$ is defined to be the least ordinal number α such that there exists an injective increasing map $f : \alpha \to \aleph$ that is *cofinal* in $\aleph$, that is such that for any ordinal number $\gamma < \aleph$ there exists an ordinal $\beta < \alpha$ with $f(\beta) \geq \gamma$. The cofinality of $\aleph$ is always a cardinal number that we will denote by $\mathrm{cof}(\aleph)$. It is clear that $\mathrm{cof}(\aleph) \leq \aleph$. A cardinal number $\aleph$ is called *regular* if $\mathrm{cof}(\aleph) = \aleph$. Otherwise, $\aleph$ is called *singular*. An uncountable cardinal $\aleph$ is said to be *inaccessible* if it is both regular and limit (i.e., it is not the successor of any other cardinal). The main result of [58] shows that $\mathrm{Gdim}(M)$ is always attained whenever it is not an inaccessible cardinal.

Lemma 11.33 *Let $\aleph$ be an infinite cardinal, M an $\aleph$-generated module, and $\aleph'$ a cardinal number with $\aleph' > \aleph$. Given an $N \in \mathrm{Add}[M]$, the following conditions are equivalent:*

1. *N can be decomposed as a direct sum of $\aleph'$ nonzero direct summands.*
2. *Every generating set of N has cardinality at least $\aleph'$.*

In particular, given an uncountable cardinal number α, a projective module P is the direct sum of α nonzero submodules if and only if every generating set of P has cardinality at least α.

Proof (1)$\Rightarrow$(2). Assume $N = \oplus_{\mathcal{I}} N_i$ with $|\mathcal{I}| = \aleph'$ and $N_i \neq 0$ for every $i \in \mathcal{I}$. Let $\{x_k\}_{\mathcal{K}}$ be a generating set of N. For every $k \in \mathcal{K}$, there exists a finite subset $\mathcal{I}_k \subseteq \mathcal{I}$ such that $x_k \in \oplus_{i \in \mathcal{I}_k} N_i$. Thus $N = \Sigma_{k \in \mathcal{K}} x_k R = \Sigma_{k \in \mathcal{K}} (\oplus_{i \in \mathcal{I}_k} N_i) = \oplus_{i \in (\cup_{k \in \mathcal{K}} \mathcal{I}_k)} N_i$. Therefore $\mathcal{I} = \cup_{k \in \mathcal{K}} \mathcal{I}_k$. Since each $\mathcal{I}_k$ is finite, we have $|\mathcal{K}| \geq |\mathcal{I}|$.

$(2)\Rightarrow(1)$. Assume that every generating set of N has cardinality at least $\aleph'$. By Kaplansky's Theorem (see [12], Theorem 26.1), N is a direct sum of $\aleph$-generated submodules, say $N = \oplus_{\mathcal{I}} N_i$. For every $i \in \mathcal{I}$, choose a generating set X_i of N_i of cardinality $\aleph$. Then $\cup_{i \in \mathcal{I}} X_i$ is a generating set of N. By hypothesis, $|\cup_{i \in \mathcal{I}} X_i|$ must be at least $\aleph'$. Since $\aleph < \aleph'$, we have $|\mathcal{I}| \geq \aleph'$. $\qquad\square$

Recall that a right R-module M is called *hereditary* if every submodule of M is projective.

Lemma 11.34 *Let P be a hereditary right R-module. Then:*

1. *P is nonsingular.*
2. *Given an infinite cardinal number $\aleph$, P is $\aleph$-generated (resp., finitely generated) if and only if it contains an essential $\aleph$-generated (resp., finitely generated) submodule.*
3. *Any submodule of a direct sum of copies of P is a direct sum of submodules of P. In particular, any direct sum of copies of P is hereditary.*

Proof

(1) Let $x\,(\neq 0) \in P$. We have $R/ann_r(x) \cong xR$. Since $xR \subseteq P$, xR is projective. Therefore $ann_r(x)$ is a direct summand of R_R and so cannot be an essential right ideal of R. Thus P is nonsingular.

(2) Let N be an essential $\aleph$-generated submodule of P. Since P is projective, there exists a splitting epimorphism $\pi : R^{(\mathcal{I})} \to P$ for some index set $\mathcal{I}$. Let $f : P \to R^{(\mathcal{I})}$ be such that $\pi f = 1_P$. Since N is $\aleph$-generated (resp., finitely generated), there exists a subset $\mathcal{K} \subseteq \mathcal{I}$ of cardinality $\aleph$ (resp., finite) such that $f(N) \subseteq R^{(\mathcal{K})}$. Let $q : R^{(\mathcal{I})} \to R^{(\mathcal{K})}$ and $g : R^{(\mathcal{K})} \to R^{(\mathcal{I})}$ be the canonical projection and injection, respectively. Then $1_P - (\pi g q f) = 0$ (because P is nonsingular) and this implies that P is a direct summand of $R^{(\mathcal{K})}$.

(3) See [235], 39.7(2). $\qquad\square$

Lemma 11.35 *Let $\{M_i : i \in \mathcal{I}\}$ be a family of modules. Then $\mathrm{Gdim}(\oplus_{i \in \mathcal{I}} M_i) = \Sigma_{i \in \mathcal{I}}\, \mathrm{Gdim}(M_i)$.*

Proof See [58], Theorem 13. $\qquad\square$

Lemma 11.36 *Let $f : M \to N$ be a splitting epimorphism of right R-modules. If L is an essential submodule of M, then $f(L)$ is an essential submodule of N. In particular, if M contains an essential $\aleph$-generated submodule, for some cardinal number $\aleph$, then so does N.*

Proof Since f is a splitting epimorphism, there exists $g : N \to M$ such that $fg = 1_N$. Let K be a nonzero submodule of N. Then $g(K)$ is a nonzero submodule of M. Since L is essential in M, $L \cap g(K) \neq 0$. Therefore $0 \neq f(L \cap g(K)) \subseteq f(L) \cap K$. This shows that $f(L)$ is essential in N. $\qquad\square$

Lemma 11.37 *Let P be a hereditary module, M be a finitely generated submodule of P, and $Q(M)$ be the P-injective hull of M. If $Q(M)$ is $\aleph$-presented for some infinite cardinal number $\aleph$, then every submodule of M has a generating set with cardinality strictly smaller than $\aleph$.*

Proof Assume to the contrary that there exists a submodule L of M such that every generating set of L has cardinality has at least $\aleph$. We claim that L contains a direct sum of $\aleph$ nonzero submodules, say $\oplus_{i \in \mathcal{I}} L_i$. If $\aleph > \aleph_0$, this is clear by Lemma 11.33. If $\aleph = \aleph_0$, it follows from the fact that hereditary modules of finite Goldie dimension are finitely generated by Lemma 11.34(2). Moreover, we can assume that each L_i is finitely generated and, therefore, $\oplus_{\mathcal{I}} L_i$ is $\aleph$-generated. Let Q be the P-injective hull of $\oplus_{\mathcal{I}} L_i$ within $Q(M)$. Then Q is a direct summand of $Q(M)$ and thus, it is $\aleph$-presented by Schanuel's Lemma (the reader may refer to [161], p. 165 for Schanuel's Lemma). Let $p : Q(M) \to Q$ be the structural projection. By Lemma 11.36, $M' = p(M)$ is an essential submodule of Q containing $\oplus_{\mathcal{I}} L_i$ and thus, Q contains a finitely generated essential submodule.

We know that P is a direct sum of countably generated projective modules by Lemma 11.33. As Q is $\aleph$-generated, with $\aleph \geq \aleph_0$, there exists a direct sum $\oplus_A P_a$ of countably generated direct summands of P and an epimorphism $\pi : \oplus_A P_a \to Q$ with $|A| \leq \aleph$. Since Q is $\aleph$-presented, there also exists an epimorphism $\pi' : R^{(\aleph)} \to Q$ such that $\mathrm{Ker}(\pi')$ is $\aleph$-generated. Again by Schanuel's Lemma, $(\oplus_A P_a) \oplus \mathrm{Ker}(\pi') \cong R^{(\aleph)} \oplus \mathrm{Ker}(\pi)$ and thus, $\mathrm{Ker}(\pi)$ is also $\aleph$-generated. Using ([94], Theorem 2.2), we deduce that there exists a submodule N of Q such that $\aleph^+$ is attained in Q/N. Moreover, the proof of ([94], Theorem 2.2) shows that it is possible to choose this N containing $\oplus_{\mathcal{I}} L_i$. Let $X = \oplus_{j \in \mathcal{J}} X_j$ be a direct sum of nonzero modules contained in Q/N with $|\mathcal{J}| = \aleph^+$. Let $q : Q \to Q/N$ be the canonical projection. Then $(q\pi)^{-1}(X)$ is a submodule of $\oplus_A P_a$ that cannot have a generating set of cardinality at most $\aleph$, since every generating set of X has cardinality at least $\aleph^+$. Moreover, $(q\pi)^{-1}(X)$ is projective, since $\oplus_A P_a$ is hereditary by Lemma 11.34. Thus, by Lemma 11.33(3), $(q\pi)^{-1}(X)$ is a direct sum of $\aleph^+$ nonzero modules. Let us write $(q\pi)^{-1}(X) = \oplus_B Y_b$ with $|B| = \aleph^+$.

Let us consider an $\aleph$-generated submodule Z of $\oplus_A P_a$ such that $\pi(Z) = \oplus_{\mathcal{I}} L_i$. Then $Z + \mathrm{Ker}(\pi)$ is a submodule of $(q\pi)^{-1}(X)$ because $\oplus_{\mathcal{I}} L_i \subseteq N = \mathrm{Ker}(q)$. We claim that it is, actually, an essential submodule. Choose a nonzero element $x \in (q\pi)^{-1}(X)$. If $\pi(x) = 0$, then $x \in \mathrm{Ker}(\pi) \subseteq Z + \mathrm{Ker}(\pi)$. Otherwise, there exists an element $r \in R$ such that $0 \neq \pi(xr) = \pi(x)r \in \oplus_{\mathcal{I}} L_i$, since $\oplus_{\mathcal{I}} L_i$ is an essential submodule of Q. Let us choose an element $z \in Z$ such that $\pi(z) = \pi(xr)$. Then $xr \in xR \cap (Z + \mathrm{Ker}(\pi))$ and $xr \neq 0$, because $\pi(xr) \neq 0$. Now since Z and $\mathrm{Ker}(\pi)$ are $\aleph$-generated, $Z + \mathrm{Ker}(\pi)$ is also $\aleph$-generated. Thus there exists a subset $B' \subseteq B$ of cardinality $\aleph < |B|$ such that $Z + \mathrm{Ker}(\pi) \subseteq \oplus_{B'} Y_b$. Choose an element $b \in B \setminus B'$. Then $Y_b \cap (Z + \mathrm{Ker}(\pi)) = 0$, a contradiction to the fact that $Z + \mathrm{Ker}(\pi)$ is an essential submodule of $(q\pi)^{-1}(X)$. $\square$

Theorem 11.38 *Let P be a finitely generated hereditary module and $\aleph$ an infinite cardinal number. If the quasi-injective hull $Q(P)$ of P is $\aleph$-generated, then $\aleph$ is not attained in P/L for any submodule L of P.*

Proof Assume to the contrary that $\aleph$ is attained in P/L for some submodule L of P. Then $\aleph$ is attained in P by the same reason as shown in the proof of Lemma 11.37. Let $\oplus_{\mathcal{I}} L_i$ be a direct sum of nonzero submodules of P with $|\mathcal{I}| = \aleph$. Let $\pi : P^{(A)} \to Q(P)$ be an epimorphism with $|A| \leq \aleph$. Clearly, $\mathrm{Ker}(\pi)$ is projective and hence a direct sum of countably generated modules, say $\mathrm{Ker}(\pi) = \oplus_{\mathcal{J}} K_j$. Let $\aleph' = \max\{\aleph, |\mathcal{J}|\}$. We claim that P contains an infinite direct sum of $\aleph'$ nonzero submodules. By Lemma 11.35, we know that

$$\mathrm{Gdim}(P^{(A)}) = \Sigma_A \, \mathrm{Gdim}(P) = |A| . \, \mathrm{Gdim}(P) = \max\{\aleph, \mathrm{Gdim}(P)\}$$

and, as $\aleph \leq \mathrm{Gdim}(P)$, we deduce that $\mathrm{Gdim}(P) = \mathrm{Gdim}(P^{(A)}) \geq \aleph'$. Now we proceed to verify that $\aleph'$ is attained in P. If $\aleph' = \aleph$, this is obvious, since P contains $\oplus_{\mathcal{I}} L_i$ with $|\mathcal{I}| = \aleph$. So let us assume that $\aleph' = |\mathcal{J}| > \aleph$. If $|\mathcal{J}|$ is not an inaccessible cardinal, then $|\mathcal{J}|$ is attained in P by ([58], Theorem 6). Otherwise, $|\mathcal{J}|$ equals its cofinality, and therefore $\mathrm{cof}(|\mathcal{J}|) = |\mathcal{J}| > \aleph$ is attained in P, since it is clearly attained in $P^{(A)}$ (see [58], Lemma 2). On the other hand, $Q(P)$ is $\aleph'$-presented. Therefore $\aleph'$ cannot be attained in P by Lemma 11.37. This yields a contradiction. Thus we conclude that $\aleph$ cannot be attained in P/L. $\qquad\square$

Corollary 11.39 *Let P be a finitely generated hereditary R-module with a countably generated quasi-injective hull. Then P is a noetherian module.*

Proof Let N be any submodule of P. Theorem 11.38 shows that N must have finite Goldie dimension. Therefore it is finitely generated by Lemma 11.34(2). $\square$

In particular, we have the following for right hereditary rings.

Theorem 11.40 *Let R be a right hereditary ring. If the injective hull $E(R_R)$ is countably generated, then R is right noetherian.*

11.2   *q*-hypercyclic rings

A ring R is called *right q-hypercyclic* if each cyclic right R-module has a cyclic quasi-injective hull. For a commutative ring R, R can be shown to be q-hypercyclic ($=qc$-ring) if R is hypercyclic. Whether a right hypercyclic ring (not necessarily commutative) is right q-hypercyclic is considered by showing that a local right hypercyclic ring R is right q-hypercyclic if and only if the Jacobson radical of R is nil. However, it is not known if there exists a local right hypercyclic ring with non-nil Jacobson radical.

An R-module M is said to have finite *Azumaya-diagram* if $M = \oplus_{i=1}^{k} M_i$ where each submodule M_i has a local endomorphism ring. We begin with considering semiperfect q-hypercyclic ring and show the following.

Lemma 11.41 *Let R be semiperfect and right q-hypercyclic. Then R is right self-injective.*

Proof Let I be a right ideal of R such that R/I is the quasi-injective hull of R. Let $\varphi : R \to R/I$ be the embedding. Since R/I contains a copy of R, R/I is injective. Let $\varphi(R) = B/I$. Then $B/I \subset_e R/I$. Hence $B \subset_e R$. Since $R \cong B/I$, B/I is projective. Thus $B = I \oplus K$ for some $K_R \subseteq B_R$. Now $R \cong B/I = (I \oplus K)/I \cong K$. Therefore $E(R) \cong E(K)$. But then $I \oplus K \subseteq_e R$ implies $E(R) = E(I) \oplus E(K) \cong E(I) \oplus E(R)$. Since $E(R) \cong R/I$, $E(R)$ is a finite direct sum of indecomposable modules, by Lemma 11.6. Thus $E(R)$ has finite Azumaya-diagram. Therefore $E(R) \cong E(R) \oplus E(I)$ implies $E(I) = 0$. Hence $I = 0$. Thus R is right self-injective. $\qquad\square$

In the next lemma we show that the property of being a q-hypercyclic ring passes to homomorphic images.

Lemma 11.42 *Let R be right q-hypercyclic. Then every homomorphic image of R is also right q-hypercyclic.*

Proof Let A be a two-sided ideal of R. Let $\overline{R} = R/A$. Let $\overline{R}/\overline{I}$ be a cyclic right $\overline{R}$-module, where $\overline{I} = I/A$. But $\overline{R}/\overline{I} \cong R/I$. Since $A \subset I$, A annihilates R/I. Let R/K be the quasi-injective hull of R/I as an R-module. Then $R/K \cong \mathrm{End}_R(E(R/I))R/I$. Then it follows that A annihilates R/K. Thus R/K may be regarded as an $\overline{R}$-module. Since R/K is quasi-injective as an R-module, R/K is quasi-injective as an $\overline{R}$-module. Since A is a two-sided ideal and annihilates R/K, $A \subset K$. Hence $R/K \cong (R/A)/(K/A)$. Clearly $\overline{R}/\overline{K}$ is the quasi-injective hull of $\overline{R}/\overline{I}$ as an $\overline{R}$-module. Hence $\overline{R}$ is right q-hypercyclic. $\qquad\square$

The following is easy to prove.

Lemma 11.43 *Let $R = \oplus_{i=1}^{k} R_i$ be a finite direct sum of rings. Then R is right q-hypercyclic if and only if each R_i is right q-hypercyclic.*

Jain and Malik [139] studied local q-hypercyclic rings and obtained the following results.

Lemma 11.44 *Let R be a local right q-hypercyclic ring. Then:*

1. *R is both left and right uniserial.*
2. *R is left bounded or right bounded.*
3. *$J(R)$ is nil.*
4. *R is a duo ring.*

Proof

1. Suppose R is a local right q-hypercyclic ring. Suppose, we have two non-comparable right ideals A and B. Then $R/A \cap B$ is not uniform. Therefore injective hull of $R/A \cap B$ decomposes, and so does its quasi-injective hull. However the quasi-injective hull of $R/A \cap B$, being local, is indecomposable,

which is a contradiction. Hence R is right uniserial. By Lemma 11.41, R is right self-injective. Now, we proceed to show that R is left uniserial. Let $x_1, x_2 \in R$. Then as we have already shown that R is right uniserial, $ann_r(x_1) \subseteq ann_r(x_2)$ or $ann_r(x_2) \subseteq ann_r(x_1)$. Now since R is right self-injective, any finitely generated left ideal is the annihilator of some right ideal. So, $ann_l(ann_r(x_1)) = Rx_1 \subseteq ann_l(ann_r(x_2)) = Rx_2$ or $Rx_1 \supseteq Rx_2$. Let I_1, I_2 be two left ideals of R such that $I_1 \not\subseteq I_2$. Let $x \in I_1 - I_2$. For $y \in I_2$, $Ry \not\supseteq Rx$ since $x \notin I_2$. Hence $Ry \subseteq Rx$ or $I_2 \subseteq I_1$.

2. If $\mathrm{Soc}(_R R) \neq 0$, then by linear ordering of left ideals, $\mathrm{Soc}(_R R)$ is a nonzero two two-sided ideal contained in each left ideal of R, and hence R is left bounded. So, now we consider the case that $\mathrm{Soc}(_R R) = 0$. It is not difficult to show that if I is a nonzero right ideal of R such that R/I is quasi-injective, then I contains a nonzero two-sided ideal of R. Using this, it may be easily deduced that R is right bounded as well.

3. Let $a \in J(R)$. Suppose $a^n \neq 0$ for any positive integer n. Let $S = \{a^n : n > 0\}$. By Zorn's Lemma there exists an ideal P of R maximal with respect to the property that $P \cap S = \emptyset$. Then P is prime. Hence R/P is a prime local right q-hypercyclic ring. Thus, by (2), R/P is either left bounded or right bounded. Then it follows that R/P is a domain. Since R/P is also local and right q-hypercyclic, R/P is a right self-injective domain and hence a division ring. Therefore P is a maximal ideal of R. Thus $P = J(R)$, a contradiction. Hence $J(R)$ is nil.

4. Let $0 \neq x, y \in R$. Suppose $xy \notin Rx$. Then $Rxy \supsetneq Rx$, so $x = zxy$ for some $z \in J(R)$. Since $J(R)$ is nil, $z^n = 0$ for some $n > 1$. Then $x = zxy = z^2 x y^2 = \cdots = z^n x y^n = 0$, a contradiction. Hence $xR \subseteq Rx$. By symmetry, $Rx \subseteq xR$. Thus, $xR = Rx$ for all $x \in R$. Hence R is a duo ring. $\qquad\square$

Theorem 11.45 *For a ring R, we have the following:*

1. *Let R be a local ring. Then R is right q-hypercyclic if and only if R is a qc-ring.*

2. *Let R be a local right hypercyclic ring. Then R is right q-hypercyclic if and only if $J(R)$ is nil.*

Proof

(1) Let R be right q-hypercyclic and let A be a nonzero right ideal of R. Then by the above lemma, A is a two-sided ideal of R. But then by Lemmas 11.41 and 11.42, R/A is a right self-injective ring. Thus R/A is a quasi-injective R-module, proving that R is a qc-ring. The converse is obvious.

(2) Let R be a local right hypercyclic ring with $J(R)$ nil. Then by Theorem 11.22, R is a duo, right self-injective, uniserial ring. Then by ([154], Theorem 2.3), R is maximal. Therefore, by Theorem 7.3, R is a qc-ring and hence a right q-hypercyclic ring. The converse is obvious. $\qquad\square$

Theorem 11.46 *For a commutative ring R, we have the following:*

1. *If R is a q-hypercyclic ring, then R must be self-injective.*
2. *R is q-hypercyclic if and only if R is a qc-ring.*
3. *If R is a hypercyclic ring, then R is q-hypercyclic.*

Proof

(1) This is obvious.

(2) This is similar to the proof of Theorem 11.45 (1).

(3) Let R be a commutative hypercyclic ring. Then by Theorem 11.21, R is a finite direct sum of commutative local hypercyclic rings. So, in view of Lemma 11.43, it suffices to show that a commutative local hypercyclic ring is q-hypercyclic. Let R be commutative local and hypercyclic. Then, by Proposition 11.5, R is uniserial and, by Lemma 11.1, R is self-injective. By Remark 11.23, $J(R)$ is nil. Then by ([154], Theorem 2.3), R is maximal. Since $J(R)$ is nil, R has rank 0. Thus R is a rank 0 maximal uniserial ring and hence a qc-ring by Theorem 7.3. This completes the proof of the theorem. $\qquad\square$

The following example of Jain and Malik [139] shows that a q-hypercyclic ring need not be hypercyclic.

Example. Let F be a field and x be an indeterminate over F. Let $W = \{\{\alpha_i\} : \{\alpha_i\}$ is a well ordered sequence of nonnegative real numbers$\}$. Let $T = \{\Sigma_{i=0}^{\infty} a_i x^{\alpha_i} : a_i \in F, \{\alpha_i\} \in W\}$ and $R = T/xJ(T)$. Let S be the socle of R. Then R/S is a q-hypercyclic ring but not a hypercyclic ring. $\qquad\square$

Jain and Malik [139] also obtained the following analog of Theorem 11.20 by considering semiperfect q-hypercyclic rings.

Theorem 11.47 *Let R be a semiperfect right q-hypercyclic ring. Then R is a finite direct sum of right q-hypercyclic matrix rings over local rings.*

Proof $R = e_1 R \oplus \cdots \oplus e_n R$, where e_i, $1 \leq i \leq n$ are primitive idempotents. By Lemma 11.41, R is right self-injective and so each $e_i R$ is an injective R-module. We will show that for $i \neq j$, either $e_i R \cong e_j R$, or $\mathrm{Hom}_R(e_i R, e_j R) = 0$. Suppose for some $i \neq j$, $\mathrm{Hom}_R(e_i R, e_j R) \neq 0$. By renumbering, if necessary, we may assume that $i = 1, j = 2$. Let $\alpha : e_1 R \to e_2 R$ be a nonzero R-homomorphism. Then $e_1 R/\mathrm{Ker}(\alpha)$ embeds in $e_2 R$. Since $e_2 R$ is indecomposable, $E(e_1 R/\mathrm{Ker}(\alpha)) \cong e_2 R$. Hence $B = e_2 R \oplus \cdots \oplus e_n R$ contains a copy of $E(e_1 R/\mathrm{Ker}(\alpha))$. Now $R/\mathrm{Ker}(\alpha) \cong (e_1 R)/\mathrm{Ker}(\alpha) \times e_2 R \times \cdots \times e_n R$. Let $A = (e_1 R)/\mathrm{Ker}(\alpha)$. Then B is injective and contains a copy of $E(A)$. Hence $\mathrm{Hom}_R(B, E(A))B = E(A)$. Since R is right q-hypercyclic, for some right ideal I, $R/I \cong Q(R/\mathrm{Ker}(\alpha)) \cong Q(A \times B) \cong E(A) \times B$. Thus $R/I \cong e_2 R \times B$. Then R/I is projective. Hence $R = I \oplus K$ for some right ideal K. Then $K \cong R/I \cong e_2 R \times e_2 R \times \cdots \times e_n R$. Thus $R = I \oplus K \cong I \times e_2 R \times e_2 R \times \cdots \times e_n R$. Hence by Azumaya-diagram, $e_1 R \cong I \times e_2 R$. Since $e_1 R$ is indecomposable, $I = 0$. Consequently, $R = K$. Then $e_1 R \times e_2 R \times \cdots \times e_n R \cong e_2 R \times e_2 R \times \cdots \times e_n R$. Again by Azumaya-diagram, $e_1 R \cong e_2 R$. Thus for $i \neq j$, either $e_i R \cong e_j R$ or $\mathrm{Hom}_R(e_i R, e_j R) = 0$. Set $[e_k R] = \Sigma e_i R$, $e_i R \cong e_k R$. Renumbering, if necessary, we may write $R = [e_1 R] \oplus \cdots \oplus [e_t R]$, $t \leq n$. Then for all $1 \leq k \leq t$, $[e_k R]$ is an ideal. Since for any k, $1 \leq k \leq n$, $e_k R$ is indecomposable, $\mathrm{End}_R(e_k R) \cong e_k R e_k$ is a local ring. Thus $[e_k R] = \oplus_i e_i R$ is the $n_k \times n_k$ matrix ring over the local ring $e_k R e_k$ where n_k is the number of $e_i R$ appearing in $\oplus_i e_i R$. Since a finite direct sum of right q-hypercyclic rings is right q-hypercyclic, the matrix ring is right q-hypercyclic. $\qquad \square$

Following the same method as for semiperfect q-hypercyclic rings (see Lemma 11.41), Jain and Saleh [143] obtained the following:

Lemma 11.48 *Let R be a right q-hypercyclic ring with finite Goldie dimension. Then R is right self-injective.*

This lemma is used in the proof of the following theorem.

Theorem 11.49 *Let R be a right q-hypercyclic ring with Krull dimension. Then R is right artinian.*

Proof There exists a prime ideal P of R such that $\mathcal{K}(R/P) = \mathcal{K}(R)$. Since R is right q-hypercyclic, $S = R/P$ is a right q-hypercyclic ring by Lemma 11.42. Therefore S is right self-injective; that is, $E(S) = S$. Since S is a prime right Goldie ring, $Q(S) = E(S) = S$. Thus S is right artinian, that is, $K(S) = 0$, which gives that R is right artinian. $\qquad \square$

11.3 π-hypercyclic rings

Goel and Jain [89] considered rings with finite uniform dimension such that any cyclic right R-module is π-injective, or more generally has cyclic π-injective hull. These results generalize the results for semiperfect rings over which each cyclic right R-module has an injective or quasi-injective hull. A ring R is called a *right π-hypercyclic ring* if each cyclic right R-module has a cyclic π-injective hull.

Recall that if $K = \mathrm{Hom}_R(E(A), E(A))$, then $Q(A) = KA$ and the π-injective hull $\pi(A)$ of A is given by VA, where V is the subring of K generated by the idempotents of K.

Lemma 11.50 *Let M be π-injective and $E(M) = \oplus A_i$ be a direct sum of submodules. Then $M = \oplus (A_i \cap M)$.*

Proof This follows from Lemma 1.19(7). $\qquad \square$

Lemma 11.51 *Let R be a ring with finite Goldie dimension. If the π-injective hull of R is cyclic, then R is π-injective.*

Proof Suppose $\pi(R) \cong R/I$ for some right ideal I of R. There exists A_R such that $R \cong A/I \subseteq_e R/I$. Thus $A \cong I \oplus K$ for some $K_R \cong R_R$, and $A \subseteq_e R$. Now $\dim R = \dim A = \dim(I \oplus K) = \dim I + \dim R$ which forces I to be 0. That is, $\pi(R) = R$. $\qquad\square$

Lemma 11.52 *A ring homomorphic image of a right π-hypercyclic ring is also a right π-hypercyclic ring.*

Proof Let A be a two-sided ideal of R. Let $\overline{R} = R/A$. Let $\overline{R}/\overline{I}$ be a cyclic right $\overline{R}$-module, where $\overline{I} = I/A$. By hypothesis, there exists K_R such that R/K is the π-injective hull of R/I as an R-module. Now $R/K \cong S(R/I)$, where S is the subring of $\operatorname{End}_R(E(R/I))$ generated by the idempotents. Since A annihilates R/I, it follows that A annihilates R/K. Thus R/K is an $\overline{R}$-module. Clearly, $\overline{R}/\overline{K} \,(\cong R/K)$ is the π-injective hull of $\overline{R}/\overline{I} \,(\cong R/I)$ as an $\overline{R}$-module. Hence $\overline{R}$ is right π-hypercyclic. $\qquad\square$

Lemma 11.53 *A right π-hypercyclic ring R is local if and only if R is right uniserial.*

Proof Let R be a local right π-hypercyclic ring. Then by Lemma 11.51, R_R is uniform. Suppose, we have two non-comparable right ideals A and B. Then $R/A \cap B$ is not uniform. Therefore the π-injective hull of $R/A \cap B$ decomposes. However the π-injective hull of $R/A \cap B$, being local, is indecomposable, which is a contradiction. Hence R is right uniserial. The converse is obvious. $\qquad\square$

The next lemma is quite useful.

Lemma 11.54 *Let R be any ring. Let A be a right R-module essential in an injective right R-module B. Then the π-injective hull $\pi(A \times B)$ of $A \times B$ is given by $E(A) \times B$.*

Proof Let $M = \pi(A \times B)$. By Lemma 1.19(7), $M = (M \cap E(A)) \times B$. Since $A \subseteq_e B$, $E(M \cap E(A)) \cong B = E(B)$. Thus it follows that $M \cap E(A) \cong B$. Therefore, M is injective, and hence $M = E(A) \times B$. $\qquad\square$

Jain and Saleh [144] studied π-hypercyclic rings with finite Goldie dimension and proved the following.

Theorem 11.55 *Let R be a right π-hypercyclic indecomposable ring with finite Goldie dimension other than 1. Then $R \cong \mathbb{M}_n(A)$, where A is a local right uniserial ring, and $n > 1$, if and only if R is right self-injective.*

Proof Let R be right self-injective. Thus there exist primitive orthogonal idempotents e_i, $1 \le i \le n$ such that $R = e_1 R \oplus \cdots \oplus e_n R$. Let $\alpha : e_1 R \to e_2 R$ be a nonzero R-homomorphism. Then $R/I \cong e_2 R \times e_2 R \times \cdots \times e_n R$ for some right ideal I of R. Thus R/I is projective and hence $R = I \oplus K$ where $K \cong e_2 R \times e_2 R \times \cdots \times e_n R$. Since R is right self-injective, Azumaya-diagram gives $e_1 R \cong I \times e_2 R$ which forces I to be zero as $e_1 R$ is indecomposable. Thus $e_1 R \cong e_2 R$. Let $[e_k R] = \Sigma e_i R, e_i R \cong e_k R$. Since R is indecomposable, $R = [e_1 R]$.

Therefore R is a matrix ring over a local ring $A \cong eRe$ where $e = e_1$. It remains to show that eRe is a right uniserial ring. We first show that for any right ideal $I \subset eR$, eR/I is uniform. Then it follows that the submodules of eR are linearly ordered. Let A and B be right ideals of eRe. Then $AeR \subset (eR)^2$ and $BeR \subset (eR)^2$. Since the submodules of eR are linearly ordered, $AeR \subset BeR$ or $BeR \subset AeR$ and so $AeRe \subset BeRe$ or $BeRe \subset AeRe$, that is, $A \subset B$ or $B \subset A$. The converse follows from the fact that an $n \times n$ matrix ring $\mathbb{M}_n(A)$, $n > 1$ is self π-injective if and only if it is self-injective. $\qquad\square$

11.56 Questions

1. Let E be a finitely generated module such that any pure quotient is pure-injective. Is E a direct sum of indecomposable pure-injective modules?

   The answer is in the affirmative when E is finitely presented, as obtained in [93]. Dinh, Guil Asensio, and López-Permouth [59] showed that if R is a ring of cardinality at most $2^{\aleph_0}$, and E is a countably generated injective right R-module such that every quotient of E is injective, then E is a direct sum of indecomposable modules.
2. Let R be a local right hypercyclic ring. Is $J(R)$ nil?
3. Is a von Neumann regular right hypercyclic ring R right self-injective? This has an affirmative answer if R is left perfect or commutative.

12 Cyclic modules essentially embeddable in free modules

A ring R is called a *right FGF ring* if each finitely generated right R-module embeds in a free module. More generally, a ring R is called a *right CF ring* if each cyclic right R-module embeds in a free module. Clearly, every right FGF ring is a right CF ring. However, the converse is not true. Björk gave an example of a left CF ring which is not a left FGF ring.

Björk Example. Let F be a field and $\overline{F}\,(\neq F)$ be a subfield of F. Let $a \mapsto \bar{a}$ be an isomorphism $F \to \overline{F}$. Let R denote the left vector space over F with basis $\{1, x\}$ and make R into an F-algebra by defining $x^2 = 0$ and $xa = \bar{a}x$ for each $a \in F$. It may be checked that $J(R) = Rx = Fx$ is the only proper left ideal of R. The only cyclic left R-modules are 0, R, and $R/J(R)$. Since each of these is embedded in $_R R$, R is a left CF ring. This ring R is not a left FGF ring as it is not left mininjective (see [26] for more details).

The following two problems are still open:

1. *FGF* **Problem:** Let R be a right FGF ring. Must R be a QF ring?
2. *CF* **Problem:** Let R be a right CF ring. Must R be right artinian?

It is well-known that a ring R is quasi-Frobenius if and only if each right R-module embeds in a projective or, equivalently, in a free module. This motivated the first problem above. Levy first asked this question for right Ore rings [166]. Later, this question was asked for any right FGF ring by Faith. It is known that if R is both left and right FGF then R is a QF ring.

Since the answer to the FGF problem is affirmative for right artinian rings, an affirmative answer to the CF problem would immediately imply the same answer to the FGF problem.

Björk [27] and Tolskaya [217] independently proved that every right self-injective right CF ring is right artinian. Gómez Pardo and Guil Asensio [92] extended this result by proving that every right CS right CF ring is right artinian. To present their proof, we start with some technical lemmas and definitions.

Lemma 12.1 *Let R be a ring, M an injective right R-module, and $S = \mathrm{End}_R(M)$. Then there exists a one-to-one correspondence between the set of isomorphism classes of indecomposable direct summands of M and the set of isomorphism classes of minimal right ideals of $S/J(S)$.*

Proof Let L_R be a direct summand of M_R. Then we have a canonical isomorphism $\operatorname{Hom}_R(M, L) \otimes_S M \cong L$. Similarly, if N_S is a direct summand of S_S, then $\operatorname{Hom}_R(M, N \otimes_S M) \cong N$. Thus the assignments $L \mapsto \operatorname{Hom}_R(M, L)$ and $N \mapsto N \otimes_S M$ define inverse bijections between the sets of isomorphism classes of direct summands of M_R and S_S. It is also clear that these bijections preserve the property of being indecomposable. Actually, if $e \in S$ is the idempotent associated to the direct summand L of M (so that $L = eM \cong eS \otimes_S M$), then the corresponding direct summand of S_S is precisely $\operatorname{Hom}_R(M, eM) \cong eS$. Since $\operatorname{End}(eM_R) \cong eSe \cong \operatorname{End}(eS_S)$, we have that if eM_R is indecomposable, then the corresponding direct summand eS of S_S is an indecomposable projective module with local endomorphism ring. It is well-known that eSe is local if and only if $eS/eJ(S)$ is a simple right S-module whose projective cover is precisely eS. But $eS/eJ(S)$ is isomorphic, as a right $S/J(S)$-module, to the minimal right ideal $\bar{e}(S/J(S))$ of $S/J(S)$, where $\bar{e} = e + J(S)$, and so we may assign to eS this minimal right ideal.

Conversely, since $S/J(S)$ is a von Neumann regular ring and idempotents lift modulo $J(S)$, each minimal right ideal of $S/J(S)$ is of the form $\bar{e}(S/J(S))$, where $e \in S$ is an idempotent and $\bar{e} = e + J(S)$. Then we can assign to $\bar{e}(S/J(S))$ the right ideal eS of S, which is, clearly, a projective cover in Mod-S of $\bar{e}(S/J(S))$, so that eSe is local. Hence we get a bijection between isomorphism classes of direct summands of S_S with local endomorphism rings and isomorphism classes of minimal right ideals of $S/J(S)$, which completes the proof. $\qquad\square$

Next, we define an *idempotent-orthogonal family* of simple modules.

Definition Let S be a ring and $\{C_k\}_{k \in K}$ a family of pairwise non-isomorphic simple right S-modules. The family is called an *idempotent-orthogonal* family of simple modules when there exists a family $\{e_k\}_{k \in K}$ of idempotents of S satisfying $C_j e_k = 0$ if $j \neq k$, and $C_k e_k \neq 0$ for each $k \in K$. More generally, the family is called *idempotent-semiorthogonal* if $C_k e_k \neq 0$ for each $k \in K$ and, if $j \neq k$, then either $C_j e_k = 0$ or $C_k e_j = 0$.

We denote by $|X|$ the cardinality of a set X.

Lemma 12.2 *Let R be a von Neumann regular right self-injective ring, and let $\{C_i\}_{i \in I}$ be a set of pairwise non-isomorphic minimal right ideals of R. If I is an infinite set, then there is an idempotent-orthogonal family $\{C_k\}_{k \in K}$ of simple right R-modules such that $|I| < |K|$.*

Proof For each subset $J \subseteq I$, we define $A_J := \Sigma\{C \subseteq R_R : C \cong C_i$ for some $i \in J\}$, and let $E(A_J)$ be its injective envelope. Since R is right self-injective, $E(A_J)$ embeds in R_R and so there exists an idempotent $e_J \in R$ such that $E(A_J) \cong e_J R$. Following the proof of [188, Lemma 8], we first show that e_J is central. It suffices to prove that, for every $x \in R$, $e_J x = e_J x e_J$ and $x e_J = e_J x e_J$. Assume that $e_J x(1 - e_J) \neq 0$. Since A_J is essential in $e_J R$, there exist an element $q \in R$ and a simple right R-module C such that $C \cong C_i$ for some $i \in J$ and

$0 \neq e_J x (1 - e_J) q \in C$. Thus the homomorphism $(1 - e_J) q R \to C$ given by left multiplication with $e_J x$ is nonzero and, since C_R is a projective module, it is actually a split epimorphism. Therefore $(1 - e_J) R$ contains a simple submodule isomorphic to C_i, for some $i \in J$, which contradicts the fact that all submodules of C_i are contained in $e_J R$. Thus we have proved that $e_J R (1 - e_J) = 0$.

Assume now that $(1 - e_J) x e_J \neq 0$ for some $x \in R$. Then the left multiplication by $(1 - e_J) x$ gives a nonzero homomorphism $e_J R \to (1 - e_J) R$. Its image $(1 - e_J) x e_J R$ is a principal right ideal of R contained in $(1 - e_J) R$, and, since R is von Neumann regular, it is a projective module. Hence $e_J R$ contains a submodule isomorphic to $(1 - e_J) x e_J R$. But the latter module has zero intersection with $e_J R$ and thus contains no submodules isomorphic to C_i, for any $i \in J$, which contradicts the fact that A_J is essential in $e_J R$. Thus we see that all the e_J are indeed central idempotents.

Next, we construct the idempotent-orthogonal family $\{C_k\}_{k \in K}$. As in the proof of [188, Theorem 1], the infinite set I can be decomposed, using a lemma from Tarski, as the union of a class K of subsets of I such that the following conditions are satisfied:

(i) $|K| > |I|$,
(ii) $|X| = |Y|$ for each $X, Y \in K$,
(iii) $|X \cap Y| < |X| = |Y|$ if $X \neq Y$ $(X, Y \in K)$.

For each subset X of I, we have $A_X = A_{(X \setminus Y)} \oplus A_{(X \cap Y)}$, so that $E(A_X) = E(A_{(X \setminus Y)}) \oplus E(A_{(X \cap Y)})$, and hence $e_X R = e_{X \setminus Y} R \oplus e_{X \cap Y} R$. Thus we see that $e_{X \cap Y} \in e_X R$ and, similarly, $e_{X \cap Y} \in e_Y R$. Consequently, $e_{X \cap Y} \in e_X R \cap e_Y R \subseteq e_X e_Y R = e_Y e_X R$. On the other hand, if C is a simple right R-module contained in $e_X e_Y R$, then $C \cong C_i$ for some $i \in X$ and, similarly, $C \cong C_j$ for some $j \in Y$. Thus $C_i \cong C_j$ and so $i = j \in X \cap Y$. Therefore $C \in A_{(X \cap Y)}$. Since $e_X e_Y R$ has essential socle, we have that $e_X e_Y R \subseteq E(A_{(X \cap Y)}) = e_{X \cap Y} R$, and so $e_{X \cap Y} R = e_X e_Y R$. Thus we see that $e_X e_Y R$ is a ring in which both $e_X e_Y$ and $e_{X \cap Y}$ are identities, and hence $e_X e_Y = e_{X \cap Y}$.

Now let $N = \Sigma\{e_Z R : Z \subseteq I \text{ and } |Z| < |X| \text{ for each } X \in K\}$ and define, for each $X \in K$, $N_X = (1 - e_X) R + N$. Since e_X and all the e_Z are central idempotents of R, both N and N_X are two-sided ideals of R. Assume that $e_X \in N_X$. Then we may write $e_X = (1 - e_X) x_0 + n$, with $x_0 \in R$, $n \in N$. Thus there exist sets $Z_1, \ldots, Z_r \subseteq I$ such that $|Z_j| < |X|$ for $j = 1, \ldots, r$, and $n = \sum_{j=1}^r e_{Z_j} x_j$, with $x_j \in R$. Since $|X|$ is infinite, we have that $|\bigcup_{j=1}^r Z_j| < |X|$, and hence there exists an element $i \in X \setminus (\bigcup_{j=1}^r Z_j)$. But, since $e_X R = (1 - e_X) x_0 R + \sum_{j=1}^r e_{Z_j} x_j R$, we then have that $e_X R$ does not contain any simple submodule isomorphic to C_i. This contradicts the fact that $i \in X$ and shows that $e_X \notin N_X$. On the other hand, if $X, Y \in K$ and $X \neq Y$, then $e_Y = (1 - e_X) e_Y + e_X e_Y = (1 - e_X) e_Y + e_{X \cap Y}$. Since $|X \cap Y| < |X|$, for each $X \in K$, we see that $e_{X \cap Y} \in N$, and so $e_Y \in N_X$.

Now let M_X be a maximal right ideal of R containing N_X. Since $e_Y \in N_X \subseteq M_X$, we see that the simple right R-module R/M_X is annihilated by e_Y, for each $Y \in K$, $Y \neq X$. It is clear, however, that $e_X \notin M_X$. Then we can define a set of simple right R-modules indexed by K, by setting $C_k = R/M_k$ for each $k \in K$, and we have that $C_k e_k \neq 0$ and $C_j e_k = 0$ for each $k, j \in K$, $j \neq k$. Furthermore, since the e_k are central idempotents for all $k \in K$, we have that $C_j \cong C_k$ implies $j = k$. $\square$

Lemma 12.3 *Let P be a finitely generated projective right R-module, $E = E(P)$, and $S = \mathrm{End}(E_R)$. If N is a right ideal of S containing $J(S)$, then $N = \{s \in S : s(P) \subseteq NE\}$.*

Proof It is clear that if $s \in N$, then $s(P) \subseteq NE$. To prove the converse inclusion, let $s \in S$ be such that $s(P) \subseteq NE$. Since $s(P)$ is finitely generated, there exist elements $h_1, \ldots, h_n \in N$ such that $s(P) \subseteq \Sigma_{i=1}^n Im(h_i)$. Let $\pi_i : E^n \to E$ denote the canonical projections and set $h = \Sigma_{i=1}^n (h_i \circ \pi_i) : E^n \to E$. Then $Im(h) = \Sigma_{i=1}^n Im(h_i)$ and hence $s(P) \subseteq Im(h)$. Thus, if we set $X = Im(h)$, with canonical projection $\beta : E^n \to X$ and canonical injection $\alpha : X \to E$, and $u : P \to E$ is the canonical inclusion, we obtain a morphism $f : P \to X$ such that $s \circ u = \alpha \circ f$. Since P is projective, there exists a morphism $g : P \to E^n$ such that $\beta \circ g = f$. Using the injectivity of E^n, we also obtain an extension $t : E \to E^n$ of g, so that $t \circ u = g$. Thus we have

$$s \circ u = \alpha \circ f = \alpha \circ \beta \circ g = h \circ g = h \circ t \circ u$$

so that $(s - h \circ t) \circ u = 0$ and hence $\mathrm{Ker}(s - h \circ t)$ is essential in E_R. This gives that $s - h \circ t \in J \subseteq N$ and, since $h \circ t = \Sigma_{i=1}^n (h_i \circ \pi_i) \circ t = \Sigma_{i=1}^n h_i \circ (\pi_i \circ t) \in N$, we see that $s \in N$. $\square$

If R is a ring, $\Omega(R)$ will denote a set of representatives of the isomorphism classes of simple right R-modules. If M is a right R-module, then $C(M)$ denotes a set of representatives of the isomorphism classes of simple submodules of M. Denote by $[X]$, the isomorphism class of a right R-module X.

Lemma 12.4 *Let P be a finitely generated projective right R-module, $E = E(P)$, and $S = \mathrm{End}_R(E)$. Assume that each direct summand of E_R has an essential finitely generated projective submodule. If $\{C_k\}_{k \in K}$ is an idempotent-semiorthogonal family of simple right $S/J(S)$-modules, then there exists an injective mapping from K to $\Omega(R)$.*

Proof Since idempotents of $S/J(S)$ lift modulo $J(S)$, there exist idempotents $\{s_k\}_{k \in K}$ of S such that $C_k s_k \neq 0$ for any $k \in K$ and either $C_j s_k = 0$ or $C_k s_j = 0$ for $k \neq j$. Let $c_k \in C_k$ be an element such that $c_k s_k \neq 0$ for each $k \in K$, and let $p_k : S_S \to C_k$ be the homomorphism defined by $p_k(1) = c_k s_k$. If $s_{k^*} = \mathrm{Hom}_R(E, s_k)$ is the endomorphism of S_S given by left multiplication with s_k, we have that $(p_k \circ s_{k^*})(1) = c_k s_k^2 = c_k s_k = p_k(1)$, and so $p_k \circ s_{k^*} = p_k$, from which it follows that

$$(p_k \otimes_S E) \circ s_k = (p_k \otimes_S E) \circ (s_{k^*} \otimes_S E) = (p_k \circ s_{k^*} \otimes_S E = p_k \otimes_S E.$$

If we set $E_k = s_k(E)$ and $E'_k = (1 - s_k)(E)$, then E_k and E'_k have finitely generated projective essential submodules P_k and P'_k, respectively, so that $P_k \oplus P'_k$ is essential in $E = E_k \oplus E'_k$. Then $(p_k \otimes_S E)(P'_k) \subseteq ((p_k \otimes_S E) \circ s_k \circ (1 - s_k))(E) = 0$ and so, if $(p_k \otimes_S E)(P_k) = 0$, we have that $(p_k \otimes_S E)(P_k \oplus P'_k) = 0$. Let $N_k = \mathrm{Ker}(p_k)$. Since C_k is simple we have that $J(S) \subseteq N_k$. Moreover, $\mathrm{Ker}(p_k \otimes_S E) = N_k E$ and so $P_k \oplus P'_k \subseteq N_k E$. Thus we obtain from Lemma 12.3 that $1_S \in N_k$, a contradiction. This proves that $(p_k \otimes_S E)(P_k) \neq 0$. Let $h_k : P_k \to E_k$, $i_k : E_k \to E$, and $t_k = i_k \circ h_k : P_k \to E$ be the inclusions, and set $L_k := Im((p_k \otimes_S E) \circ t_k)$, with canonical projection $q_k : P_k \to L_k$ and inclusion $w_k : L_k \to C_k \otimes_S E$. As we have just seen, L_k is a finitely generated nonzero module, and hence we can choose for each $k \in K$ a simple quotient U_k of L_k with canonical projection $\pi_k : L_k \to U_k$. We define a map from K to the set of isomorphism class of simple right R-modules by assigning $k \mapsto [U_k]$. Now we proceed to show that this map is injective. Suppose that $[U_j] = [U_k]$ for $j, k \in K$. Since $\{C_k\}_{k \in K}$ is an idempotent-semiorthogonal family of simple right $S/J(S)$-modules, we can assume that, say, $C_k s_j = 0$. Let $\varphi : U_j \to U_k$ be an isomorphism. Suppose $\alpha_k : U_k \to E(U_k)$ denotes the inclusion, for each $k \in K$. We obtain by injectivity an R-homomorphism $\phi : E(U_j) \to E(U_k)$ satisfying that $\phi \circ \alpha_j = \alpha_k \circ \varphi$. Also, $\alpha_k \circ \pi_k$ has an extension π'_k to $C_k \otimes_S E$, so that $\alpha_k \circ \pi_k = \pi'_k \circ w_k$. On the other hand, using the projectivity of P_j, we obtain a homomorphism $\psi : P_j \to P_k$ such that $\varphi \circ \pi_j \circ q_j = \pi_k \circ q_k \circ \psi$. Now, from the injectivity of E, we obtain an endomorphism $\tau : E \to E$, that is, an element $\tau \in S$, such that $\tau \circ t_j = t_k \circ \psi$. Observe then that $\phi \circ \alpha_j = \alpha_k \circ \varphi$ is a monomorphism, and hence the morphism $\phi \circ \alpha_j \circ \pi_j \circ q_j : P_j \to E(U_k)$ is nonzero (with image isomorphic to U_j). Thus we see that:

$$0 \neq \phi \circ \alpha_j \circ \pi_j \circ q_j = \alpha_k \circ \varphi \circ \pi_j \circ q_j = \alpha_k \circ \pi_k \circ q_k \circ \psi = \pi'_k \circ w_k \circ q_k \circ \psi$$

$$= \pi'_k \circ (p_k \otimes_S E) \circ t_k \circ \psi = \pi'_k \circ (p \otimes_S E) \circ \tau \circ t_j$$

$$= \pi'_k \circ (p \otimes_S E) \circ \tau \circ i_j \circ h_j.$$

Assume now that $j \neq k$ and consider the homomorphism $p_k \circ \tau_* \circ i_{j_*} : s_j S \to C_k$, where $\tau_* = \mathrm{Hom}_R(E, \tau)$ and $i_{j_*} = \mathrm{Hom}_R(E, i_j)$. If we set $x := (p_k \circ \tau_*)(1) \in C_k$, we have that $(p_k \circ \tau_* \circ i_{j_*})(s_j) = (p_k \circ \tau_*)(s_j) = x s_j \in C_k s_j = 0$. Tensoring with $_S E$, we then see that $(p_k \otimes_S E) \circ (\tau_* \otimes_S E) \circ (i_{j_*} \otimes_S E) = 0$ and, since $\tau_* \otimes_S E \cong \tau$ and $i_{j_*} \otimes_S E \cong i_j$, that $(p_k \otimes_S E) \circ \tau \circ i_j = 0$, which gives a contradiction and shows that we must have $j = k$. $\qquad \square$

Theorem 12.5 *Let P be a finitely generated projective right R-module such that each direct summand of $E(P)$ has an essential finitely generated projective submodule and $|\Omega(R)| \leq |C(P)|$. Then P cogenerates the simple right R-modules and has finite essential socle.*

Proof Let $E_R = E(P_R)$, $S = \text{End}(E_R)$, and $J = J(S)$. By Lemma 12.1, there exists a bijection between the set $\mathcal{E}$ of isomorphism classes of indecomposable direct summands of E_R and the set $\mathcal{M}$ of isomorphism classes of minimal right ideals of S/J. Clearly, the assignment $[C] \mapsto [E(C)]$ defines an injection from $C(P)$ to $\mathcal{E}$, and so we have that $|\Omega(R)| = |C(P)| \leq |\mathcal{E}| = |\mathcal{M}|$. Assume $|\mathcal{M}|$ is infinite. Then, by Lemma 12.2, there exists an idempotent-orthogonal family $\{C_k\}_{k \in K}$ of simple right S/J-modules, such that $|\mathcal{M}| < |K|$. By Lemma 12.4, there is an injective mapping from K to $\Omega(R)$, and so we get a chain of inequalities:

$$|\Omega(R)| \leq |\mathcal{M}| < |K| \leq |\Omega(R)|,$$

which gives a contradiction. This shows that $|\mathcal{M}|$ is finite, say r, and that $|\Omega(R)| = n \leq r$ is also finite. Thus we have $\Omega(R) = C(P)$. This means that P_R cogenerates all the simple right R-modules.

We claim that $|\mathcal{M}| = n$. Let $C_1, \ldots, C_r$ be a set of representatives of the elements of $\mathcal{M}$. Suppose that there exists a simple right S-module $C = C_{r+1}$ which is not isomorphic to any of the $C_1, \ldots, C_r$. For each $1 \leq i \leq r$, there exist idempotent elements $e_1, e_2, \ldots, e_r \in S$ such that, if $\bar{e}_i = e_i + J$, then $C_i = \bar{e}_i(S/J)$. Since $\bar{e}_i(S/J)\bar{e}_j = \text{Hom}_{S/J}(\bar{e}_j(S/J), \bar{e}_i(S/J))$, we see that $\bar{e}_i(S/J)\bar{e}_j = 0$ for $i, j \leq r$, $i \neq j$ and $\bar{e}_i(S/J)\bar{e}_i \neq 0$ for all $i = 1, \ldots, r$. Thus the family $\{C_i\}$, $i = 1, \ldots, r+1$ is an idempotent-semiorthogonal family of simple right S/J-modules with respect to the idempotents $\{\bar{e}_1, \ldots, \bar{e}_r, 1\}$. From Lemma 12.4, we obtain that $r + 1 \leq n$, a contradiction that shows that the simple module C cannot exist, and hence that S/J is a semisimple artinian ring. Therefore S is a semiperfect ring and E_R is a finite-dimensional module. Thus E_R is a finite direct sum of indecomposable submodules. From the preceding argument it also follows that $r \leq n$ and hence that $r = n$. Since the number of isomorphism classes of indecomposable direct summands of E_R is exactly n, we see that each of them is an injective envelope of a simple right R-module, so that E_R, and hence P_R, has finite essential socle. $\qquad\square$

If R is a right CS ring, then every direct summand of $E(R_R)$ has an essential finitely generated projective submodule. As a consequence, we have:

Corollary 12.6 *Let R be a ring such that R is right CS and cogenerates the simple right R-modules. Then R_R has finite essential socle.*

Since a ring R such that every cyclic right R-module has finite essential socle is right artinian, we have the following:

Corollary 12.7 *Let R be a right CS right CF ring. Then R is right artinian.*

Thus it follows immediately that:

Corollary 12.8 *Let R be a right CS right FGF ring. Then R is QF.*

Chen and Li [48] studied right CF strongly right C2-rings. Here a ring R is called a *strongly right C2-ring* if R^n satisfies condition C2 as a right R-module for every $n \geq 1$.

The following is well known.

Lemma 12.9 *A ring R is right π-injective if and only if, for any two right ideals A and B of R with $A \cap B = 0$, we have $ann_l(A) + ann_l(B) = R$.*

Proof Suppose R_R is π-injective. Let A and B be right ideals of R with $A \cap B = 0$. Let $P \supseteq B$ be a complement of A and let $Q \supseteq A$ be a complement of P. Then P is also a complement of Q. Since R is right π-injective, it follows that $R = P \oplus Q$. Thus there exists an idempotent e in R such that $Q = eR$ and $P = (1 - e)R$. So we have $A \subseteq eR$ and $B \subseteq (1 - e)R$. Clearly, then $ann_l(A) \supseteq R(1 - e)$ and $ann_l(B) \supseteq Re$. Hence $R = ann_l(A) + ann_l(B)$.

Conversely, suppose for any two right ideals A and B of R with $A \cap B = 0$, we have $ann_l(A) + ann_l(B) = R$. Let C and D be right ideals of R that are the complement of each other. To prove that R is right π-injective, it suffices to show that $R = C \oplus D$ (see [179], Theorem 2.8). Since $C \cap D = 0$, by assumption, $R = ann_l(C) + ann_l(D)$. Thus we have $1 = e + f$ where $e \in ann_l(C)$ and $f \in ann_l(D)$. Then $1 - f = e \in ann_l(C)$. So $fc = c$ for each $c \in C$. Similarly, we have $ed = d$ for each $d \in D$. So $C = fC, D = eD$, and $fD = 0 = eC$. Clearly, $D \subseteq ann_r(f) \subseteq ann_r(f^2)$ and $ann_r(f^2) \cap C = 0$. Now, since D is a complement of C, we have $D = ann_r(f) = ann_r(f^2)$. Similarly, we can show that $C = ann_r(e) = ann_r(e^2)$. Suppose $fer \in C \oplus D$, where $r \in R$. Then we have $f^2 e^2 r = 0$ because $fe = ef$ and $fD = 0 = eC$. Since $ann_r(e) = ann_r(e^2)$ and $ann_r(f) = ann_r(f^2)$, this yields $fer = 0$. Therefore $feR \cap (C \oplus D) = 0$. But, $C \oplus D \subseteq_e R_R$. Thus $fe = 0 = ef$. Now $eR \subseteq ann_r(f) = D$ and $fR \subseteq ann_r(e) = C$. This yields $R = eR + fR = C + D = C \oplus D$. Hence R_R is π-injective. $\qquad \square$

The following is a useful lemma.

Lemma 12.10 *Let R be a strongly right C2-ring. If I is a finitely generated proper left ideal of R, then $ann_r(I) \neq 0$.*

Proof Let $I = Rx_1 + \cdots + Rx_n$ be a finitely generated proper left ideal of R. We wish to show that $ann_r(I) \neq 0$. Assume to the contrary that $ann_r(I) = 0$. Then $(x_1, \ldots, x_n)^T R \cong R_R$. So $(x_1, \ldots, x_n)^T R$ is a direct summand of R^n. Let $(y_1, \ldots, y_n)^T \in ann_l^{R^n} ann_r^R(x_1, \ldots, x_n)^T$. Then we get an isomorphism $f : (x_1, \ldots, x_n)^T R \to (y_1, \ldots, y_n)^T R$ defined by $f((x_1, \ldots, x_n)^T r) = (y_1, \ldots, y_n)^T r$. Since $(x_1, \ldots, x_n)^T R$ is a direct summand of R^n, there exists a homomorphism $g : R^n \to (y_1, \ldots, y_n)^T R$ such that $gi = f$ where $i : (x_1, \ldots, x_n)^T R \to R^n$ is the inclusion. So, there exists an element $M \in M_n(R)$ such that $M(x_1, \ldots, x_n)^T = (y_1, \ldots, y_n)^T$, which means $(y_1, \ldots, y_n)^T \in M_n(R)(x_1, \ldots, x_n)^T$. So,

$$ann_l^{R^n} ann_r^R(x_1, \ldots, x_n)^T = M_n(R)(x_1, \ldots, x_n)^T.$$

Since $ann_r(I) = 0$, $M_n(R)(x_1, \ldots, x_n)^T = R^n$ and hence $I = Rx_1 + \cdots + Rx_n = R$. This yields a contradiction to the fact that I is a proper left ideal of R and thus the proof is complete. $\square$

Next, we state an important characterization of right CF rings.

Lemma 12.11 *A ring R is a right CF ring if and only if every right ideal of R is a right annihilator of some finite subset of R.*

Proof Let R be a right CF ring. Suppose I is a right ideal of R and $\varphi : R/I \to R^n$ is an embedding given by $\varphi(1 + I) = (r_1, \ldots, r_n)$. Then $I = ann_r(S)$ where $S = \{r_1, \ldots, r_n\}$ is a finite subset of R. Conversely, suppose I is a right ideal of R such that $I = ann_r(S)$ where $S = \{r_1, \ldots, r_n\}$ is a finite subset of R. Define $\varphi : R \to R^n$ by $\varphi(r) = (a_1 r, \ldots, a_n r)$. Clearly this map φ is an R-homomorphism with kernel I. Thus R/I is embedded in R^n and hence R is a right CF ring. $\square$

Chen and Li showed the following.

Theorem 12.12 *If R is a right CF and strongly right C2-ring, then R is right artinian.*

Proof Let R be a right CF and strongly right C2-ring. Thus, by Lemma 12.11, for any two right ideals A and B with $A \cap B = 0$, we have $A = ann_r(I_1)$, $B = ann_r(I_2)$ for finitely generated left ideals I_1 and I_2 of R. So $ann_r(I_1 + I_2) = ann_r(I_1) \cap ann_r(I_2) = 0$. Since $I_1 + I_2$ is finitely generated and R is a strongly right C2-ring, by Lemma 12.10, $I_1 + I_2 = R$. Therefore $I_1 \subseteq ann_l(ann_r(I_1)) = ann_l(A)$, $I_2 \subseteq ann_l(ann_r(I_2)) = ann_l(B)$. So $ann_l(A) + ann_l(B) = R$. Thus, by Lemma 12.9, R is right π-injective and hence a right CS ring. Therefore, by Corollary 12.7, it follows that R is right artinian. $\square$

Chen and Li further studied left perfect right CF rings. In order to prove their result, we first give some lemmas.

Lemma 12.13 *Let R be a semiregular right CF ring and let M be a cyclic right R-module. If no nonzero direct summands of M embed in the Jacobson radical of a finitely generated free module, then M is projective and π-injective.*

Proof Since $M + f(M) \subseteq (1 - f)(M) \oplus f(M) \subseteq M + f(M)$ for any $0 \neq f = f^2 \in \text{End}(E(M_R))$, $M + f(M) = (1 - f)(M) \oplus f(M)$. Since R is right CF and both $f(M)$ and $(1 - f)(M)$ are both cyclic, $M + f(M) = (1 - f)(M) \oplus f(M)$ embeds in a finitely generated free module F. Since R is semiregular, M admits a decomposition $M = P \oplus N$, where P is a direct summand of F, and $N \subseteq J(R)F$ (see [183]). By hypothesis, $N = 0$ and $M = P$ is projective. Since $M + f(M) \subseteq F$ and P is a direct summand of F, $M (= P)$ is a direct summand of $M + f(M)$. Note that $M \subseteq_e M + f(M)$. Thus $f(M) \subseteq M$ and so M is π-injective. $\square$

The next two lemmas are straightforward.

Lemma 12.14 *Let R be a ring and let X be a finite left or right T-nilpotent subset of R. Then there exists a positive integer n such that every product of n elements of X is zero.*

Lemma 12.15 *For a ring R, we have the following:*

1. *If R is right semiartinian, then the socle length of $J(R)$ is less than the socle length of R_R.*
2. *If R is right CF, then R is right semiartinian if and only if $\operatorname{Soc}(R_R) \subseteq_e R_R$.*

Theorem 12.16 *If R is a right CF ring, then the following are equivalent:*

1. *R is semilocal and $\operatorname{Soc}(R_R) \subseteq_e R_R$.*
2. *R is left perfect.*
3. *R is right artinian.*

Proof (1)$\Leftrightarrow$(2). It is well-known that a ring is left perfect if and only if it is right semiartinian and semilocal. Suppose R is semilocal and $\operatorname{Soc}(R_R) \subseteq_e R_R$. By Lemma 12.15, R is right semiartinian. Therefore R is left perfect. The converse follows easily from the first assertion.

(3)$\Rightarrow$(2) is clear.

(2)$\Rightarrow$(3). Let $R = \oplus_{i=1}^n e_i R$, where $\{e_1, \ldots, e_n\}$ is a set of orthogonal local idempotents. If none of the $e_i R$ is π-injective, then by Lemma 12.13, R embeds in $(J(R))^m$ for some positive integer m. Then the socle length of R_R is less than or equal to the socle length of $J(R)$. This contradicts Lemma 12.15. Thus we may assume that $\{e_1, \ldots, e_n\}$ is ordered in such a way that $e_i R$ embeds in $(J(R))^k$, $k \in \mathbb{N}$ for $i = 1, \ldots, r$, while for $i = r+1, \ldots, n$, $e_i R$ is π-injective. Thus $\operatorname{Soc}(R_R) \subseteq_e R_R$ implies that $\operatorname{Soc}(e_i R)$ is simple as a right ideal and $\operatorname{Soc}(e_i R) \subseteq_e e_i R$ for $i = r+1, \ldots, n$. So $e_i R$ is finitely cogenerated for $i = r+1, \ldots, n$. Thus $\oplus_{i=r+1}^n e_i R$ is finitely cogenerated. Set $e = e_1 + \ldots + e_r$, and $1 - e = e_{r+1} + \ldots + e_n$. Then there exists a monomorphism $\varphi : \oplus_{i=1}^r e_i R \to (J(R))^k$, where k is a positive integer. Hence $eR \hookrightarrow R^k = eR^k \oplus (1-e)R^k$. Suppose that the first component of $Im(\varphi) \subseteq e(J(R))^k \oplus (1-e)R^k$ takes e to $(x_1, \ldots, x_k) \in e(J(R))^k$. Now take the monomorphism $\varphi^k : eR^k \to eR^{k^2} \oplus (1-e)R^{k^2}$ and consider the composition

$$eR \to eR^k \oplus (1-e)R^k \to eR^{k^2} \oplus (1-e)R^{k^2} \oplus (1-e)R^k.$$

This composition is a monomorphism whose first component $eR \to eR^{k^2}$ maps e onto the element $(x_{i_2}, x_{i_1})_{(i_2, i_1) \in \mathbb{N}^2}$ of eR^{k^2}. By recursively continuing in this way, we get for each $t > 0$, a monomorphism

$$f = (\varphi^{k^t} \oplus 1)(\varphi^{k^{t-1}} \oplus 1) \cdots \varphi : eR \to eR^{k^t} \oplus (1-e)R^{k^t + \cdots + k}$$

whose first component $eR \to eR^{k^t}$ maps e onto the element $(x_{i_t}, \ldots, x_{i_1})_{(i_t, \ldots, i_1) \in \mathbb{N}^t}$. Lemma 12.14 ensures that for a large enough t, that component is zero. As a consequence, $eR \hookrightarrow (1-e)R^s$. Since $(1-e)R^s$ is finitely cogenerated, so is the submodule eR. Thus R is finitely cogenerated, and so R is right artinian. $\qquad\square$

The following result of Rada and Saorin [197] follows immediately from the above theorem.

Corollary 12.17 *Every right FGF left perfect ring is QF.*

Jain and López-Permouth studied rings under a tighter embedding hypothesis, more specifically, rings whose cyclic modules are essentially embeddable in projective modules (direct summands of R_R) [135]. A ring R is called a *right CEP ring* if each cyclic right R-module is essentially embeddable in a projective module. A ring R is called a *right CES ring* if each cyclic right R-module is essentially embeddable in a direct summand of R_R. Examples of right CEP rings include QF-rings and right uniserial rings.

It is easy to see that a ring R is a QF-ring if and only if R is both a right and left CEP-ring. Jain and López-Permouth [135] provided the following characterization of QF-rings.

Theorem 12.18 *For a ring R, the following are equivalent:*

1. *R is QF.*
2. *R is CEP and QF-3.*
3. *Every cyclic R-module has a projective injective hull.*

The following example due to Jain and López-Permouth [135] is an example of a local right CEP ring which is neither right uniserial nor quasi-Frobenius.

Example 12.19 *Let S be a ring having only three right ideals, namely, (0), $J(S)$, and S and not right self-injective. Let $R = S \propto S$ be the trivial extension of S by itself. Then R is a local right CEP ring which is neither right uniserial nor quasi-Frobenius.*

In [10] Al-Huzali, Jain, and López-Permouth asked if each right CEP ring (and hence right CES ring) is semiperfect. This question was answered in the affirmative by Gómez Pardo and Guil Asensio [91].

Corollary 12.20 *Every right CEP ring is right artinian. In particular, every right CES ring is right artinian.*

Proof Let R be a right CEP ring. From Theorem 12.5 we know that R_R has finite essential socle. Since every cyclic right R-module embeds in a finitely generated free module, it follows that every cyclic right R-module has finite essential socle. Hence R is right artinian. $\qquad\square$

We have the following structure for right CES rings.

Corollary 12.21 *Let R be a ring. Then the following conditions are equivalent:*

1. *R is right CES.*
2. *R is of one of the following types:*
   (a) R is (artinian) uniserial as a right R-module;
   (b) R is an $n \times n$ matrix ring over a right self-injective ring of type (a); or
   (c) R is a direct sum of rings of types (a) or (b).

12.22 Questions

1. Is every right FGF ring a QF ring?
2. Is every right CF ring right artinian?

13 Serial and distributive modules

We begin this chapter with the following theorem from Warfield [234].

Theorem 13.1 *For a ring R, the following conditions are equivalent:*

1. *R is a serial ring.*
2. *All finitely presented right R-modules are serial.*
3. *All finitely presented left R-modules are serial.*
4. *R is a semiperfect ring, all finitely presented indecomposable right R-modules are local, and all finitely presented indecomposable left R-modules are local.*

We would like to caution the reader here that the proof of the above theorem presented by Drozd [60] seems to have a gap in the diagonalization argument.

As a consequence of the Warfield Theorem, we have the following:

Corollary 13.2 *A ring R is a serial right noetherian ring if and only if all finitely generated right R-modules are serial noetherian modules.*

A module M is said to be *distributive* if the lattice of all submodules of M is distributive; that is, $(X + Y) \cap Z = X \cap Z + Y \cap Z$ for any three submodules X, Y, Z of M. A direct sum of distributive modules is called a *semidistributive* module.

It is well-known that any uniserial module is distributive, and any serial module is semidistributive. The ring of integers $\mathbb{Z}$ is a distributive $\mathbb{Z}$-module which is not serial.

If R is a ring and X, Y are two subsets of a right R-module M, then we set $(X : Y) = \{a \in R \mid Xa \subseteq Y\}$.

The next two lemmas are combinations of some results from [172], [173], [214], [40], and [221].

Lemma 13.3 *For a right R-module M, the following conditions are equivalent.*

1. *M is a distributive module.*
2. *All 2-generated submodules in M are distributive.*
3. *In M, every 2-generated submodule is contained in some distributive submodule.*
4. *$X \cap (Y + Z) = X \cap Y + X \cap Z$ for all cyclic submodules X, Y, Z in M.*

5. $(\sum_{i \in I} X_i) \cap (\sum_{j \in J} Y_j) = \sum_{i \in I, j \in J}(X_i \cap Y_j)$ *for any two sets* $\{X_i\}_{i \in I}$ *and* $\{Y_j\}_{j \in J}$ *of submodules in* M.

6. *All subfactors of* M *are distributive.*

7. R *contains a unitary subring* R' *such that the induced* R'-*module* M *is distributive.*

8. *For any* $x, y \in M$, *there exists an element* $a \in R$ *with* $xaR + y(1 - a)R \subseteq xR \cap yR$.

9. *For any* $x, y \in M$, *there exist* $a, b, c, d \in R$ *such that* $1 = a + b$, $xa = yc$, *and* $yb = xd$.

10. $R = (x : yR) + (y : xR)$ *for all* $x, y \in M$.

11. *For any* $x, y \in M$, *there exists a right ideal* B *of* R *with* $(x + y)R = xB + yB$.

Proof The equivalences $(1) \Leftrightarrow (6)$ and $(8) \Leftrightarrow (9) \Leftrightarrow (10)$ are easily verified.

$(1) \Rightarrow (8)$. Let $T = xR \cap yR$. Since $(x + y)R = (x + y)R \cap xR + (x + y)R \cap yR$, there exist $b, d \in R$ such that

$$(x + y)b \in xR, \quad (x + y)d \in yR, \quad x + y = (x + y)b + (x + y)d.$$

Therefore $yb = (x + y)b - xb \in T$ and $xd = (x + y)d - yd \in T$. Let $a \equiv 1 - b$ and $z \equiv a - d = 1 - b - d$. Then

$$1 = a + b, \quad (x + y)z = (x + y) - (x + y)b - (x + y)d = 0,$$

$$xa = xd + xz = xd + (x + y)z - yz = xd - yz,$$

$$yz = -xz \in T, \quad xa \in T$$

and a is the required element.

$(8) \Rightarrow (1)$. Let $X, Y, Z \in M$ and let $z = x + y \in (X + Y) \cap Z$, where $x \in X$ and $y \in Y$. By assumption, there exist $a, b \in R$ such that $1 = a + b$, $xa \in yR$, and $yb \in xR$. Then $zb = xb + yb \in xR \cap zR$,

$$za = xa + ya \in yR \cap zR, \quad z = zb + za \in X \cap Z + Y \cap Z,$$

$$(X + Y) \cap Z \subseteq X \cap Z + Y \cap Z \subseteq (X + Y) \cap Z.$$

$(1) \Rightarrow (11)$. We set $B = (x : (x + y)R)$. Then $(x + y)R \cap xR = xB$. If $b \in B$, then $yb = (x + y)b - xb \in (x + y)R$. Therefore:

$$B = (y : (x + y)R), \quad (x + y)R \cap yR = yB,$$

$$(x + y)R = (x + y)R \cap xR + (x + y)R \cap yR = xB + yB.$$

$(11) \Rightarrow (1)$. Let $x, y \in M$. By assumption, $(x + y)R = xB + yB$, where B is a right ideal of R. It follows from the modular law that:

$$xR \cap (x + y)R = xR \cap (xB + yB) = xB,$$

$$yR \cap (x + y)R = yR \cap (xB + yB) = yB,$$

$$(x + y)R = xR \cap (x + y)R + yR \cap (x + y)R$$

and M is distributive.

$(1) \Leftrightarrow (7)$ follows from the proved equivalence $(1) \Leftrightarrow (9)$. $\square$

A ring R is said to be an *abelian ring* if all idempotents of R are central. We will denote the lattice of submodules of a module M by $\mathrm{Lat}(M)$.

Lemma 13.4 *For a module M, the following conditions are equivalent.*

1. *M is distributive.*
2. *$\mathrm{Hom}(X, Y) = 0$ for any subfactor $X \oplus Y$ of M.*
3. *M does not have subfactors which are direct sums of two isomorphic nonzero modules.*
4. *M does not have subfactors which are direct sums of two isomorphic simple modules.*
5. *In every subfactor of M, any direct summand is fully invariant.*
6. *The endomorphism ring of any subfactor of M is an abelian ring.*
7. *$\mathrm{Hom}(X/(X \cap Y), Y/(X \cap Y)) = 0$ for any submodules X and Y in M.*
8. *$\mathrm{Hom}((X + Y)/Y, (X + Y)/X) = 0$ for any submodules X and Y in M.*
9. *$\mathrm{Hom}(M'/Y, M'/X) = 0$ for any $X, Y, M' \in \mathrm{Lat}(M)$ with $X + Y = M'$.*
10. *For any subfactor $\overline{M}$ of M and each submodule X in $\overline{M}$, we have that $X + f(X)$ is an essential extension of X for any homomorphism $f \colon X \to \overline{M}$.*
11. *For any subfactor $\overline{M}$ of M, every closed submodule in $\overline{M}$ is fully invariant in $\overline{M}$.*

Proof The implications $(2) \Rightarrow (3)$, $(3) \Rightarrow (4)$, $(10) \Rightarrow (11)$, and $(11) \Rightarrow (5)$, and equivalences $(2) \Leftrightarrow (5)$, $(5) \Leftrightarrow (6)$, $(2) \Leftrightarrow (7)$, and $(8) \Leftrightarrow (9)$ are easily verified. The equivalence $(7) \Leftrightarrow (8)$ follows from the natural isomorphisms $X/(X \cap Y) \cong (X + Y)/Y$ and $Y/(X \cap Y) \cong (X + Y)/X$.

$(2) \Rightarrow (10)$. Let Y be a submodule in $X + f(X)$ with $Y \cap X = 0$. Then $Y = Y \cap (X + f(X)) = Y \cap X + Y \cap f(X) = Y \cap f(X)$. Therefore there exists a submodule X' in X such that $Y = f(X')$ and $X' \cap Y = 0$. By (2), $Y = 0$. Thus $X + f(X)$ is an essential extension of X.

$(4) \Rightarrow (3)$. We assume that M has a subfactor $X \oplus Y$ such that X and Y are nonzero modules and there exists an isomorphism $f \colon X \to Y$. The nonzero module X has a nonzero cyclic submodule N. The module N has a simple factor module N/T. Then Y has a simple subfactor $f(N)/f(T)$. Therefore M has a subfactor isomorphic to $N/T \oplus N/T$.

$(3) \Rightarrow (1)$. We assume that M is not distributive. There exist $X, Y, Z \in \mathrm{Lat}(M)$ such that $X \subseteq Y + Z$ and $X/(X \cap Y + X \cap Z)$ is a nonzero submodule in $(Y + Z)/(X \cap Y + X \cap Z)$. Let $h\colon Y + Z \to (Y + Z)/(X \cap Y + X \cap Z)$ be the natural epimorphism. Then:

$$0 \neq h(X) \subseteq h(Y + Z) = h(Y) \oplus h(Z), \quad h(X) \cap h(Y) = 0, \quad h(X) \cap h(Z) = 0.$$

Let $f\colon h(Y) \oplus h(Z) \to h(Y)$ and $g\colon h(Y) \oplus h(Z) \to h(Z)$ be the natural projections. Since $h(X) \cap h(Y) = 0$ and $h(X) \cap h(Z) = 0$, we have $f(h(X)) \cong h(X) \cong g(h(X)) \neq 0$. This is a contradiction.

$(1) \Rightarrow (2)$. Let $f \in \mathrm{Hom}(X, Y)$, $x \in X$, and let $y = f(x) \in Y$. By Lemma 13.3 there exists an element $a \in R$ such that:

$$xaR + y(1 - a)R \subseteq xR \cap yR \subseteq X \cap Y = 0,$$

$$xa = y(1 - a) = 0, \quad y = ya = f(x)a = f(xa) = f(0) = 0.$$

Therefore $f \equiv 0$ and $\mathrm{Hom}(X, Y) = 0$. $\qquad\square$

Lemma 13.5 *For any ring R, any nonzero right R-module M is an essential extension of a direct sum of nonzero cyclic modules.*

Proof Let $\mathcal{E}$ be the nonempty set of all submodules in M that are direct sums of nonzero cyclic modules. In $\mathcal{E}$, there exists a partial order $\leq$ such that for any $S, T \in \mathcal{E}$, the inequality $S \leq T$ means that $S \oplus E = T$ for some $E \in \mathcal{E}$. Since $\mathcal{E}$ contains the union of any ascending (with respect to $\leq$) chain of elements of $\mathcal{E}$, Zorn's Lemma implies that $\mathcal{E}$ contains at least one maximal element X. It is clear that M is an essential extension of $X = \oplus_{i \in I} X_i$, where all X_i are nonzero cyclic modules. $\qquad\square$

The following result is from Tuganbaev [218].

Theorem 13.6 *For a ring R, there exists a cardinal number $\alpha = \alpha(R)$ such that every distributive right R-module has cardinality $\leq \alpha$.*

Proof Let $\{X_i\}_{i \in I}$ be the set of all nonisomorphic cyclic R-modules, M be the injective hull of $\oplus_{i \in I} X_i$, and let α be the cardinality of M. Let Y be a distributive right R-module. By Lemma 13.5, the module Y is an essential extension of a module $\oplus_{j \in J} Y_j$, where Y_j are cyclic modules. By Lemma 13.4 the modules Y_j are nonisomorphic modules. Therefore there exists a monomorphism $f\colon \oplus_{j \in J} Y_j \to \oplus_{i \in I} X_i$. Since the module M is injective, f can be extended to a monomorphism $Y \to M$. Therefore the cardinality of Y does not exceed α. $\qquad\square$

Theorem 13.7 *(see [75, 20.23]) Let R be a ring and let α be a cardinal number. If every right R-module is a direct sum of modules with generator systems of cardinality $\leq \alpha$, then the ring R is right artinian.*

Tuganbaev [218] proved the following:

Theorem 13.8 *If R is a ring such that all right R-modules are semidistributive, then R is right artinian.*

Proof By Theorem 13.6, there exists a cardinal number α such that every right R-module is a direct sum of modules with generator systems of cardinality $\leq \alpha$. By Theorem 13.7, the ring R is right artinian. $\qquad\square$

A module M is called *completely cyclic* if each submodule of M is cyclic. The next lemma is easy to prove.

Lemma 13.9 *If R is a left or right artinian ring, then every uniserial module is completely cyclic. Therefore every left serial right artinian ring is artinian.*

The following result is from Skornyakov [205].

Theorem 13.10 *If R is a ring such that all right R-modules are serial, then R is an artinian serial ring.*

Proof By Theorem 13.8, the ring R is right artinian. By Theorem 13.1, the ring R is serial. By Lemma 13.9, the left serial right artinian ring R is artinian. $\quad\square$

Lemma 13.11 *For a ring R, the following conditions are equivalent:*

1. *R is a right noetherian ring.*
2. *For every right R-module M, there exists a decomposition $M = X \oplus Y$ such that X is an injective module and the module Y does not have nonzero injective submodules.*

Proof $(1) \Rightarrow (2)$. Since R is a right noetherian ring, all direct sums of injective right R-modules are injective. Let $\{X_i\}_{i\in\mathcal{I}}$ be the set of all submodules in M, which are direct sums of injective modules. On this set, we define a partial order $\leq$ such that for arbitrary $i, j \in \mathcal{I}$, $X_i \leq X_j$ if and only if $X_i \oplus X_k = X_j$ for some $k \in \mathcal{I}$. In this set, every ascending chain has an upper bound, since all direct sums of injective right R-modules are injective. By Zorn's Lemma, the set $\{X_i\}_{i\in\mathcal{I}}$ has a maximal element X. Then there exists a decomposition $M = X \oplus Y$, which is the required decomposition.

$(2) \Rightarrow (1)$. Let $M = \oplus_{i\in I} M_i$, where all M_i are injective right R-modules. It is sufficient to prove that the module M is injective. By assumption, there exists a decomposition $M = X \oplus Y$ such that X is injective and Y does not have nonzero injective submodules. If $Y = 0$, then $M = X$, and so M is injective. We assume that $Y \neq 0$. Since X is not an essential submodule in M, we have that $X \cap M_i$ is not an essential submodule in M_i for some $i \in I$. Then the injective module M_i has a nonzero direct summand N with $X \cap N = 0$. Let $\pi\colon M \to Y$ be the projection with kernel X. Since $X \cap N = 0$, we have that $\pi(N) \cong N$ and the module Y has a nonzero injective submodule $\pi(N)$. This is a contradiction. $\quad\square$

Lemma 13.12 *Let R be an artinian serial ring with Jacobson radical $J(R) = J$ and let n be the nilpotence index of J.*

1. *If X is a right R-module which is either finitely generated or injective, then X is a direct sum of completely cyclic uniserial modules of length $\leq n$.*

2. *If Y is a right R-module with $Y J^{n-1} \neq 0$, then Y has a nonzero injective uniserial direct summand.*

3. *For every right R-module M, there exists a direct decomposition $M = X \oplus Y$ such that X is an injective serial module and $Y J^{n-1} = 0$.*

Proof

(1) By Lemma 13.9, $X = \oplus_{i \in I} X_i$ and for any i, we have that X_i is a completely cyclic uniserial module with $X_i J^n = 0$. Let $i \in I$. Since X_i is uniserial, we have that for any positive integer $k \leq n$, the module $X_i J^{k-1}/X_i J^k$ either is simple or it is equal to zero. Therefore X_i is a module of length $\leq n$.

(2) Since $Y J^{n-1} \neq 0$, we have that Y has a cyclic submodule P with $P J^{n-1} \neq 0$. By (1), we can assume that P is a uniserial module of length $\leq n$. Then P is a module of length n, since otherwise $P J^{n-1} = 0$. The injective hull Q of P is indecomposable. By (1), Q is a cyclic uniserial module of length $\leq n$, and Q contains the submodule P of length n. Therefore $P = Q$ is an injective direct summand in Y.

(3) By Lemma 13.11, there exists a decomposition $M = X \oplus Y$ such that X is an injective module and Y does not have nonzero injective submodules. By (2), $Y J^{n-1} = 0$. By (1), X is serial. $\qquad\square$

The following result is due to Nakayama.

Theorem 13.13 *([182], see also [75, 25.4.2])*

1. *If R is an artinian serial ring and n is the nilpotence index of $J(R)$, then every R-module is a direct sum of completely cyclic uniserial modules of length $\leq n$.*

2. *Let R be a ring such that R_R is a direct sum of indecomposable modules. If every finitely generated indecomposable right R-module is a uniserial artinian module, then R is an artinian serial ring.*

Proof

(1) We use induction on n. For $n = 1$, R is semisimple, every R-module is semisimple, and the assertion holds. We assume that $n > 1$, $S = R/J^{n-1}$, and every S-module is a direct sum of completely cyclic uniserial modules of length $\leq n - 1$. Let M be a right R-module. By Lemma 13.12(3), there exists a decomposition $M = X \oplus Y$ such that X is an injective serial module and $Y J^{n-1} = 0$. By Lemma 13.12(1), X is a direct sum of completely cyclic uniserial modules of length $\leq n$. Since Y is an S-module, it follows from the induction hypothesis that Y is a direct sum of completely cyclic uniserial modules of length $\leq n - 1$.

(2) Since R_R is a finite direct sum of artinian modules, R is right artinian. Since R is right artinian, every finitely generated right R-module M is a

direct sum of indecomposable modules, which are serial by assumption. Then M is serial. By Theorem 13.1, R is a serial ring. By Lemma 13.9, the ring R is artinian. $\qquad\qquad\square$

Example Let F be a field and let R be the Five-dimensional F-algebra consisting of all 3×3 matrices of the form $\begin{pmatrix} f_{11} & f_{12} & f_{13} \\ 0 & f_{22} & 0 \\ 0 & 0 & f_{33} \end{pmatrix}$, where $f_{ij} \in F$. Then $e_{11}R = e_{11}F + e_{12}F + e_{13}F$ is an indecomposable distributive noetherian artinian nonuniserial completely cyclic R-module. It may be checked that R is a right semidistributive, left serial, right and left artinian ring which is not right serial. $\qquad\qquad\square$

13.14 Questions

1. Let R be a (right and left) serial ring. Does every cyclic right R-module have finite Goldie dimension?
2. Characterize rings over which all (right) finitely presented modules are semidistributive.
3. Characterize rings over which all (right) finitely generated modules are semidistributive.

14 Rings characterized by decompositions of their cyclic modules

The study of rings via decomposition properties of cyclic modules has been considered by many authors. Chatters [45] studied rings where each cyclic module is a direct sum of a projective module and a noetherian module and proved the following.

Theorem 14.1 *A ring R is right noetherian if and only if every cyclic module is a direct sum of a projective module and a noetherian module.*

In [117], Huynh and Dung considered the rings whose cyclic modules are the direct sum of an injective module and a finitely cogenerated module and showed the following.

Theorem 14.2 *A ring R is right artinian if and only if each cyclic right R-module is the direct sum of an injective module and a finitely cogenerated module.*

Huynh [112] extended this result in the next theorem. Recall that a ring R is called *hereditarily artinian* if every ideal of R is a right artinian ring.

Theorem 14.3 *A ring R is hereditarily artinian if and only if each cyclic right R-module is the direct sum of an injective module and a finite module.*

The next lemma is a simple observation.

Lemma 14.4 *Let R be a ring such that every cyclic right R-module is either semisimple or has a nonzero injective submodule. Then:*

1. *For any right R-module M, the Jacobson radical $J(M)$ is semisimple. Consequently, $J(M) \subseteq \mathrm{Soc}(M)$ and $J(M) \cdot J(R) = 0$.*
2. *$(J(R))^2 = 0$ and $Z(R_R) \subseteq N(R) = J(R) \subseteq \mathrm{Soc}(R_R)$.*
3. *If M is an indecomposable nonsimple right R-module, then M is a cyclic injective uniserial module of length 2.*
4. *If R_R is a direct sum of indecomposable modules, then R is an artinian serial ring and $(J(R))^2 = 0$.*

Proof In any module, each injective submodule is a direct summand. Therefore it follows from the assumption that every cyclic nonsemisimple right R-module has a nonzero injective direct summand.

(1) Since $J(M)$ is the sum of all small submodules of M, every finitely generated submodule X of $J(M)$ is small in M. Therefore X does not contain nonzero injective submodules. It follows from the assumption that every cyclic module in $J(M)$ is semisimple. Therefore $J(M)$ is semisimple, since $J(M)$ is a sum of semisimple submodules.

(2) By (1), $(J(R))^2 = 0$. Therefore $J(R) \subseteq N(R)$, and the inclusion $J(R) \supseteq N(R)$ holds for all rings. We assume that $Z(R_R) \nsubseteq J(R)$. Then there exists an element $x \in Z(R_R) \setminus J(R)$. If the cyclic module xR is not semisimple, then it follows from the assumption that $Z(R_R)$ contains a nonzero idempotent e; this contradicts the relation $eR \cap (1 - e)R = 0$. Therefore the module xR is semisimple. Thus $xR = X_1 \oplus \cdots \oplus X_n$, where all modules X_i are simple. Since $x \notin J(R)$, we have that $xR + B = R$ for some B. Therefore $X_i + B = R$ for some i. Then $X_i \cap B = 0$ and $X_i \oplus B = R_R$. Therefore $X_i = e'R$ for some nonzero idempotent $e' \in X_i \subseteq Z(R)$; this is a contradiction.

(3) Since the indecomposable nonsimple module M cannot be semisimple, $M \neq \operatorname{Soc}(M)$. It follows from the assumption that for any $x \in M \setminus \operatorname{Soc}(M)$, the cyclic nonsemisimple module xR contains a nonzero injective submodule X. Then X is a nonzero direct summand of the indecomposable module M. Therefore $M = X = xR$ is a cyclic indecomposable injective nonsimple module, and M is generated by any element $x \in M \setminus \operatorname{Soc}(M)$. Then $M/\operatorname{Soc}(M)$ is a simple module. Since the indecomposable injective module M is uniform, $\operatorname{Soc}(M)$ is an indecomposable semisimple module. Therefore $\operatorname{Soc}(M)$ is a simple proper submodule of M and M is a uniserial module of length 2.

(4) By (2), $(J(R))^2 = 0$. By (3), every finitely generated indecomposable right R-module is a uniserial artinian module. By Theorem 13.13(2), R is an artinian serial ring. $\qquad\qquad\square$

Lemma 14.5 *Let R be a semiprime ring. Then:*

1. *If X is a finitely generated semisimple right ideal of R, then any submodule of X_R is a direct summand of R_R. Therefore it is a cyclic projective right R-module generated by an idempotent of R.*

2. *If X is a finitely generated semisimple right ideal of R, then X is an injective right R-module.*

3. *If every cyclic right R-module is a direct sum of a semisimple module and an injective module, then R is a semisimple artinian ring.*

Proof

(1) Since every submodule of a finitely generated semisimple module is a finitely generated semisimple module, it is sufficient to prove that X is a direct summand of R_R. Let $X = X_1 \oplus \cdots \oplus X_n$, where all X_i are simple modules.

We assume that $n = 1$; the module X_R is simple. Since R is semiprime, $X^2 \neq 0$. Then $xX \neq 0$ for some $x \in X$. Therefore $X = xX$, since X is a minimal right ideal. Therefore $x = xe \neq 0$, where $0 \neq e \in X$. Since X is a minimal right ideal, $X = eR$ and $e^2 - e \in X \cap ann_r(x)$. But $X \cap ann_r(x) = 0$, since $X \cap ann_r(x) \neq X$. Therefore $e^2 - e = 0$ and X_R is a direct summand of R_R.

Now we assume that $n > 1$ and for any $k < n$, any right ideal, which is a direct sum of k simple modules, is a direct summand of R_R. There exists an idempotent $e \in R$ with $eR = X_1 \oplus \cdots \oplus X_{n-1}$. Then $X = eR \oplus Y$, where $Y = X \cap (1 - e)R$, $X_n \cong X/eR \cong Y$, and Y_R is a simple module. By the above, Y is a direct summand of R_R with $Y \subseteq (1 - e)R$. Then there exists a decomposition $(1 - e)R = Y \oplus Z$. Therefore $R_R = eR \oplus Y \oplus Z = X \oplus Z$.

(2) Let B be a right ideal of R and let $f \in \mathrm{Hom}(B, X)$. By Baer's criterion, it is sufficient to prove that f can be extended to $g \in \mathrm{Hom}(R_R, X)$. By (1), the module $f(B)$ is projective. Therefore $B = C \oplus \mathrm{Ker}(f)$, where $C \cong f(B)$, and $f(B)$ is a finitely generated semisimple module by (1). By (1), $C = \pi(R_R)$, where $\pi = \pi^2 \in \mathrm{End}(R_R)$. We set $g = f\pi \in \mathrm{Hom}(R_R, X)$. Then g coincides with f on $B = C \oplus \mathrm{Ker}(f)$.

(3) By assumption, $R_R = X \oplus Y$, where X is a semisimple module and Y is an injective module. By (2), X is injective. Therefore R is right self-injective. By Lemma 14.4(2), $Z(R_R) \subseteq N(R) = 0$. Therefore R is a von Neumann regular ring. Now let C be any cyclic right R-module. Then, by assumption, $C = M \oplus N$, where M is a semisimple module and N is an injective module. Since M is cyclic semisimple, it has finite Goldie dimension. Therefore, by Theorem 4.11, R has finite right Goldie dimension. Hence R is semisimple artinian. $\qquad\square$

Theorem 14.6 *([65, 13.5], [226]) For a ring R, the following conditions are equivalent:*

1. *Every right R-module is a direct sum of a semisimple module and an injective module.*

2. *Every cyclic right R-module is a direct sum of a semisimple module and an injective module.*

3. *R_R is a finite direct sum of indecomposable modules and every cyclic right R-module either is semisimple or has a nonzero injective submodule.*

4. *R_R is a finite direct sum of indecomposable modules and every indecomposable right R-module is a uniserial module of length ≤ 2.*

5. *R is an artinian serial ring and $(J(R))^2 = 0$.*

Proof The implication $(5) \Rightarrow (1)$ follows from Lemma 13.12(3). The implication $(1) \Rightarrow (2)$ is obvious.

$(2) \Rightarrow (3)$. Let $P = N(R)$ be the prime radical. The ring R/P is semiprime and every cyclic right R/P-module is a direct sum of a semisimple module and an injective module. By Lemma 14.5(3), R/P is a semisimple ring, and $P^2 = 0$ by Lemma 14.4(2). Therefore R is semiprimary. Then R_R is a finite direct sum of indecomposable modules. Now, let C be any cyclic right R-module. Then, by assumption, $C = X \oplus Y$, where X is semisimple and Y is injective. If $Y = 0$, then $C = X$ and so C is semisimple, otherwise C has a nonzero injective submodule Y.

The implication $(3) \Rightarrow (4)$ follows from Lemma 14.4(3).

$(4) \Rightarrow (5)$. It follows from (4) that R is a right artinian right serial ring and $J(R)$ is a semisimple right R-module. Then $(J(R))^2 = 0$. By Theorem 13.13(2), R is an artinian serial ring. $\qquad\square$

Goel, Jain, and Singh studied rings for which each cyclic module is injective or projective [87]. More generally, Osofsky and Smith [193] studied rings where each cyclic module is a direct sum of a projective module and an injective module and obtained the following.

Theorem 14.7 *A ring R is right noetherian if every cyclic right R-module is a direct sum of a projective module and an injective module.*

Proof Let R be a ring such that each cyclic right R-module is a direct sum of a projective module and an injective module. First, we will show that each cyclic singular right R-module is injective. Let R/I be a cyclic singular module. Then I is essential in R. Therefore R/I cannot contain a nonzero projective submodule. This implies R/I is injective.

Next, we claim that every singular right R-module is injective. Since submodules and quotient modules of a singular module are singular, it follows that every singular right R-module is semisimple. Since the singular submodule of R cannot contain a direct summand of R, R is right nonsingular. Since the injective hull of a singular module is singular (see, e.g., [96], Proposition 1.2.3), it follows that every singular module is a direct summand of its (singular) injective hull and is therefore injective.

Now we will show that $R/\operatorname{Soc}(R_R)$ is right noetherian. A simple $R/\operatorname{Soc}(R_R)$-module is a simple R-module annihilated by the socle of R and hence is singular as an R-module. If E is a direct sum of simple $R/\operatorname{Soc}(R_R)$-modules then E is singular and hence injective as an R-module and so injective as an $R/\operatorname{Soc}(R_R)$-module. Thus $R/\operatorname{Soc}(R_R)$ is a ring such that every direct sum of injective hulls of simple $R/\operatorname{Soc}(R_R)$-modules is injective and hence $R/\operatorname{Soc}(R_R)$ is right noetherian (see [159] or [202, Theorem 4.1]).

Now let xR be any cyclic injective right R-module. Let $S = \operatorname{Soc}(R_R)$. Then xR/xS is a cyclic right R/S-module and hence noetherian. Now, by Lemma 3.7, xS is of finite length. Therefore xR is noetherian. Thus we have shown that every cyclic injective right R-module is noetherian. Hence, by Theorem 14.1, it follows that R is right noetherian. $\qquad\square$

As a consequence, it follows that if each cyclic right R-module is injective or projective then the ring R is right noetherian. The structure of rings where each cyclic module is a direct sum of a projective module and an injective module was completely described by Huynh in [115].

In 1991, Smith asked the question whether a ring is right noetherian if each cyclic module is a direct sum of a projective module and a module that is either injective or noetherian? This question has been recently answered in the affirmative by Huynh and Rizvi [123].

To present their proof, let us fix some notation and terminology. A module M is called *completely injective* if every factor module of M is injective.

A cyclic module M is called a *special module* if the following conditions are satisfied:

(i) M is uniserial.

(ii) M is non-noetherian but every proper submodule of the maximal submodule N of M is noetherian.

(iii) For any nonzero proper submodule $K \subset M$, M/K is injective.

We call a ring R a *right PIN-ring* if every cyclic right R-module $C = P \oplus Q$ where P is a projective module and Q is either injective or noetherian.

First, notice that if R is a right PIN-ring and I is a two-sided ideal of R then R/I is also a right PIN-ring. We will start with a very useful lemma.

Lemma 14.8 *Let R be a right PIN-ring. Then R is a right q.f.d. ring.*

Proof Let R be a right PIN-ring and let C be an arbitrary cyclic right R-module. Let E be an essential submodule of C. Then $M = C/E$ is a singular module. Hence M and each cyclic subfactor of M cannot contain nonzero cyclic projective submodules. By hypothesis, every cyclic subfactor of M is injective or noetherian. By [63], M has Krull dimension. In particular M has finite Goldie dimension. Hence $C/\operatorname{Soc}(C)$ has finite Goldie dimension by ([65], 5.14). Now we will show that $\operatorname{Soc}(C)$ is finitely generated. Assume to the contrary that $\operatorname{Soc}(C)$ is not finitely generated. Then we may write $\operatorname{Soc}(C) = S_1 \oplus S_2$ where S_1 and S_2 are infinitely generated. Since S_1 cannot be a direct summand of C, the factor module C/S_1 is not projective. Hence by assumption, $C/S_1 = \overline{P} \oplus \overline{Q}$ where $\overline{P}$ is projective and $\overline{Q} \neq 0$ and is injective or noetherian. If $\overline{Q}$ is injective then $\overline{Q}$ has finite Goldie dimension by [63] because $\overline{Q}/\operatorname{Soc}(\overline{Q})$ has finite Goldie dimension as concluded above. Hence, in any case, $\overline{Q}$ has finite Goldie dimension. This implies that $\operatorname{Soc}(\overline{P})$ is infinitely generated. Let Q be the inverse image of $\overline{Q}$ in C. Then $C = P \oplus Q$ for some submodule $P \subseteq C$ with $P_R \cong \overline{P}_R$. Hence both P_R and Q_R have infinitely generated socles. Notice that Q contains S_1 and S_1 is an infinitely generated semisimple module. It follows that $P \neq \operatorname{Soc}(P)$ and $Q \neq \operatorname{Soc}(Q)$ due to the fact that C is cyclic. Hence $C/\operatorname{Soc}(C)$ has Goldie dimension at least 2. Further, by a similar argument, we can show that $P = P_1 \oplus P_2$ where $\operatorname{Soc}(P_1)$ and $\operatorname{Soc}(P_2)$ are both infinitely generated. As P_R is cyclic we have $P_i \neq \operatorname{Soc}(P_i)$ where $i = 1, 2$. This implies that $P/\operatorname{Soc}(P)$ has Goldie dimension at least 2, and hence the Goldie dimension of $C/\operatorname{Soc}(C)$ is at least 3. Continuing in this way we can prove that the Goldie dimension of $C/\operatorname{Soc}(C)$ can be greater than or equal to any given positive integer n. This is a contradiction to the fact that $C/\operatorname{Soc}(C)$ has finite Goldie dimension. Hence $\operatorname{Soc}(C)$ has finite composition length, and so C has finite Goldie dimension. Thus every cyclic right R-module has finite Goldie dimension, that is, R is a right $q.f.d.$ ring. $\qquad\square$

Corollary 14.9 *Let R be a right PIN-ring. If M is a cyclic right R-module such that every simple subfactor of M is M-injective, then M is noetherian.*

Proof By Lemma 14.8, M is $q.f.d.$. Hence M is noetherian by ([65], 16.4). $\qquad\square$

Lemma 14.10 *For any ring R, the module R_R is not a special module.*

Proof Suppose R_R is a special module with the maximal right ideal M. Since M_R is not noetherian, there exists $x\,(\neq 0) \in M$ such that xR is not minimal and $xR \subset M$. As $xR \cong R/ann_r(x)$, $ann_r(x)$ is not maximal in R_R. Hence $ann_r(x) \subset M$. But, being proper submodules of M, both xR and $ann_r(x)$ are noetherian. This implies that R is right noetherian, a contradiction. Thus R is not a special module right R-module. $\qquad\square$

Lemma 14.11 *Let R be a right PIN-ring. Suppose there is a cyclic non-noetherian indecomposable right R-module M such that every proper submodule of M is noetherian. Then*

  1. *If M is injective, then M is a special module.*
  2. *$M/\operatorname{Soc}(M)$ is a special module.*

Proof

  (1) Let N be the sum of all noetherian submodules of M. Since M is cyclic and non-noetherian, $M \neq N$. Also, since every proper submodule of M is noetherian, N is the unique maximal submodule of M, that is, M is a local module. Let U be a nonzero noetherian submodule of M. By assumption, $M/U = \overline{P} \oplus \overline{Q}$ where $\overline{P}_R$ is projective and $\overline{Q}_R$ is either injective or noetherian. Let Q be the inverse image of $\overline{Q}$ in M. Then $M/Q \cong \overline{P}$, and hence $M \cong \overline{P} \oplus Q$. Since M is indecomposable, we must have $\overline{P} = 0$, that is, M/U is either injective or noetherian. (Note that Q contains U and so Q is nonzero.) If M/U is noetherian, then M is noetherian, a contradiction. Hence M/U is injective. Now we can apply the last part of the proof of ([122], Lemma 6) to obtain that M is special if M is injective.

  (2) Let $S = \operatorname{Soc}(M)$. By Lemma 14.8, S_R has finite composition length. Suppose that $S \neq 0$. Then $\overline{M} = M/S$ is injective by the previous argument. Then the image $\overline{N}$ of N in $\overline{M}$ is the unique maximal submodule of $\overline{M}$. Since for every noetherian submodule $\overline{T}$ of $\overline{M}$, $\overline{M}/\overline{T}$ is injective, we can use the argument used at the end of the last paragraph to show that $\overline{M}$ is a special module. This completes the proof of (2) in case $\operatorname{Soc}(M) \neq 0$. Now consider the case when $\operatorname{Soc}(M) = 0$. Let H and K be two nonzero finitely generated submodules of N. Then H and

K are noetherian. If $H \cap K = 0$, then as $\mathrm{Soc}(M) = 0$, we can pick two nonzero elements x, y with $x \in H$, $y \in K$ such that $xR \neq H$, $yR \neq K$, and consider the factor module $M/(xR + yR)$. By the above, $M/(xR + yR)$ is an injective local module. But $M/(xR + yR)$ contains a nontrivial direct sum $H/xR \oplus K/yR$, a contradiction. Hence $I = H \cap K \neq 0$. By the same observation about M/I we obtain $I = H$ or $I = K$, that is, $H \subseteq K$ or $K \subseteq H$. From this it follows easily that M is a uniserial module. Let V be a proper submodule of N. Then there exists an $x \in N$ with $x \notin V$. Hence $xR \supset V$, and therefore, V is noetherian. This completes the proof of (2). $\qquad\square$

Lemma 14.12 *Let R be a right PIN-ring. If R_R is indecomposable and not noetherian, then:*

1. *For any cyclic non-noetherian submodule C of R_R, $C \cong R_R$.*
2. *If there is a nonzero noetherian submodule A of R_R, then R/A is the direct sum of an injective special module and a completely injective noetherian module. Moreover, in this case, there exists a uniform noetherian submodule U of R_R such that R/U is a special module.*

Proof Let C be a cyclic non-noetherian submodule of R_R. If C is injective, then $C = R$ as R_R is indecomposable and then clearly (1) holds. Assume that C is not injective. Then by assumption, $C = P \oplus Q$ where P is a nonzero projective module and Q is either injective or noetherian. Since P is cyclic, there is a right ideal P' of R such that $R/P' \cong P$. Hence P' splits in R_R. But, as R_R is indecomposable we must have $P' = 0$ and so $R_R \cong P_R$. This, together with the fact that R_R has finite Goldie dimension (by Lemma 14.8), gives us $Q = 0$. Thus $C = P \cong R_R$, proving (1).

Now assume that R_R contains a nonzero noetherian submodule A. Then, as R_R is indecomposable, the non-noetherian module R/A does not contain a nonzero projective direct summand. Hence by assumption, R/A is injective. Since by Lemma 14.8, R/A has finite Goldie dimension, we have $R/A = X_1 \oplus \cdots \oplus X_n$, where each X_i is a cyclic injective indecomposable module. From this decomposition, we see at least one direct summand, say X_1, is not noetherian. We aim to show first that this X_1 is special. Let Y be an arbitrary nonzero cyclic proper submodule of X_1. Then Y is not injective, and as X_1 is uniform, Y is projective or noetherian. Suppose that Y is projective. Since R_R is indecomposable, we have $Y_R \cong R_R$, and so R_R is uniform. Let $u \in R$ such that $u + A$ generates Y. Then:

$$R_R \cong Y_R = (uR + A)/A \cong uR/(uR \cap A).$$

It follows that $uR = L \oplus (uR \cap A)$ where L is a submodule of uR with $L \cong R_R$. Since R_R is uniform we must have $uR \cap A = 0$. But this is impossible because $uR \neq 0$, $A \neq 0$. Thus Y_R cannot be projective, and hence Y_R is noetherian. Then by Lemma 14.11(2), X_1 is a special module, as desired.

Let $y \in R$ such that $y + A$ generates X_1. Then yR has a noetherian submodule $H = yR \cap A$. Because $X_1 \cong (yR + A)/A \cong yR/H$, yR/H is a special module. In particular, yR is not noetherian. Hence by (1) of this lemma proved above, $yR \cong R_R$. It follows that R_R contains a noetherian submodule K such that R/K is a special module. As $(A + K)/K$ is a noetherian submodule of R/K, $R/(A + K)(\cong (R/K)/[(A + K)/K])$ is special. On the other hand,

$$R/(A + K) \cong (R/A)/[(A + K)/A],$$

whence R'/K' is special, where $R' := R/A$ and K' is the image of K in R'.

Moreover, from the fact that $X_1 \cap K'$ is noetherian we see that $(X_1 + K')/K'$ is a special submodule of R'/K'. Now $R'/K' = (X_1 + K')/K'$ and hence $R' = X_1 + K'$. It follows that R'/X_1 is noetherian. This means that in the decomposition of R/A, $X := X_2 \oplus \cdots \oplus X_n$ is noetherian. Let Z be any submodule of X, and denote by C the inverse image of Z in R. Then C_R is noetherian. Hence as before, we see that R/C is also injective. This together with

$$R/C \cong (R/A)/(C/A) = (R/A)/Z \cong X_1 \oplus (X/Z)$$

shows that X/Z is injective. Thus X is a completely injective module. Therefore $R/A = X_1 \oplus X$, and so the first statement of (2) is verified. If we choose A to be uniform, then $H = yR \cap A$ is also uniform. Since $yR \cong R_R$, R_R contains a uniform noetherian submodule U such that R/U is special. This completes the proof of (2). $\qquad\qquad\qquad\qquad\qquad\qquad\qquad\qquad\qquad\qquad\qquad\qquad\qquad\square$

Now we are ready to prove the following theorem due to Huynh and Rizvi.

Theorem 14.13 *A ring R is right noetherian if and only if R is a right PIN-ring.*

Proof Let R be a right PIN-ring. Assume to the contrary that R is not right noetherian. By Lemma 14.8, R has finite right Goldie dimension. Hence

$$R_R = R_1 \oplus \cdots \oplus R_k \oplus R_{k+1} \oplus \cdots \oplus R_n, \qquad (14.1)$$

where each R_i is an indecomposable right R-module; $R_1, \ldots R_k$ are not noetherian, and $R_{k+1}, \ldots R_n$ are noetherian. By assumption, $k \geq 1$.

We will first show that $k \leq 1$. Since each R_i is indecomposable, by assumption, we have that for each nonzero submodule $V \subseteq R_1$, R_1/V is noetherian or injective. For $i = 1, 2, \ldots, k$ there exists a nonzero submodule $U_i \subseteq R_i$ such that R_i/U_i is not noetherian, and hence injective. Note that we may choose each U_i to be an essential submodule of R_i because otherwise $R_i/\operatorname{Soc}(R_i)$ will be noetherian (cf. [121]) with finite composition length $\operatorname{Soc}(R_i)$, and so R_i is noetherian, a contradiction. Therefore each R_i/U_i is singular. For $j \neq 1$, we consider the factor module R_j/W where W is an arbitrary essential submodule of R_j. By (14.1),

$$R/(U_1 \oplus W) \cong (R_1/U_1) \oplus R_2 \oplus \cdots \oplus R_{j-1} \oplus (R_j/W) \oplus R_{j+1} \oplus \cdots \oplus R_n.$$

From this, $(R_1/U_1) \oplus (R_j/W)$ is cyclic and singular. Hence by assumption, $(R_1/U_1) \oplus (R_j/W)$ is either injective or noetherian. But R_1/U_1 is not noetherian, hence $(R_1/U_1) \oplus (R_j/W)$ must be injective. In particular, $(R_j/W)_R$ is injective. Thus we have shown that

for $j \neq 1$, $(R_j/W)_R$ is injective for each essential submodule W of R_j. $\qquad (*)$

If $k \geq 2$, we take W to be an arbitrary essential submodule of R_1 and consider the isomorphisms

$$R/(W \oplus U_2) \cong (R_1/W) \oplus (R_2/U_2) \oplus R_3 \oplus \cdots \oplus R_n,$$

where R_2/U_2 is as above, that is, it is singular, injective and non-noetherian. By the same reasons as before, R_1/W is injective. Hence we have proved that, in this case, for each essential submodule W of R_l $(l = 1, 2, \ldots, n)$, R_l/W is injective. Now let M be an essential, maximal right ideal of R. Then there exists an R_t in (14.1) such that $M \not\supseteq R_t$, and so $R_t + M = R$. Hence:

$$R/M = (R_t + M)/M \cong R_t/(R_t \cap M).$$

Therefore R/M is injective. This shows that every simple singular right R-module is injective. By Corollary 14.9, for every essential submodule $E \subseteq R_R$, R/E is noetherian. By ([65], 16.13(2)), $R/\mathrm{Soc}(R_R)$ is right noetherian. But $\mathrm{Soc}(R_R)$ has finite length (cf. Lemma 14.8), and this implies that R is right noetherian. This is a contradiction to our assumption. Thus $k \leq 1$.

So the only possible non-noetherian indecomposable summand of R_R in (1) is R_1. Let M_i be a maximal submodule of R_i. If the simple module R_i/M_i is nonsingular, then R_i/M_i is projective. Hence M_i splits in R_i, a contradiction. Therefore each maximal submodule of R_i must be essential in R_i. By $(*)$, for any maximal submodule $M_j \subseteq R_j$ with $j \geq 2$, R_j/M_j is injective. Let $A = \Sigma_{\alpha \in \Omega} x_\alpha R$ where each $x_\alpha R$ is a right ideal of R which is isomorphic to a homomorphic image of some R_j in (1) with $2 \leq j \leq n$. Then A is a two-sided ideal of R. Hence, as noticed for R_j before, for each maximal submodule $M_\alpha \subseteq x_\alpha R$, $x_\alpha R/M_\alpha$ is injective. Assume there are countably infinitely many $x_\alpha R$, say $x_1 R, x_2 R, \ldots$, which form a strictly ascending chain

$$x_1 R \subset x_1 R + x_2 R \subset \cdots \subset x_1 R + x_2 R + \cdots + x_m R \subset \cdots$$

Set $B_i = x_1 R + \cdots + x_i R$ and $B = \cup_{i=1}^{\infty} B_i$. Let M_1 be a maximal submodule of B_1. Then B_1/M_1 is injective. Hence $B_2/M_1 \cong (B_1/M_1) \oplus (X_2/M_1)$ where $X_2 = M_1 + x_2 R$. Let M_2 be a maximal submodule of X_2 with $M_1 \subset M_2$. Then X_2/M_2 is injective. Hence $B_3/M_2 \cong (B_1/M_1) \oplus (X_2/M_2) \oplus (X_3/M_2)$ where $X_3 = M_2 + x_3 R$. By induction we can prove that for any $i \geq 2$, there exists a maximal submodule M_i of X_i with $M_{i-1} \subset M_i \subset X_i = M_{i-1} + x_i R$ such that X_i/M_i is injective, and hence

$$B_i/M_i \cong (B_1/M_1) \oplus \cdots \oplus (X_i/M_i) \oplus (X_{i+1}/M_i),$$

where $X_{i+1} = M_i + x_{i+1}R$. Let $U = \cup_{i=1}^{\infty} M_i$. Then we have:

$$B/U \cong (B_1/M_1) \oplus [\oplus_{i=1}^{\infty}(X_i/M_i)].$$

This is a contradiction to Lemma 14.8 because B/U is a submodule of R/U that has finite uniform dimension. Therefore there is a finite subset $F \subseteq \Omega$ such that $A = \Sigma_{\alpha \in F} x_\alpha R$. Hence A_R is noetherian.

Now we consider the following two cases.

Case 1. $A' := R_1 \cap A \neq 0$.

Since A' is noetherian, and R_1 is indecomposable, it follows from assumption that R_1/A' is injective. Hence R/A is a right self-injective non-noetherian ring. Moreover, R/A satisfies (A). Hence, instead of R/A, we may assume that our ring R is right self-injective and not right noetherian. Then we can use the arguments as before to show that R has a decomposition (1) with $k = 1$, but this time all R_i's are uniform because of the injectivity. Hence $R_2, R_3, \ldots, R_n$ are all noetherian and completely injective.

For each $0 \neq x \in R_1$, consider the R-homomorphism $\varphi_j : R_j \to R_1$ defined by $\varphi_j(r_j) = xr_j$ $(r_j \in R_j,\ j = 2, 3, \ldots, n)$. Then $R_j/\operatorname{Ker}\varphi_j$ is isomorphic to a proper submodule of R_1. Since $R_j/\operatorname{Ker}\varphi_j$ is injective, we must have $R_j = \operatorname{Ker}\varphi_j$, that is, $xR_j = 0$. This shows $(R_1)(R_2 \oplus \cdots \oplus R_n) = 0$, or equivalently, $R_2 \oplus \cdots \oplus R_n$ is a two-sided ideal of R. Hence we may restrict ourselves on the factor ring $R/(R_2 \oplus \cdots \oplus R_n)$ which is R-isomorphic to R_1. This means, again, we may assume that R is a right self-injective right PIN-ring, R_R is indecomposable, and not noetherian. From this, it follows that every cyclic proper submodule of R_R is noetherian. Hence by Lemma 14.11(b), R is itself a special right R-module which is impossible by Lemma 14.10.

Case 2. $R_1 \cap A = 0$.

Since $R/A \cong R_1$, we may assume that R is a right PIN-ring, R_R is indecomposable, but not noetherian. However, in this case, R may not be right self-injective. By Lemma 14.12, the Goldie dimension d of R_R is at most 2. If $S = \operatorname{Soc}(R_R) \neq 0$, then S_R is noetherian by Lemma 14.8. Hence R/S is a right self-injective right PIN-ring. Using the arguments as in Case 1, we shall arrive at a contradiction. Hence $\operatorname{Soc}(R_R) = 0$.

Suppose $d = 2$. By Lemma 14.12, R_R contains a uniform noetherian submodule U such that R/U is special. Let V denote the complement of U in R. Then by assumption, $V \neq 0$ and since V_R embeds in R/U, V_R contains a nonzero noetherian submodule V_1. By Lemma 14.12 we have $R/V_1 = \overline{S} \oplus \overline{T}$, where $\overline{S}$ is injective and special, $\overline{T}$ is noetherian and completely injective. As $\operatorname{Soc}(R_R) = 0$ we may assume that $V_1 \neq V$, and therefore the uniform dimension of R/V_1 is at least 2. This means that in the above decomposition, $\overline{T} \neq 0$. Let $\overline{U}$ be the image of U in R/V_1, and $\overline{V}_1$ be the image of V_1 in R/U. Since $\overline{V}_1$ is noetherian and

$$(R/V_1)/\overline{U} \cong R/(U + V_1) \cong (R/U)/\overline{V}_1,$$

$(R/V_1)/\overline{U}$ must be special. Since $(\overline{T} + \overline{U})/\overline{U} \cong \overline{T}/(\overline{T} \cap \overline{U})$, and $\overline{T}$ is completely injective, it follows that $(\overline{T} + \overline{U})/\overline{U}$ is an injective noetherian submodule of the special module $(R/V_1)/\overline{U}$. Hence $(\overline{T} + \overline{U})/\overline{U} = 0$, or equivalently $\overline{T} \subseteq \overline{U}$. Since $\overline{T}$ is injective and $\overline{U}$ is uniform, we must have $\overline{T} = \overline{U}$. But $\overline{U} \cong U$, and so U_R is injective. This is a contradiction because R_R is indecomposable. Thus $d = 1$, that is, R_R is uniform.

Now there are again two possibilities to consider: either

(i) there is a nonzero noetherian submodule in R_R; or
(ii) there is no nonzero noetherian submodule in R_R.

First, assume that there is $0 \neq a \in R$ such that aR is noetherian. We know that noetherian modules have Krull dimension [100]. Since R_R is uniform, every cyclic subfactor of R/aR does not contain a nonzero projective direct summand. Hence, by assumption, they are either noetherian or injective. Hence by [63] R/aR has Krull dimension. It follows that R has right Krull dimension. If the prime radical N of R is zero, then R is semiprime, and since R has right Krull dimension, R is right Goldie (see [100]). It follows that R is a right Ore domain, in particular, R does not contain nonzero zerodivisors. Since $aR \cong R/ann_r(a)$ where $ann_r(a) = 0$, R is right noetherian, a contradiction. Thus we must have $N \neq 0$. In this case the Jacobson radical J of R is nonzero. Recall that J is the intersection of all maximal right ideals of R. Since R_R is uniform, J_R contains a nonzero noetherian submodule $H = J \cap aR$. By Lemma 14.12, $R/H = S' \oplus T'$ where S'_R is special and T'_R is noetherian. Since $H \subseteq J$, we easily see that J/H is also the intersection of all maximal submodules of R/H, that is, J/H is the Jacobson radical of the right R-module R/H. Therefore $J/H = ((J/H) \cap S') \oplus ((J/H) \cap T')$. Hence every finitely generated submodule of J_R is noetherian and $(J/H) \cap S'$ is exactly the maximal submodule of the special module S'. Let X be a cyclic submodule of R_R such that $(X + H)/H = S'$. Then $X_1 := X \cap J$ is not noetherian and X/X_1 is simple. Let Y be an arbitrary nonzero cyclic submodule of X_1. Then Y_R is noetherian. By Lemma 14.10, $X \cong R_R$ and, if we denote the factor module X/Y by $\overline{X}$, then $\overline{X} = \overline{H} \oplus \overline{K}$ where $\overline{H}_R$ is injective and special, and $\overline{K}_R$ is injective and noetherian. Let K be the inverse image of $\overline{K}$ in X. Then K_R is noetherian. If $K \not\subseteq X_1$, then $K + X_1 = X$, implying the fact that $X_1/(X_1 \cap K)$ is cyclic. Since $X_1 \cap K$ is noetherian, in particular finitely generated, it follows that X_1 is finitely generated, too. This together with $X_1 \subseteq J$ shows that X_1 is noetherian, a contradiction. Hence $K \subset X_1$. If $K \not\subseteq Y$ then, since $K/Y (= \overline{K})$ is injective, $R/Y = (K/Y) \oplus (L/Y)$ for some right ideal L of R containing Y. From this we easily see the existence of a maximal right ideal L' of R with $L' \not\supseteq K$, a contradiction to $K \subset X_1 \subseteq J$. Hence $K \subseteq Y$, or equivalently $\overline{K}$ is zero, i.e., X/Y is a special module. Thus we have shown that for each nonzero cyclic submodule Y of X_1, X/Y is special. From this and since X_R is uniform it is easy to see that every cyclic proper submodule of X is noetherian. By Lemma 14.11, X_R or $X/\operatorname{Soc}(X_R)$ is special. But $\operatorname{Soc}(X_R) \subseteq \operatorname{Soc}(R_R) = 0$.

Hence R_R being isomorphic to X_R, is a special module, a contradiction to Lemma 14.10. This last contradiction shows that (i) cannot happen.

Now we consider the second possibility, that is, for each $0 \neq x \in R$, xR is not noetherian. In this case, by Lemma 14.12, $xR \cong R_R$ for any $0 \neq x \in R$.

Let M be an arbitrary maximal submodule of R_R. We will show that R/M is an injective right R-module. There is a nonzero cyclic submodule $yR \subset M$ such that R/yR is not noetherian. By assumption, R/yR is injective because R/yR cannot contain a nonzero projective direct summand. Write $R/yR = I_1 \oplus \cdots \oplus I_t$, where each I_i is indecomposable and injective. We may assume, without loss of generality, that I_1 is not noetherian. Since I_1 is indecomposable and singular, by assumption, each cyclic proper submodule of I_1 is noetherian. By Lemma 14.11, I_1 is special. Now, since $yR \cong R_R$ there exists a nonzero submodule $V \subseteq yR$ such that yR/V is not noetherian. By the same arguments as for R/yR we obtain that yR/V is injective and yR/V contains an injective special submodule, say J_1. Write

$$R/V = yR/V \oplus T$$

for some submodule $T \subseteq R/V$. This yields $R/yR \cong T$. Hence T_R is injective and

$$R/V = J_1 \oplus J_2 \oplus \cdots \oplus J_h$$

where each J_i is injective, indecomposable, and J_1, J_2 are special modules. Let M_i be a maximal submodule of J_i for each $i = 1, 2, \ldots, h$. Then for $i \neq 1$ we have

$$(R/V)/M_i \cong J_1 \oplus \cdots \oplus J_{i-1} \oplus (J_i/M_i) \oplus J_{i+1} \oplus \cdots \oplus J_h$$

The singular module $(R/V)/M_i$ is not noetherian, and it does not contain a nonzero projective direct summand. By assumption, $(R/V)/M_i$ must be injective. Therefore J_i/M_i is injective for any $i \neq 1$. Furthermore, as J_2 is a special module, the same observation for

$$(R/V)/M_1 \cong (J_1/M_1) \oplus J_2 \oplus \cdots \oplus J_h$$

shows that J_1/M_1 is also injective. From these facts it is clear that for each maximal submodule $\overline{H} \subset R/V$ the factor module $(R/V)/\overline{H}$ is injective. Since M was a maximal submodule of R_R containing V, it follows that R/M is injective as desired. In fact, we have shown that R is a right V-ring, and hence R is right noetherian by Corollary 14.9. This is a contradiction to our assumption that R is not right noetherian. Therefore R must be right noetherian. The converse is obvious. $\qquad\qquad\square$

This clearly extends the above two results of Chatters [45] and Osofsky and Smith [193].

The theorem of Huynh and Rizvi also implies that a ring R is right noetherian if and only if every cyclic right R-module is either injective or noetherian.

Holston, Jain, and Leroy [110] have studied rings where each cyclic module is a direct sum of a projective module and a module that is either CS or noetherian.

We will say that a right R-module M satisfies property $(\mathcal{Q})$ if we can write $M = A \oplus B$, where A is either a CS-module or a noetherian module, and B is a projective module.

It was shown in [194] that a ring R is right noetherian if and only if every 2-generated right R-module satisfies property $(\mathcal{Q})$. We remark that it is not sufficient to assume that every cyclic satisfies property $(\mathcal{Q})$ in order for R to be right noetherian, as may be seen from the example of Levy. The ring R in Levy's example (see the example on p. 24) is not noetherian, but every homomorphic image is self-injective, and so R satisfies property $(\mathcal{Q})$. We now present some results on property $(\mathcal{Q})$, beginning with a useful proposition.

Proposition 14.14 *Let C be a cyclic right R-module such that each cyclic subfactor of C satisfies property $(\mathcal{Q})$, and let $S = \mathrm{Soc}(C)$. Then C/S has finite Goldie dimension. Furthermore, if R is a right WV-ring, then C/S is noetherian.*

Proof Let $E \subseteq_e C$ and X/D be a cyclic subfactor of C/E, where $E \subseteq D \subseteq X \subseteq C$. Then, by $(\mathcal{Q})$, $X/D = B/D \oplus A/D$ where B/D is projective and A/D is CS or noetherian. Since D splits from B, essentiality shows that $B = D$. Theorem 4.11 then applies to give that C/E has finite Goldie dimension. Since E was arbitrary such that $E \subseteq_e C$, this implies in particular that C/E has $q.f.d.$ Then $C/\mathrm{Soc}(C)$ has finite Goldie dimension by Lemma 3.6.

Now assume that R is a right WV-ring and let $Z \subset Y \subseteq C/E$, where, as above, $E \subseteq_e C$. If $0 \neq x \in J(Y/Z)$ let K be maximal in xR. Since C/E is singular it is proper cyclic and the simple module xR/K is C/E-injective, so it splits in $(Y/Z)/K$, a contradiction since $xR/K \subseteq J((Y/Z)/K)$. We conclude that $J(Y/Z) = 0$. Then by [203, Theorem 3.8], C/E is noetherian. Hence, by [65, Theorem 5.15 (1)], $C/\mathrm{Soc}(C)$ is noetherian. $\qquad\square$

Under a stronger assumption on a cyclic module C than property $(\mathcal{Q})$, namely, if every cyclic subfactor of C is projective, CS, or noetherian, we next show that C is noetherian when R is a right WV-ring.

Theorem 14.15 *Let C be a cyclic right R-module such that each cyclic subfactor of C is either CS, noetherian, or projective.*

1. *Then C has finite Goldie dimension.*
2. *If R is a right WV-ring, then C is noetherian.*

Proof

(1) Let $S = \mathrm{Soc}(C)$. By Proposition 14.14, C/S has finite Goldie dimension. We show S is finitely generated. Suppose S is infinitely generated. Write $S = S_1 \oplus S_2$, where S_1 and S_2 are both infinitely generated. Now by hypothesis, C/S_1 is either CS or noetherian or projective. If projective, then $S_1 \subset_\oplus C$ and hence S_1 is cyclic, a contradiction as S_1 is infinitely

generated. If noetherian, then so is $S/S_1 \cong S_2$, a contradiction as S_2 is infinitely generated. So C/S_1 is CS. Furthermore, $(S_1 \oplus S_2)/S_1 \cong S_2$ is infinitely generated. Since $C/S \cong \dfrac{C/S_1}{(S_1 \oplus S_2)/S_1}$, we get a contradiction by invoking Lemma 3.7. Hence S is finitely generated and so C has finite Goldie dimension.

(2) Since C/S is noetherian, and S has finite Goldie dimension, it follows that C is noetherian. $\qquad\square$

The next lemma is straightforward.

Lemma 14.16 *Let C be a right R-module and $S = \mathrm{Soc}(C)$. If C/S is a uniform R-module, then for any two submodules A and B of C with $A \cap B = 0$, either A or B is semisimple.*

Proof Let K be a complement submodule of A in C containing B. Then $A \oplus K \subseteq_e C$. This yields $\mathrm{Soc}(A \oplus K) = S$. Thus $(A \oplus K)/(\mathrm{Soc}(A \oplus K)) \subseteq C/S$. Since $(A \oplus K)/(\mathrm{Soc}(A \oplus K)) \cong A/\mathrm{Soc}(A) \times K/\mathrm{Soc}(K)$ and C/S is uniform as an R-module, either $A/\mathrm{Soc}(A)$ or $K/\mathrm{Soc}(K)$ is zero. So $A = \mathrm{Soc}(A)$ or $K = \mathrm{Soc}(K)$. In other words, either A or K (and hence B) is semisimple. $\qquad\square$

Lemma 14.17 *If C is a right R-module, and if $C/I = A/I \oplus B/I$ is a direct sum with B/I a projective module, then $C = A \oplus B'$, where $B = B' \oplus I$.*

Proof From the decomposition $C/I = A/I \oplus B/I$, we have $C = A + B$, where $A \cap B = I$. Since B/I is projective, $B = B' \oplus I$ for some B'. Then $C = A + (B' \oplus I) = A + B'$. We claim that $A \cap B' = 0$.

Let $x \in A \cap B' \subseteq A \cap B = I$. Then $x \in B' \cap I = 0$. Thus $C = A \oplus B'$. $\qquad\square$

Lemma 14.18 *Let R be a right WV-ring. Let C be a cyclic right R-module with a projective socle S (equivalently, $S = \mathrm{Soc}(C)$ is embeddable in R). If C/S is a uniform right R-module and each cyclic subfactor of C satisfies the property $(\mathcal{Q})$, then C is noetherian.*

Proof First assume R is a right V-ring. Let C'/I be a cyclic subfactor of C and write $C'/I = A/I \oplus B/I$ as a direct sum of a CS or noetherian module and a projective module, respectively. Then, by Lemma 14.17, $C' = A \oplus B'$, where $B = B' \oplus I$. Since $C'/\mathrm{Soc}(C') \cong (C' + S)/S \subseteq C/S$ is uniform, either A or B' is semisimple (Lemma 14.16). Note both A and B' are cyclic.

Case 1: A is semisimple. Since A/I is semisimple cyclic and R is a right V-ring, A/I is injective. Moreover A/I embeds in $A \subseteq \mathrm{Soc}(C') \subseteq \mathrm{Soc}(C) = S$. The hypothesis that S embeds in R yields that A/I and hence C'/I are projective.

Case 2: B' is semisimple. Since $B/I \cong B'$ is semisimple and cyclic, it is a finite direct sum of simple injective modules. If A/I is noetherian, then clearly C/I will be also. Now if A/I is CS then, as A/I and B/I are relatively injective, $A/I \oplus B/I$ is also CS (see [65, Proposition 7.10]). Thus any cyclic subfactor of the cyclic module C is either CS or noetherian or projective. Therefore C is noetherian by Theorem 14.15.

Now, if R is not a right V-ring, then by Theorem 6.6, R is right uniform. Since S embeds in R, S is trivially noetherian. By Proposition 14.14, C/S is noetherian. Therefore C is noetherian. $\square$

Proposition 14.19 *Let R be a right V-ring. Let C be a cyclic right R-module with essential and projective socle. Suppose each cyclic subfactor of C satisfies the property $(\mathcal{Q})$. Then C is semisimple.*

Proof Let $S = \mathrm{Soc}(C)$. We know by Proposition 14.14 that C/S has finite Goldie dimension. Suppose $C/S \neq 0$. Then C/S contains a nonzero cyclic uniform submodule. Thus we can find $u \in C$ with $U = (uR + S)/S \cong uR/\mathrm{Soc}(uR)$ uniform. Since $\mathrm{Soc}(uR) \subseteq_\oplus S$, we know that $\mathrm{Soc}(uR)$ is projective. Moreover, every cyclic subfactor of uR also satisfies the property $(\mathcal{Q})$, hence Lemma 14.18 implies that uR is noetherian. Then $\mathrm{Soc}(uR)$ is a finite direct sum of simple modules, and hence it is injective. Since $\mathrm{Soc}(uR) \subset_e uR$, $uR = \mathrm{Soc}(uR)$. This yields $U = 0$, a contradiction. Thus $C/S = 0$, that is, $C = S$, completing the proof. $\square$

Theorem 14.20 *Let R be a von Neumann regular right WV-ring such that each cyclic right R-module satisfies $(\mathcal{Q})$. Then R is semisimple artinian.*

Proof By Proposition 6.9 (4), a von Neumann regular right WV-ring is a right V-ring. Let $S = \mathrm{Soc}(R)$. Then R/S has finite right Goldie dimension and hence it is semisimple artinian. Let T be a complement of S. Then T embeds essentially in R/S. Thus $T = 0$. Hence $S \subset_e R$. So by Proposition 14.19, R is semisimple artinian. $\square$

Finally, we prove the following general result.

Theorem 14.21 *For a ring R, we have the following:*

1. *Let R be a right V-ring. Let M be a finitely generated R-module with projective socle. Suppose each cyclic subfactor of M satisfies the property $(\mathcal{Q})$. Then M is noetherian, and $M = X \oplus T$, where X is semisimple and T is noetherian with zero socle.*
   In particular, if R is a right V-ring such that each cyclic module satisfies the property $(\mathcal{Q})$, then $R = S \oplus T$, where S is semisimple artinian and T is a finite direct sum of simple noetherian rings with zero socle.

2. *If R is a right WV-ring, then R is right noetherian if and only if each cyclic right R-module satisfies the property $(\mathcal{Q})$.*

Proof

(1) First, assume M is cyclic. Let $S_0 = \mathrm{Soc}(M)$ and let T_0 be a complement of S_0 in M. Consider the cyclic module $X_0 = M/T_0$. Then S_0 is essentially embeddable in X_0. Since $\mathrm{Soc}(X_0) \cong S_0$, X_0 is semisimple by Proposition 14.19. So X_0, and hence S_0, is a finite direct sum of simples. In particular S_0 is injective and we have $M = S_0 \oplus T_0$. Since M/S_0 is noetherian (Proposition 14.14), T_0 is noetherian and obviously it has zero socle. In

general, $M = \sum_{i=1}^{n} x_i R$. By above, each $x_i R$ is noetherian, and hence M is noetherian. $X = \mathrm{Soc}(M)$ is finitely generated and injective by hypothesis. Therefore $M = X \oplus T$, where X is semisimple and T is noetherian with zero socle.

Finally, let $S = \mathrm{Soc}(R_R)$ which is clearly projective in a right V-ring and let T be its complement. Then, as shown above, R is a right noetherian right V-ring. Therefore R is a direct sum of simple noetherian rings ([78], p. 70). So $R = S \oplus T$, where S is semisimple artinian and T is a finite direct sum of simple noetherian rings with zero socle.

(2) Note that if R is a right WV-ring and not a right V-ring, then R is right uniform. In this case, $\mathrm{Soc}(R_R)$ is either zero or a minimal right ideal. Since $R/\mathrm{Soc}(R_R)$ is noetherian (Proposition 14.14), we conclude that R is right noetherian. The converse is obvious. $\qquad\square$

15 Rings each of whose modules is a direct sum of cyclic modules

Köthe was first to study rings each of whose modules is a direct sum of cyclic modules. He proved the following:

Theorem 15.1 (*Köthe, [158]*). *Let R be an artinian principal ideal ring. Then each R-module is a direct sum of cyclic modules.*

Later Cohen and Kaplansky obtained the following:

Theorem 15.2 (*Cohen and Kaplansky, [52]*). *If R is a commutative ring such that each R-module is a direct sum of cyclic modules, then R must be an artinian principal ideal ring.*

Asano [15] proved that a commutative artinian ring R is a principal ideal ring if and only if R is isomorphic to a finite direct product of uniserial rings. Thus, it follows that for a commutative ring R, each R-module is a direct sum of cyclic modules if and only if R is an artinian principal ideal ring, if and only if R is isomorphic to a finite direct product of artinian uniserial rings.

However, finding the structure of noncommutative rings each of whose modules is a direct sum of cyclic modules is still an open question. Chase [43] proved the following.

Theorem 15.3 *If R is a ring such that each left R-module is a direct sum of finitely generated modules, then R is left artinian.*

Thus, in particular, if R is a ring such that each left R-module is a direct sum of cyclic modules, then R is left artinian. The result of Köthe was generalized by Nakayama [182] who proved that if R is an artinian serial ring then each R-module is a direct sum of cyclic modules. Further, Nakayama [181] gave an example of a noncommutative right artinian ring R where each right module is a direct sum of cyclic modules but R is not a principal right ideal ring.

Example Let R be the ring of 2×2 lower triangular matrices over a field K. The ring R is an artinian serial ring and therefore each right R-module is a direct sum of cyclic modules. The injective hull of R_R is $M_2(K)$. Thus R is not right self-injective and hence R is not quasi-Frobenius. Since every artinian serial principal right ideal ring is quasi-Frobenius (see [75, Proposition 25.4.6B]), it follows that R is not a principal right ideal ring. $\square$

Recently Behboodi et al. [22] have given noncommutative analogue of the Cohen–Kaplansky Theorem. To present their proof, we start with the following basic lemma.

Lemma 15.4 *Let R be a local ring with maximal ideal M such that M is nilpotent. Then:*

>   *(i) R is a principal left ideal ring if and only if M is a principal left ideal.*
>
>   *(ii) R is a left artinian principal left ideal ring if and only if the ring R/M^2 is a principal left ideal ring.*

Proof

(i) Let $M = Rm = RmR$ and let X be a nonzero left ideal in R. Assume that X is not principal. Since R is local and X is not principal, X is contained in M. There is a positive integer n such that X is contained in M^n and X is not contained in M^{n+1}. Since $M = Rm = RmR$, we have that $M^n = Rm^n$ is a principal left ideal. Therefore $X = Bm^n$ for some left ideal B, where B is not equal to R, since X is not principal. Thus B is contained in M and $X = Bm^n$ is contained in M^{n+1}. This is a contradiction. Therefore X is principal and hence R is a principal left ideal ring. The converse is obvious.

(ii) Assume that R/M^2 is a principal left ideal ring. Suppose $M/M^2 = R(a + M^2)$ where $a \in M$. Then $M = Ra + M^2$ and so $M(M/Ra) = M/Ra$. Therefore $M = Ra$. Hence by (i), R is a principal left ideal ring. The converse is obvious. $\qquad\square$

Theorem 15.5 *Let $R = \prod_{i=1}^{n} R_i$ be a finite product of rings R_i such that each R_i has no nontrivial idempotent elements. If each left R-module is a direct sum of cyclic modules, then R is an artinian principal right ideal ring.*

Proof By Theorem 15.3, R is left artinian. It suffices to assume that $n = 1$, that is, R has no nontrivial idempotents. Since R is left artinian with no nontrivial idempotents, R is a local ring. Let M be the maximal ideal of R. Then $M^k = 0$ for some $k \geq 1$. By Lemma 15.4(ii), we may assume that $M^2 = 0$.

We first consider the case that R is a principal left ideal ring. Then R is a left uniserial ring. Consequently, $_R R$ is uniform. Thus $E(_R R)$ is indecomposable. Since $E(_R R)$ is a direct sum of cyclic R-modules, $E(_R R)$ is cyclic and so by Lemma 11.12, R is left self-injective. Thus R is a quasi-Frobenius ring and hence R is an artinian principal ideal ring by [71, Section 4, Theorem 1].

Next, we consider the case that R is not a principal left ideal ring. Thus by Lemma 15.4(i), M is not a principal left ideal. Since R is left artinian and $M^2 = 0$, M on each side is a module over the division ring R/M. Therefore $M = \mathrm{Soc}(R_R) = \mathrm{Soc}(_R R)$. Thus $_R M = Ry_1 \oplus \cdots \oplus Ry_t$ such that $t \geq 2$ and each Ry_i is a simple left R-module. It follows that $l(_R R) = t + 1$. We claim that $M_R = xR$. Because, if not, then we can assume that $M_R = \oplus_{i \in \mathcal{I}} x_i R$ where $|\mathcal{I}| \geq 2$ and each $x_i R$ is a simple right R-module. We set

$_R N = (R \oplus R)/R(x_1, x_2)$. Since $ann_l^R(R(x_1, x_2)) = M$, $R(x_1, x_2)$ is simple and hence $l(_R N) = 2 \times l(_R R) - l(_R R(x_1, x_2)) = 2(t+1) - 1$. We claim that every nonzero cyclic submodule Ra of N has length 1 or $t+1$. If $Ma = 0$, then $l(Ra) = 1$ since $Ra \cong R/M$. Suppose that $Ma \neq 0$, then there exist $c_1, c_2 \in R$ such that $a = (c_1, c_2) + R(x_1, x_2)$. If $c_1, c_2 \in M$, then $Ma = 0$, since $M^2 = 0$. Thus without loss of generality, we can assume that $a = (1, c_2) + R(x_1, x_2)$ (since if $c_1 \notin M$, then c_1 is a unit). Now let $r \in ann_l^R(a)$. Then $r(1, c_2) = t(x_1, x_2)$ for some $t \in R$. It follows that $r = tx_1$ and $rc_2 = tx_2$. Thus $tx_2 = tx_1 c_2$. If $t \notin M$, then t is a unit and so $x_2 = x_1 c_2$ which is a contradiction as $x_1 R \cap x_2 R = 0$. Thus $t \in M$ and so $r = tx_1 = 0$. Therefore, $ann_l^R(a) = 0$ and so $Ra \cong R$. It follows that $l(Ra) = t + 1$. Now since each left R-module is a direct sum of cyclic modules, we have

$$N = Rw_1 \oplus \cdots \oplus Rw_k \oplus Rv_1 \oplus \cdots \oplus Rv_l,$$

where $l, k \geq 0$, and each Rw_i is of length $t+1$ and each Rv_j is of length 1. Clearly $M \oplus M$ is not a simple left R-module. Since $R(x_1, x_2)$ is simple, $MN = (M \oplus M)/R(x_1, x_2) \neq 0$. It follows that $k \geq 1$. Also, $l(_R N) = 2(t+1) - 1 = k(t+1) + l$ and this implies that $k = 1$ and $l = t$. Since $Mv_i = 0$ for each i, $MN = Mw_1$ and hence $N/MN \cong Rw_1/Mw_1 \oplus Rv_1 \oplus \cdots \oplus Rv_t$. It follows that $l(_R N/MN) = 1 + t$. On the other hand, we have $N/MN \cong R/M \oplus R/M$ and so $l(_R N/MN) = 2$ and so $t = 1$, a contradiction.

Thus M is simple as a right R-module. Thus R is right artinian and so R is a principal right ideal ring. Since R is also left artinian, R is an artinian principal right ideal ring. $\square$

As a consequence, we have the following

Corollary 15.6 *Let* $R = \prod_{i=1}^n R_i$ *be a finite product of rings* R_i *such that each* R_i *has no nontrivial idempotent elements. Then the following statements are equivalent;*

1. *Each left, and each right, R-module is a direct sum of cyclic modules.*
2. *R is an artinian principal ideal ring.*
3. *R is isomorphic to a finite product of artinian uniserial rings.*

If $_R R$ is uniform, then R has no nontrivial idempotent elements. Hence, we have the following:

Corollary 15.7 *Let* $R = \prod_{i=1}^n R_i$ *be a finite product of rings* R_i *where each* $_{R_i} R_i$ *is uniform. Then the following statements are equivalent:*

1. *Each left R-module is a direct sum of cyclic modules.*
2. *Each left, and each right, R-module is a direct sum of cyclic modules.*
3. *R is an artinian principal ideal ring.*
4. *R is isomorphic to a finite product of artinian uniserial rings.*

Proof We only need to show that $(1) \implies (3)$. Without loss of generality, we may assume that $_RR$ is uniform. Then $E(_RR)$ is indecomposable. Since $E(_RR)$ is a direct sum of cyclic modules, $E(_RR)$ is cyclic. As the composition length of $E(_RR)$ is not less than that of $_RR$, we get $E(_RR)=_RR$. Hence R is left self-injective. Since R is already right uniserial, we get that R is also left uniserial. Hence R is an artinian principal ideal ring. $\qquad\square$

15.8 Question

1. Describe noncommutative rings where each right module is a direct sum of cyclic modules.

16 Rings each of whose modules is an I_0-module

For a module M, we say that a submodule X of M *lies over a direct summand* of M if there exists a decomposition $M = M_1 \oplus M_2$ such that $M_1 \subseteq X$ and $M_2 \cap X$ is small in M_2. In this case, $M_2 \cap X$ is small in M and $M_2 \cap X \subseteq J(M)$.

A module M is said to be *semiregular* if every finitely generated submodule of M lies over a direct summand of M. A module M is called a *lifting* module if every submodule of M lies over a direct summand of M. Clearly, every lifting module is semiregular. However, there are examples of semiregular modules which are not lifting. For example, the direct product R of an infinite number of fields is an example of a semiregular R-module which is not a lifting module. This is because the semiprimitive lifting modules are precisely the semisimple modules, which follows from the fact that $J(M)$ is the sum of all small submodules of M for any module M and a module M is called a *semiprimitive module* if $J(M) = 0$.

We will denote the lattice of submodules of a module M by $\mathrm{Lat}(M)$. Most of the results of this chapter are from Tuganbaev [226] and Abyzov and Tuganbaev [3].

Theorem 16.1 *For a ring R, the following conditions are equivalent:*

1. *Every right R-module is semiregular.*
2. *Every right R-module is a lifting module.*
3. *Every right R-module is a direct sum of a semisimple module and an injective module Q with $\mathrm{Soc}(Q) \subseteq J(Q)$.*
4. *Every right R-module is a direct sum of a semisimple module and an injective module which is a direct sum of uniserial modules of length 2.*
5. *R is an artinian serial ring with $(J(R))^2 = 0$.*

Proof The implication $(5) \Rightarrow (4)$ follows from Theorem 14.6. The implications $(4) \Rightarrow (3)$ and $(2) \Rightarrow (1)$ are easy.

$(3) \Rightarrow (2)$. Let M be a right R-module and let $X \in \mathrm{Lat}(M)$. By (3), X/Y is a semisimple module for some injective $Y \in \mathrm{Lat}(X)$. There exist decompositions $M = Y \oplus M_1$ and $X = Y \oplus X_1$, where X_1 is a semisimple submodule of M_1. By (3), $M_1 = P \oplus Q$, where P is a semisimple module and Q is an injective module with $\mathrm{Soc}(Q) \subseteq J(Q)$. Since the module X_1 is semisimple, there exists a decomposition $X_1 = X_2 \oplus (X_1 \cap Q)$, where $X_2 \cap Q = 0$. Since the submodule $X_2 \oplus Q$ of M_1 contains the direct summand Q of M_1, there exists a decomposition $X_2 \oplus Q = P_1 \oplus Q$, where P_1 is a submodule of the semisimple module P.

Since P is semisimple, there exists a decomposition $P = P_1 \oplus P_2$. Then

$$M = Y \oplus M_1 = Y \oplus P_1 \oplus P_2 \oplus Q = Y \oplus P_2 \oplus X_2 \oplus Q,$$

$$X = Y \oplus X_1 = Y \oplus X_2 \oplus (X_1 \cap Q),$$

where the semisimple module $X_1 \cap Q$ is contained in the small submodule $\mathrm{Soc}(Q)$ of Q. Therefore X lies over the direct summand $Y \oplus X_2$ of M and condition (2) holds.

$(1) \Rightarrow (5)$. By Theorem 14.6, it is sufficient to prove that every cyclic right R-module X is the direct sum of a semisimple module and an injective module. Let M be the injective hull of X. Since M is semiregular, there exists a decomposition $M = M_1 \oplus M_2$ such that $M_1 \subseteq X$ and $M_2 \cap X$ is a small submodule of M_2. Then $X = M_1 \oplus (M_2 \cap X)$, where M_1 is injective. We know that $J(M_2)$ is semisimple. Since $M_2 \cap X \subseteq J(M_2)$, the module $M_2 \cap X$ is semisimple. $\qquad\square$

Lemma 16.2 *Let M be a module such that, for any nonzero cyclic small submodule X of M, every maximal submodule Y of X is a direct summand of X. Then the Jacobson radical $J(M)$ is semisimple.*

Proof Without loss of generality, we can assume that $J(M) \neq 0$. It is sufficient to prove that every nonzero cyclic submodule X of $J(M)$ is semisimple. Let X_1 be a proper submodule of X. There exists a submodule X_2 of X such that $X_1 \cap X_2 = 0$ and $X_1 \oplus X_2$ is an essential submodule of X. If $X = X_1 \oplus X_2$, then X is semisimple.

We assume that $X \neq X_1 \oplus X_2$. Since X is cyclic, there exists a maximal submodule Y of X with $X_1 \oplus X_2 \subseteq Y$. Since Y contains an essential submodule $X_1 \oplus X_2$ of X, we have that Y is an essential submodule of X. Since $J(M)$ is the sum of all small submodules of M, we have that X is the sum of a finite number of small submodules of M. Therefore X is small in M. By assumption, Y is a direct summand of X. In addition, Y is an essential submodule of X. Therefore $Y = M$; this is a contradiction. $\qquad\square$

A module M is called an I_0-*module* if every (cyclic) submodule X of M with $X \nsubseteq J(M)$ contains a nonzero direct summand of M. Since $J(M)$ is the sum of all small submodules of M, the module M is an I_0-module if and only if every finitely generated (cyclic) nonsmall submodule of M contains a nonzero direct summand of M.

Remark 16.3 *Every semiregular module is an I_0-module. However, the converse is not true. Let $\mathbb{Q}$ be the field of rational numbers, $\mathbb{Z}$ be the ring of integers, $\{\mathbb{Q}_i\}_{i=1}^{\infty}$ be an infinite countable set of copies of $\mathbb{Q}$, $D = \prod_{i=1}^{\infty} \mathbb{Q}_i$, and let R be the subring of D generated by the ideal $\oplus_{i=1}^{\infty}\mathbb{Q}_i$ and by the subring $\{(z, z, z, \ldots) \mid z \in \mathbb{Z}\}$. Then it can be verified that R_R is an I_0-module, R is a commutative semiprimitive ring, R has the factor ring $R/(\oplus_{i=1}^{\infty}\mathbb{Q}_i)$ which is isomorphic to $\mathbb{Z}$, and $R/(\oplus_{i=1}^{\infty}\mathbb{Z}_i)$ is not an I_0-ring. Therefore R_R is not semiregular.*

Lemma 16.4 *Let M be a module and let X be a small submodule of M. If there exists an epimorphism $f\colon X \to S$ such that S is a simple module and $M \oplus S$ is an I_0-module, then $\mathrm{Ker}(f)$ is a direct summand of X.*

Proof We consider the submodule $X' = \{x + f(x) \mid x \in X\}$ in $M \oplus S$. There exists an isomorphism $\varphi\colon X \to X'$ such that $\varphi(x) = x + f(x)$ for all $x \in X$. We note that

$$\mathrm{Ker}(f) = \varphi(\mathrm{Ker}(f)) = \varphi^{-1}(\mathrm{Ker}(f)) = X' \cap M.$$

Since $M \oplus S = M + X'$ and $M \neq M \oplus S$, we have that X' is not small in $M \oplus S$. Therefore $X' \not\subseteq J(M \oplus S)$. By assumption, $M \oplus S$ is an I_0-module. Therefore there exists a decomposition $M \oplus S = Y' \oplus M'$ such that $Y' \neq 0$ and $Y' \subseteq X'$.

Let $\pi\colon M \oplus S \to M$ be the projection with kernel S. Then

$$M = \pi(Y' \oplus M') = \pi(Y') + \pi(M') \subseteq X + \pi(M').$$

Then $\pi(M') = M$, since X is small in M. Since $\pi(M') = M$, we have $M \oplus S = M' + S$. The following two cases are possible: (1) $M' \cap S \neq 0$; (2) $M' \cap S = 0$.

(1) We assume that $M' \cap S \neq 0$. Then $S \subseteq M'$, since S is simple. Therefore

$$0 \neq Y' = Y' \cap (M \oplus S) = Y' \cap (M' + S) = Y' \cap M' = 0.$$

This is a contradiction.

(2) We assume that $M' \cap S = 0$. Then

$$M \oplus S = M' \oplus S = M' \oplus Y', \quad Y' \cong (M' \oplus S)/M' \cong S.$$

Therefore Y' is simple. Thus either $Y' \subseteq \mathrm{Ker}(f) \subseteq X$ or $Y' \cap \mathrm{Ker}(f) = 0$. The first case is impossible, since the small submodule X of M contains a nonzero direct summand Y' of M in this case.

Therefore $Y' \cap \mathrm{Ker}(f) = 0$. Since $\mathrm{Ker}(f) = \varphi(\mathrm{Ker}(f)) = \varphi^{-1}(\mathrm{Ker}(f))$ is a maximal submodule of X and $\varphi\colon X \to X'$ is an isomorphism, $\mathrm{Ker}(f) = \varphi(\mathrm{Ker}(f))$ is a maximal submodule of X'. Therefore $X' = Y' \oplus \mathrm{Ker}(f)$. The isomorphism $\varphi^{-1}\colon X' \to X$ induces the relation $X = \varphi^{-1}(Y') \oplus \mathrm{Ker}(f)$. $\square$

Theorem 16.5 ([227]) *For a ring R, the following conditions (1)–(4) are equivalent:*

1. *Every right R-module is an I_0-module.*
2. *For every right R-module M, the module $J(M)$ is semisimple and, if $J(M) = 0$, every nonzero submodule of M contains a nonzero direct summand of M.*
3. *For every right R-module M, either M has a nonzero injective direct summand or M is semisimple and contained in the Jacobson radical of its injective hull.*

4. *For every right R-module M, either M has a nonzero injective direct summand or M is semisimple.*

Furthermore, if any of the conditions (1)–(4) hold, then the following conditions (5)–(9) hold.

5. *For every indecomposable right R-module M, either M is injective or M is a simple small submodule of the injective hull of M.*
6. *Every indecomposable nonzero right R-module M is either a simple small submodule of its injective hull or a cyclic uniserial injective module of length 2.*
7. *Every indecomposable right R-module is a cyclic uniserial module of length at most 2.*
8. *Every right R-module is a subdirect product of cyclic uniserial modules of finite length at most 2. In particular, every nonzero right R-module has a maximal submodule, whence M does not coincide with its Jacobson radical.*
9. *For every right R-module M, the Jacobson radical $J(M)$ is a semisimple small submodule of M.*

Proof

(1) $\Rightarrow$ (2). Let M be a nonzero right R-module. If $J(M) = 0$, then every nonzero submodule of the semiprimitive I_0-module M contains a nonzero direct summand of M.

We assume that $J(M) \neq 0$ and X is a nonzero cyclic submodule of $J(M)$. Then X is a cyclic small submodule of M. By Lemma 16.4, every maximal submodule of X is a direct summand of X. By Lemma 16.2, $J(M)$ is a semisimple module.

(2) $\Rightarrow$ (3). Let N be a nonzero right module with injective hull M. We assume that $N \subseteq J(M)$. Then $J(M) \neq 0$. By assumption, $J(M)$ is semisimple. The submodule N of the semisimple module $J(M)$ is semisimple.

We assume that $J(M) = 0$. By (2), the nonzero submodule N of the injective module M has a nonzero injective direct summand.

The implication (3) $\Rightarrow$ (4) is obvious.

(4) $\Rightarrow$ (1). Let M be a right R-module and let X be a cyclic submodule of M such that X is not small in M. If X has a nonzero injective direct summand Y, then Y is a direct summand of M, and the proof is complete.

We assume that the nonzero module X does not have nonzero injective direct summands. By (4), X is semisimple. Therefore there exists a decomposition $X = \oplus_{i=1}^n X_i$ such that all modules X_i are simple. Since X is not small in M, there exists an integer $i \in \{1, \ldots, n\}$ such that X_i is not small in M. Therefore there exists a proper submodule Y of M with $X_i + Y = M$. Then $X_i \cap Y \neq X_i$, since X_i is simple. Thus $X_i \cap Y = 0$, $X_i \oplus Y = M$, and X contains a nonzero direct summand of M.

(5) The assertion follows from (4).

(6) Let M be an indecomposable nonzero right R-module. Assume that M is not a simple small submodule of its injective hull. By (5), M is an injective nonsimple module. It suffices to prove that any proper nonzero submodule N of M is simple. Since M is indecomposable and $N \neq M$, the module N is not injective. By (5), N is simple.

(7) The assertion follows from (6).

(8) The assertion follows from (7) and the property that every module is a subdirect product of subdirectly irreducible modules.

(9) By (2), $J(M)$ is a semisimple module. We assume that $J(M)$ is not small in M. There exists a proper submodule X of M with $M = X + J(M)$. Then M/X is a nonzero module coinciding with Jacobson radical of M/X. This contradicts (8). $\qquad\square$

Corollary 16.6 ([227]) *For a ring R, the following conditions are equivalent:*

1. *Every right R-module is an I_0-module and R_R is a direct sum of indecomposable modules.*
2. *R is an artinian serial ring with $J(R)^2 = 0$.*

Proof Since every semiregular module is an I_0-module, the implication $(1) \Rightarrow (2)$ follows from Theorem 16.1.

$(2) \Rightarrow (1)$. By Theorem 16.5(7), every indecomposable right R-module is a cyclic uniserial module of length ≤ 2. In addition, R_R is a finite direct sum of indecomposable modules. Therefore R is a right artinian right serial ring with $J(R)^2 = 0$. By Theorem 13.13(2), R is an artinian serial ring. $\qquad\square$

Lemma 16.7 *Let R be a ring, e be a nonzero idempotent of R, and let M be a right R-module which is an I_0-module.*

1. *If N is a submodule of M, then there exists a natural isomorphism of right eRe-modules $(M/N)e \cong Me/Ne$.*
2. *If M is a semisimple right R-module, then Me is a semisimple right eRe-module.*
3. *If eRe is a von Neumann regular ring, then the injectivity of the right R-module M implies the injectivity of the right eRe-module Me.*

Proof The assertions (1) and (2) are well-known; they are easily verified. Since every eRe-module is flat, the assertion (3) follows from [74, 11.35] and the eRe-isomorphism $\mathrm{Hom}_R(eR, M) \cong Me$. $\qquad\square$

Lemma 16.8 *Let R be a ring, M be a right R-module which is an I_0-module, and let N be a submodule of M such that $(N + J(M))/J(M)$ is a simple submodule of $M/J(M)$. Then N contains a local direct summand mR of M with $(N + J(M))/J(M) = (m + J(M))R$.*

Proof Let n be an element of the submodule N such that $(N + J(M))/J(M) = (n + J(M))R$. Since M is an I_0-module, there exists a cyclic submodule mR

in M such that $mR \not\subset J(M)$, $mR \subset nR$, and mR is a direct summand in M. Then

$$(N + J(M))/J(M) = (m + J(M))R \cong mR/(J(M) \cap mR) \cong mR/J(mR);$$

therefore, the module mR is local. $\qquad\qquad\square$

In any ring R, we now define two ideals $I^{(1)}(R)$ and $I^{(2)}(R)$ using transfinite induction.

We first define the ideal $I^{(1)}(R)$ as follows. For $\alpha = 0$, set $I_\alpha^{(1)}(R) = 0$. If $\alpha = \beta + 1$, then $I_{\beta+1}^{(1)}(R)/I_\beta^{(1)}(R)$ is the sum of all simple injective submodules of the right R-module $R/I_\beta^{(1)}(R)$. For a limit ordinal number α, we set $I_\alpha^{(1)}(R) = \bigcup_{\beta < \alpha} I_\beta^{(1)}(R)$. It is clear that for some ordinal number τ, we have $I_\tau^{(2)}(R) = I_{\tau+1}^{(1)}(R)$. Further, we denote the right ideal $I_\tau^{(1)}(R)$ by $I^{(1)}(R)$; it is clear that $I^{(1)}(R)$ is an ideal.

Next, we define the ideal $I^{(2)}(R)$. For $\alpha = 0$, set $I_\alpha^{(2)}(R) = 0$. If $\alpha = \beta + 1$, then $I_{\beta+1}^{(2)}(R)/I_\beta^{(2)}(R)$ is the sum of all local injective submodules of the right R-module $R/I_\beta^{(2)}(R)$ of length ≤ 2 such that their factor modules with respect to their Jacobson radicals are injective modules. For a limit ordinal number α, we set $I_\alpha^{(2)}(R) = \bigcup_{\beta < \alpha} I_\beta^{(2)}(R)$. It is clear that for some ordinal number γ, we have $I_\gamma^{(2)}(R) = I_{\gamma+1}^{(2)}(R)$. Further, we denote the right ideal $I_\gamma^{(2)}(R)$ by $I^{(2)}(R)$; it is clear that $I^{(2)}(R)$ is an ideal.

Lemma 16.9 *Let R be a ring.*

1. *For any ordinal number α, every injective simple right $R/I_\alpha^{(1)}$-module M is an injective right R-module.*
2. *If e is a central idempotent of the ring R and eR is a semisimple R-module, then eR is an injective R-module.*

Proof

(1) We use transfinite induction. If $\alpha = 0$, then the assertion is trivial. We assume that α is an ordinal number and for every $\beta < \alpha$, each injective simple right $R/I_\beta^{(1)}$-module M is an injective right R-module. We consider an arbitrary injective simple right $R/I_\alpha^{(1)}$-module M. We assume that $E(M) \neq M$, where $E(M)$ is the injective hull of M. If $E(M)I_\alpha^{(1)} \neq 0$, then we denote by γ the least ordinal number with $E(M)I_\gamma^{(1)} \neq 0$. It is clear that $\gamma = \gamma_0 + 1$ for some ordinal number γ_0. Then $E(M)$ is an $R/I_{\gamma_0}^{(1)}$-module and for some primitive idempotent e in the ideal $I_\gamma^{(1)}/I_{\gamma_0}^{(1)}$, we have $E(M)e \neq 0$. By the induction hypothesis, the simple module

$e(R/I_{\gamma_0}^{(1)})$ is an injective R-module. It follows from [153, 6.6.3] that $E(M)$ is a nonsimple module and any two nonzero submodules of $E(M)$ have nonzero intersection. On the other hand, $E(M)$ contains a simple injective R-module. This is a contradiction; therefore $E(M)I_\alpha^{(1)} = 0$. It follows that $E(M)$ is an $R/I_\alpha^{(1)}$-module and since M is an injective simple $R/I_\alpha^{(1)}$-module, then $E(M) = M$.

(2) We have $eR = eRe$ is a division ring. Note that for any two rings A, B, if $R = A \oplus B$, then any A-module is injective if and only if it is injective as an R-module. This gives that eR is an injective R-module. $\qquad\square$

Lemma 16.10 *Let R be a right semiartinian ring where each right module is an I_0-module. Then every nonsemisimple right R-module N contains an injective local submodule of length at most 2. In particular, every local module of length 2 is injective.*

Proof Since N is not semisimple, it follows from Theorem 16.5 that N contains a nonzero injective submodule N_0. Since R is a right semiartinian ring, N_0 is semiartinian. Consequently, $N_0/J(N_0)$ contains a simple submodule. It follows from Lemma 16.8 and Theorem 16.5(7) that N_0 contains a direct summand which is an injective local module of length at most 2. $\qquad\square$

A right semiartinian right V-ring is called a *right SV-ring*.

Lemma 16.11 *Let R be a ring and let M be a right R-module.*

1. *If $MI^{(2)}(R) \neq 0$, then M contains a nonzero local injective submodule of length at most 2.*
2. *If $R = I^{(2)}(R)$, then R is a right SV-ring.*

Proof

(1) Let γ be the least ordinal number with $MI_\gamma^{(2)}(R) \neq 0$. It is clear that γ is a nonlimit ordinal number. Then the module M contains a nonzero homomorphic image of the module $I_\gamma^{(2)}(R)/I_{\gamma-1}^{(2)}(R)$. Consequently, M contains a local injective submodule of length at most 2.

(2) The assertion follows from (1). $\qquad\square$

Lemma 16.12 *For a ring R, the following conditions are equivalent:*

1. *R is a right semiartinian ring where each right module is an I_0-module.*
2. *Either R is a right SV-ring or $R/I^{(2)}(R)$ is a serial artinian ring and $J^2(R/I^{(2)}(R)) = 0$.*

Proof $(1) \Rightarrow (2)$. We denote by S the factor ring $R/I^{(2)}(R)$ which can be considered as a right R-module. We assume that $S/J(S)$ is not a semisimple ring. By Theorem 16.5, the right R-module $S/J(S)$ contains a nonzero injective submodule. Consequently, the right R-module $S/J(S)$ contains a simple injective submodule. It follows from Lemma 16.8 and Lemma 16.10 that the right R-module S contains an injective local submodule X of length ≤ 2 such that the factor module of X with respect to the Jacobson radical of X is an injective module. By the construction of the ideal $I^{(2)}(R)$, the right R-module $R/I^{(2)}(R)$ does not contain local submodules of length ≤ 2 whose factor modules with respect to their Jacobson radicals are injective modules. This is a contradiction; therefore $S/J(S)$ does not contain a nonzero injective submodule. Consequently, it follows from Theorem 16.5 that $S/J(S)$ is a semisimple module. Then S is a semilocal ring and the implication follows from Theorem 16.5 and Corollary 16.6.

$(2) \Rightarrow (1)$. We prove that every local right $R/I^{(2)}(R)$-module N of length 2 is an injective R-module. Let $E(N)$ be the injective hull of the R-module N. If $E(N)I^{(2)}(R) = 0$, then we can consider $E(N)$ as an $R/I^{(2)}(R)$-module. It follows from the injectivity of the $R/I^{(2)}(R)$-module N that $N = E(N)$. In the case $E(N)I^{(2)}(R) \neq 0$, it follows from Lemma 16.11 that $E(N)$ contains an injective local submodule of length ≤ 2; then the relation $E(N) = N$ is easily verified.

We consider an arbitrary nonsemisimple right R-module N. If $NI^{(2)}(R) = 0$, then we can consider N as a right $R/I^{(2)}(R)$-module. Then N contains a local submodule of length 2 which is an injective R-module. If $NI^{(2)}(R) \neq 0$, then it follows from Lemma 16.11 that N contains a nonzero local injective submodule. It follows from the above argument that an arbitrary nonsemisimple module N contains a nonzero injective submodule. Then the implication follows from Theorem 16.5. $\qquad\qquad\square$

Lemma 16.13 *Let R be a ring where each right module is an I_0-module, and let e be a nonzero idempotent of R such that the ring eRe is von Neumann regular. Then each right module over eRe is an I_0-module.*

Proof We assume the contrary. It follows from Theorem 16.5 that there exists a nonsemisimple right eRe-module N which does not contain a nonzero injective submodule. We consider the right R-module $M = N \underset{eRe}{\bigotimes} eR$. It is clear that $Me \cong N$. Using transfinite induction, we define submodules M_α in M for every ordinal number α, as follows. For $\alpha = 0$, we set $M_\alpha = 0$. If $\alpha = \beta + 1$, then $M_{\beta+1}/M_\beta$ is the sum of all injective submodules of the module M/M_β. If α is a limit ordinal number, we set $M_\alpha = \underset{\beta < \alpha}{\bigcup} M_\beta$. We denote by M_0 the union of all such modules. Using transfinite induction, we prove that $M_\alpha e = 0$ for every ordinal number α. If $\alpha = 0$, then the assertion is trivial. Let α be an ordinal number and let $M_\beta e = 0$ for every $\beta < \alpha$. If α is a limit ordinal number, then the relation $M_\alpha e = 0$ is trivial. We assume that α is a nonlimit ordinal number and $\alpha = \alpha_0 + 1$. By the induction hypothesis, $M_{\alpha_0} e = 0$. By Lemma 16.7, we have

$$(M_\alpha/M_{\alpha_0})e \cong (M_\alpha e)/(M_{\alpha_0} e) \cong M_\alpha e.$$

If $M_\alpha e \neq 0$, then the module M_α/M_{α_0} contains an injective submodule L with $Le \neq 0$. Since it follows from Lemma 16.7 that Le is an injective eRe-module,

$M_\alpha e$ has a nonzero injective submodule. Consequently, Me has a nonzero injective submodule; this contradicts the original assumption. It follows that $M_\alpha e = 0$ for every ordinal number α. Consequently, $M_0 e = 0$. Since M/M_0 does not contain an injective submodule, it follows from Theorem 16.5 that M/M_0 is semisimple. By Lemma 16.7, $(M/M_0)e$ is also a semisimple module. It then follows from $(M/M_0)e \cong Me$ that Me is semisimple; this contradicts the choice of the module N. $\qquad\square$

Theorem 16.14 *For a ring R, the following conditions are equivalent:*

1. *R is a right SV-ring.*
2. *R is a von Neumann regular ring and every right R-module is an I_0-module.*

Proof The implication $(1) \Rightarrow (2)$ is easy to see.

$(2) \Rightarrow (1)$. Let R be a von Neumann regular ring such that every right R-module is an I_0-module. We assume that $R \neq I^{(1)}(R)$ and we denote by S the ring $R/I^{(1)}(R)$. It follows from Lemma 16.9 that the module S_S does not contain a simple injective S-submodule; in particular, S_S is not semisimple. Since every S-module is an I_0-module, it follows from Theorem 16.5 that S_S contains a nonzero injective submodule of the form eS, where e is some idempotent of the ring S. Since the ring eSe is isomorphic to the endomorphism ring of the injective module eS and $J(eSe) = 0$, it follows from [235, 22.1] that eSe is a von Neumann regular right self-injective ring. Then it follows from Lemma 16.13 that each right module over eSe is an I_0-module. Since eS does not contain a simple submodule, the von Neumann regular ring eSe does not contain primitive idempotents, that is, $\mathrm{Soc}(eSe) = 0$. The ring eSe contains an infinite set of orthogonal nonzero idempotents of the form $\{e_{ij}\}_{i,j=1}^{\infty}$. For every i, the module $\bigoplus_{j=1}^{\infty} e_{ij}eSe$ is an essential submodule in $f_i eSe$, where f_i is some idempotent of the ring eSe. It is clear that the family $\{f_i eSe\}_{i=1}^{\infty}$ of right ideals is independent and, for some idempotent f of eSe, the right ideal $\bigoplus_{i=1}^{\infty} f_i eSe$ is essential in $feSe$. The right ideal $\bigoplus_{i,j=1}^{\infty} e_{ij}eSe$ is an essential submodule in $feSe$ and the right eSe-module $eSe/(\bigoplus_{i,j=1}^{\infty} e_{ij}eSe \oplus (e-f)eSe)$ contains a submodule which is isomorphic to $\bigoplus_{i=1}^{\infty} (f_i eSe/(\bigoplus_{j=1}^{\infty} e_{ij}eSe))$. The module $eSe/(\bigoplus_{i,j=1}^{\infty} e_{ij}eSe \oplus (e-f)eSe)$ is not semisimple. Consequently, it follows from Theorem 16.5 that it contains a nonzero injective submodule; this contradicts Lemma 4.3. $\qquad\square$

Lemma 16.15 *Let R be a ring where each right module is an I_0-module. Then R is a right semiartinian ring.*

Proof We assume that $R \neq L(R)$. We let S denote the ring $R/L(R)$ where each right module is an I_0-module. It is clear that $\mathrm{Soc}(S_S) = 0$. Consequently, it follows from Theorem 16.5(9) that $J(S) = 0$. Since the module S_S is not semisimple, it follows from Theorem 16.5(4) that S_S contains a nonzero injective submodule of the form eS, where e is an idempotent of the ring S. It follows from [235, 22.1] that the ring eSe is von Neumann regular. By Lemma 16.13 and Theorem 16.14, eSe is a right SV-ring. Consequently, eSe contains some primitive idempotent f. The module feS is simple; this contradicts the relation $\mathrm{Soc}(S_S) = 0$.   □

Theorem 16.16 *For a ring R, the following conditions are equivalent:*

1. *All right R-modules are I_0-modules.*
2. *Either R is a right SV-ring or $R/I^{(2)}(R)$ is an artinian serial ring and $J^2(R/I^{(2)}(R)) = 0$.*

Proof This follows directly from Lemmas 16.11, 16.12, and 16.15.   □

Corollary 16.17 *Let R be a ring where each right module is an I_0-module, and let τ be the cardinality of isomorphism classes of indecomposable right R-modules whose factor modules with respect to the Jacobson radical are not injective. Then τ is finite.*

Proof Let M be an indecomposable right R-module such that the factor module of M with respect to $J(M)$ is not injective. It follows from Theorem 16.5(7) that M is a local module of length two. If $MI^{(2)}(R) \neq 0$, then it follows from the proof of Lemma 16.11 that the module $M/J(M)$ is injective. This is a contradiction; therefore $MI^{(2)}(R) = 0$. Consequently, it follows from Theorem 16.16 that the module M is isomorphic to an indecomposable direct summand of the module $(R/I^{(2)}(R))_R$.   □

Remark 16.18 [229] For a ring R without infinite sets of orthogonal noncentral idempotents, the following conditions are equivalent:

1. Each right R-module is an I_0-module.
2. Either R is a right SV-ring or $R/I^{(1)}(R)$ is an artinian serial ring and $J^2(R/I^{(1)}(R)) = 0$.

16.19 Questions

1. Let R be a ring such that each right R-module is an I_0-module. Is it true that either R is a right SV-ring or $R/I^{(1)}(R)$ is an artinian serial ring and $J^2(R/I^{(1)}(R)) = 0$?
2. Let R be a ring such that each right R-module is an I_0-module. Suppose, in addition, R is a ring with polynomial identity. Is it true that either R is a right SV-ring or $R/I^{(1)}(R)$ is an artinian serial ring and $J^2(R/I^{(1)}(R)) = 0$?

17 Completely integrally closed modules and rings

A subring R of Q is said to be *classically right completely integrally closed* in Q if R contains every element $q \in Q$ such that $q^n a \in R$ for some nonzero-divisor $a \in R$ and all $n \in \mathbb{N}$. A subring R of Q is said to be *right completely integrally closed* in Q if R contains every element $q \in Q$ such that $qB \subseteq B$ for some essential right ideal B of the ring R.

If X is a module, Y is a submodule of the module X, and $f\colon Y \to X$ is a homomorphism, which preserves some essential submodule of the module Y, then f is called a Y-*essential* homomorphism. A module X is said to be *completely integrally closed* if for every submodule Y of X, every Y-essential homomorphism $Y \to X$ can be extended to a homomorphism $X \to X$. As we will see below, the class of all completely integrally closed modules contains all (quasi-)injective modules and all classically completely integrally closed commutative domains (each considered as a module over itself).

Remark 17.1 *The class of all completely integrally closed rings[1] contains all right self-injective rings. In addition, if R is a commutative domain with field of quotients Q, then it is known that the domain R is classically completely integrally closed in Q if and only if R is a completely integrally closed subring of Q (see also Proposition 17.18 below).*

For a ring Q, a subring R of Q is said to be *right classically integrally closed* in Q if R contains every element $q \in Q$ such that $q^{n+1} = q^n a_n + \cdots + q a_1 + a_0$ for some $a_0, \ldots, a_n \in R$. A module M is said to be *integrally closed* if all endomorphisms of any finitely generated submodule of M can be extended to endomorphisms of M.

Remark 17.2 *The class of all rings, that are integrally closed as right modules over themselves, contains all right self-injective and all von Neumann regular rings.*

In addition, let R be a commutative domain with field of quotients Q. The following properties are well-known:

1. *The domain R is classically integrally closed in Q if and only if R is an integrally closed R-module (see also Remark 17.25 below).*

---

[1] As right modules over themselves.

    2. If R is noetherian, then R is classically integrally closed in Q if and only if R is classically completely integrally closed in Q (see also Theorem 17.22 below).

Remarks 17.1 and 17.2 will be generalized in Theorems 17.15, 17.21, and 17.22, due to Tuganbaev [230]; these theorems are the main results of this chapter. The next result is an easy observation about skew-injective modules.

Lemma 17.3 *Let X be a skew-injective module, $\overline{X}$ be an essential extension of X, and let f be an endomorphism of $\overline{X}$.*

    1. If $f^2(x) \in X$ for any element $x \in X$ with $f(x) \in X$, then $f(X) \subseteq X$.
    2. Let $N = \{x \in X \mid f(x) \in X\}$. If $(f^2 - f)(N) \subseteq X$, then $f(X) \subseteq X$.
    3. If $(f^2 - f)(X) \subseteq X$, then $f(X) \subseteq X$.

Proof

    (1) Since $f^2(x) \in X$ for any element $x \in X$ with $f(x) \in X$, we have $f(N) \subseteq N$. The restriction of f to N can be extended to an endomorphism of X. By Lemma 1.19(5), we have $f(X) \subseteq X$.

    (2) Let $x \in N$. Then $f(x) \in X$. By assumption, $(f^2 - f)(x) \in X$. Therefore $f^2(x) = (f^2 - f)(x) + f(x) \in X$. By (1), $f(X) \subseteq X$.

    (3) This follows from (2).   □

Next, we show the relationship between skew-injective modules and π-injective modules.

Proposition 17.4 *Each skew-injective module is π-injective, and indecomposable π-injective modules coincide with uniform modules.*

Proof We assume that M is a skew-injective module, E is the injective hull of the module M, and f is an idempotent endomorphism of the module E. Since $(f^2 - f)(M) = 0 \in M$, it follows from Lemma 17.3(3) that $f(M) \subseteq M$. Thus M is π-injective. It is easily verified that indecomposable π-injective modules coincide with uniform modules.   □

Lemma 17.5 *Let M be an essential extension of a completely integrally closed module X. Then:*

    1. $f(X) \subseteq X$ for every X-essential endomorphism f of M.
    2. If f is an endomorphism of M and $\mathrm{Ker}(f)$ is an essential submodule of M, then $f(X) \subseteq X$.
    3. If X is an essential extension of some module Y which is fully invariant in M, then X is a fully invariant submodule of M.
    4. If X is an essential extension of a quasi-injective module Y, then X is a quasi-injective fully invariant submodule of M.
    5. If X is an essential extension of a semisimple module, then X is a quasi-injective fully invariant submodule of M.

Proof

(1) By the assumption, $f(Y) \subseteq Y$ for some essential submodule Y in X. The module $X \cap f^{-1}(X)$ is denoted by X_1 and the restriction of the homomorphism f to the module X_1 is denoted by $f_1 \in \mathrm{Hom}(X_1, X)$. Since $Y \subseteq X_1$ and $f_1(Y) = f(Y) \subseteq Y$, we have that f_1 is an X_1-essential homomorphism from X_1 into X. Since X is a completely integrally closed module, f_1 can be extended to the homomorphism $g \colon X \to X$. By Lemma 1.19(5), $f(X) \subseteq X$.

(2) We set $Y = X \cap \mathrm{Ker}(f)$. Then Y is an essential submodule of X and $f(Y) = 0 \subseteq Y$. By (1), $f(X) \subseteq X$.

(3) This follows from (1).

(4) Let M' be an injective hull of the module M. Then M' is an essential extension of Y, since M' is an essential extension of M, M is an essential extension of X and X is an essential extension of Y. By Lemma 1.19(3), Y is a fully invariant submodule of M and M'. By (3), X is a fully invariant submodule of M and M'. By Lemma 1.19(4), X is quasi-injective.

(5) Since all semisimple modules are quasi-injective, this follows from (4). $\square$

Proposition 17.6 *Let X be a module with injective hull M and let $\Delta(M)$ be the ideal of the ring $\mathrm{End}(M)$ consisting of all endomorphisms whose kernels are essential submodules of M. The following conditions are equivalent:*

1. X is a completely integrally closed module.

2. $f(X) \subseteq X$ for every X-essential endomorphism f of the module M.

3. X is a skew-injective module and $h(X) \subseteq X$ for every $h \in \Delta(M)$.

4. X is a skew-injective module and $X = Y \oplus Z$, where $\Delta(Y) = 0$, Z is a quasi-injective X-injective module and Z is an essential extension of the module $\sum_{h \in \Delta(M)} h(X)$.

Proof The implication $(1) \Rightarrow (2)$ follows from Lemma 17.5(1).

$(2) \Rightarrow (3)$. Let Y_1 be a submodule of X and let $f_1 \in \mathrm{End}(Y_1)$. By Lemma 1.19(6), there exists a submodule Z of X such that $Y_1 \cap Z = 0$ and X is an essential extension of $Z \oplus Y_1$. Let f_2 denote the endomorphism of $Z \oplus Y_1$ given by $f_2(z + y_1) = f_1(y_1)$ where $z \in Z$ and $y_1 \in Y_1$. The endomorphism f_2 can be extended to an endomorphism f of the injective module M. Since $f(Z \oplus Y_1) \subseteq Z \oplus Y_1$, we have that f is an X-essential endomorphism of M. By the assumption, $f(X) \subseteq X$. Then X is skew-injective, since f_1 can be extended to an endomorphism of X. Let h be an endomorphism of M such that $\mathrm{Ker}(h)$ is an essential submodule of M. We set $Y = X \cap \mathrm{Ker}(h)$. Since Y is an essential submodule of X and $h(Y) = 0 \subseteq Y$, we have that h is an X-essential endomorphism of M. By the assumption, $h(X) \subseteq X$.

$(3) \Rightarrow (1)$. Let Y_1 be a submodule of X and let f_1 be an Y_1-essential homomorphism from Y_1 to X. Then $f_1(Y_2) \subseteq Y_2$ for some essential submodule Y_2 of the module Y_1. By Lemma 1.19(6), there exists a submodule Z of X such that

$Y_1 \cap Z = 0$ and X is an essential extension of $Z \oplus Y_1$. It is easily verified that X is an essential extension of the module $Z \oplus Y_2$, since Y_1 is an essential extension of Y_2. Therefore M is an essential extension of $Z \oplus Y_2$. Let f_2 denote the homomorphism from $Z \oplus Y_1$ into X given by $f_2(z + y_1) = f_1(y_1)$ for $z \in Z$ and $y_1 \in Y_1$. We denote by f_3 the endomorphism of $Z \oplus Y_2$ such that $f_3(z + y_2) = f_1(y_2) \in Y_2 \subseteq Z \oplus Y_2$ for any two elements $z \in Z$ and $y_2 \in Y_2$. Since M is skew-injective, f_3 can be extended to an endomorphism f_4 of X. The homomorphisms f_2 and f_4 can be extended to endomorphisms f and f' of the injective module M. We set $h = f - f' \in \mathrm{End}(M)$. Since $h(Z \oplus Y_2) = (f - f')(Z \oplus Y_2) = 0$ and M is an essential extension of $Z \oplus Y_2$, we have $h(X) \subseteq X$ by assumption. Then $f(X) \subseteq h(X) + f'(X) = h(X) + f_4(X) \subseteq X$, and f coincides with f_1 on the module Y_1. Therefore f_1 can be extended to an endomorphism of X and X is a completely integrally closed module.

$(3) \Rightarrow (4)$. Let Z_1 denote the submodule $\sum_{h \in \Delta(M)} h(X)$ of M. Since $\Delta(M)$ is an ideal of the ring $\mathrm{End}(M)$, we have that Z_1 is a fully invariant submodule of the injective module M. Therefore Z_1 is quasi-injective. By the assumption, $Z_1 \subseteq X$. By Lemma 1.19(8), there exists a decomposition $X = Y \oplus Z$ such that Z is an essential extension of the quasi-injective module Z_1. Lemma 17.5(4) then shows that the completely integrally closed module is quasi-injective, that is, the module Z is Z-injective. By Lemma 1.19(9), Z is Y-injective. Clearly, then Z is $Y \oplus Z$-injective, that is, Z is X-injective. Let $h_1 \in \Delta(Y)$. We denote by h_2 the endomorphism of $X = Y \oplus Z$ such that $h_2(y + z) = h_1(y)$ for all $y \in Y$ and $z \in Z$. Since $h_1 \in \Delta(Y)$, we have $h_2 \in \Delta(X)$. The endomorphism h_2 can be extended to an endomorphism h of the injective module M. Since $\mathrm{Ker}(h_2) \subseteq \mathrm{Ker}(h)$ and $h_2 \in \Delta(X)$, we have $h \in \Delta(M)$. Therefore $h_1(Y) \subseteq h(X) \subseteq Z_1$, $h_1(Y) \subseteq Y \cap Z = 0$, $\Delta(Y) = 0$.

The implication $(4) \Rightarrow (3)$ is obvious. $\qquad\qquad\qquad\qquad\qquad\square$

Lemma 17.7 *If X is a submodule of the module M and $E(M)$ is the injective hull of M, then the following conditions are equivalent:*

1. *M is a rational extension of X.*
2. *$f(M) = 0$ for every $f \in \mathrm{End}(E(M))$ with $f(X) = 0$.*
3. *M is an essential extension of X and for any two nonzero elements $m \in M$ and $x \in X$, there exists an element $a \in R$ such that $xa \neq 0$ and $ma \in X$.*

Proof $(1) \Rightarrow (2)$. We assume that $f(M) \neq 0$. Then $M \cap f(M) \neq 0$. Let $0 \neq x = f(y) \in M \cap f(M)$, $y \in M$. Since M is a rational extension of X, there exists an element $a \in R$ such that $xa \neq 0$ and $ya \in X$. Therefore $0 = f(ya) = xa \neq 0$. This is a contradiction.

$(2) \Rightarrow (1)$. We assume that M is not a rational extension of X. Then there exist elements $x, y \in M$ with $x \neq 0$ and $(y : X) \subseteq ann_r(x)$. Since $ann_r(y) \subseteq (y : X) \subseteq ann_r(x)$, there exists an epimorphism $g \colon yR \to xR$ such that $g(ya) = xa$ for all $a \in R$. Then $X \cap yR \subseteq \mathrm{Ker}(g)$, since $(y : X) \subseteq ann_r(x)$. Therefore there exists an epimorphism $h \colon (yR + X) \to xR$ such that $X \subseteq \mathrm{Ker}(h)$ and

$h(ya) = xa$ for all $a \in R$. Since $E(M)$ is injective, h can be extended to $f \in$ End$(E(M))$, and $f(X) = h(X) = 0$. Since $f(y) = x \neq 0$, we have $f(M) \neq 0$.

$(1) \Rightarrow (3)$. Let $0 \neq m \in M$. By substituting $m_1 = m_2 = m$ into the definition of a rational extension, we see that there exists an element $a \in R$ such that $ma \in X$ and $ma \neq 0$. Therefore M is an essential extension X. The remaining part of the assertion follows from the definition of a rational extension.

$(3) \Rightarrow (1)$. Let $0 \neq m_1, m_2 \in M$. We have to prove that there exists an element $a \in R$ such that $m_1 a \neq 0$ and $m_2 a \in X$. Since M is an essential extension X, we have that $0 \neq m_1 a_1 \in X$ for some $a_1 \in R$. If $m_2 a_1 = 0$, then $m_2 a_1 \in X$ and we can set $a = a_1$. We assume that $m_2 a_1 \neq 0$. Let x and m denote the nonzero elements $m_1 a_1 \in X$ and $m_2 a_1 \in M$, respectively. By assumption, there exists an element $b \in R$ such that $xb \neq 0$ and $mb \in X$. We set $a = a_1 b$. Then $m_1 a = m_1 a_1 b = xb \neq 0$ and $m_2 a = m_2 a_1 b = mb \in X$. $\qquad \square$

Lemma 17.8 *Let X be a right module over the ring R, M be an injective hull of X, and let $X = Y \oplus Z$, where the module Y is nonsingular and the module Z is injective. Then M is a rational extension of X.*

Proof Now let $0 \neq m = y' + z_1 \in M$ and let $0 \neq x = y + z_2 \in X$, where $y' \in Y'$, $y \in Y$ and $z_1, z_2 \in Z$. Then $y' \neq 0$ and $y \neq 0$. By Lemma 1.20(2), $(y' : Y)$ is an essential right ideal of the ring R. Since Y is nonsingular, $ya \neq 0$ for some element $a \in (y' : Y)$, and $y'(y' : Y) \in Y$. Then $xa \neq 0$ and $ma \in X$. By Lemma 17.7, M is a rational extension of X. $\qquad \square$

A module M is said to be *locally noetherian* if every cyclic submodule of M is noetherian. An endomorphism f of the module X is said to be *locally nilpotent* if $X = \bigcup_{n=1}^{\infty} \text{Ker}(f^n)$, that is, if for every cyclic (finitely generated) submodule Y of X, there exists an integer $n \in \mathbb{N}$ with $f^n(Y) = 0$.

Lemma 17.9 *Let X be a locally noetherian module and f be an endomorphism of the module M. If $\text{Ker}(f)$ is an essential submodule of X, then f is locally nilpotent.*

Proof Let Y be a cyclic submodule of X and let $Y_i = Y \cap \text{Ker}(f^i)$. Since $Y_i \subseteq Y_{i+1}$ and Y is a noetherian module by assumption, we have $Y_{n+1} = Y_n$ for some $n \in \mathbb{N}$. Let $x \in \text{Ker}(f) \cap f^n(Y)$. Then $x = f^n(y)$ for some $y \in Y$. Thus $0 = f(x) = f^{n+1}(y)$. So $y \in Y_{n+1} = Y_n$. Hence $x = 0$ and consequently, $f^n(Y) = 0$. Therefore f is locally nilpotent. $\qquad \square$

Lemma 17.10 *Let X be a skew-injective module and M be an injective hull of X.*

1. *If $\overline{X}$ is an essential extension of X and f is an endomorphism of $\overline{X}$ with $f(X \cap f^{-1}(X)) \subseteq X \cap f^{-1}(X)$, then $f(X) \subseteq X$. In particular, if f is an endomorphism of M with $f(X \cap f^{-1}(X)) \subseteq X \cap f^{-1}(X)$, then $f(X) \subseteq X$.*
2. *If $X = Y \oplus Z$, where the module Y is nonsingular and the module Z is injective, then X is completely integrally closed and M is a rational extension of X.*

3. *If for every $x \in X$ and each $g \in \Delta(X)$, there exists an integer $n \in \mathbb{N}$ with $g^{n+1}(x) \in \sum_{i=0}^{n} g^i(x)R$, then X is a completely integrally closed module.*

4. *If every endomorphism $g \in \Delta(X)$ is locally nilpotent, then X is completely integrally closed.*

5. *If $\Delta(X)$ is a nil ideal in $\mathrm{End}(X)$, then X is completely integrally closed.*

6. *If X is a locally Noetherian module, then X is completely integrally closed.*

Proof

(1) Since $f(X \cap f^{-1}(X)) \subseteq X \cap f^{-1}(X)$ and X is skew-injective, $(f - g)(X \cap f^{-1}(X)) = 0$ for some endomorphism g of X. Now (1) follows from Lemma 1.19(5).

(2) There exists a decomposition $M = Y' \oplus Z$, where Y' is the injective hull of the nonsingular module Y. By Lemma 17.8, M is a rational extension of X. Let $h \in \Delta(M) = \Delta(Y' \oplus Z)$. By Lemma 1.20(3), $Z(Y') = 0$. Let π be the projection of $M = Y' \oplus Z$ onto Y' with kernel Z. By Lemma 1.20(2), $\pi h(X) \subseteq Z(Y') = 0$. Since $h(X) \subseteq M = Y' \oplus Z$, we have $h(X) \subseteq Z$. By Proposition 17.6, X is a completely integrally closed module.

(3) By Proposition 17.6, it is sufficient to prove that $h(X) \subseteq X$ for every $h \in \Delta(M)$. Let $P = X \cap h^{-1}(X)$ and $Y = \{x \in X \mid h^n(x) \in X \text{ for all } n \in \mathbb{N}\}$. Since $Y \supseteq X \cap \mathrm{Ker}(h)$, we have that M and X are essential extensions of Y. In addition, Y is the largest submodule of X with $h(Y) \subseteq Y$. Since X is skew-injective and $h(Y) \subseteq Y$, we have that $(h - g)(Y) = 0$ for some $g \in \mathrm{End}(X)$. Since $g(X \cap \mathrm{Ker}(h)) = 0$, we have $g \in \Delta(X)$. We set $\overline{X} = X/Y$. If $\overline{X} = 0$, then $X = Y$, $h(X) \subseteq X$, and the assertion has been proved in this case.

     Now we assume that $\overline{X} \neq 0$. Then $(h - g)(Y) = 0$ and the homomorphism $t \colon \overline{X} \to M$ is well defined by the relation $t(x + Y) = (h - g)(x)$. We set $\overline{Z} = t^{-1}(Y) \subseteq \overline{X}$. Let Z be the submodule of X such that $Z \supseteq Y$ and $Z/Y = \overline{Z}$. As $Y \subseteq_e X$, by Lemma 1.20(1), $\overline{Z} = t^{-1}(Y) \subseteq_e \overline{X}$. Therefore $\overline{Z} \neq 0$. In addition, $h(Z) \subseteq t(Z) + g(Z) \subseteq Y + g(Z) \subseteq X$. Since $Z \supseteq Y$, we have that X is an essential extension of Z.

     Let $x \in X \setminus Y$, $0 \neq \overline{x} = x + Y \in \overline{X}$. By the assumption, $g^{n+1}(x) \in \sum_{i=0}^{n} g^i(x)R$ for some n. Since $(h^i - g^i)(Y) = 0$ for all i, the homomorphisms $t_i \colon \overline{x}R \to M$ are well defined by the relations $t_i(\overline{x}a) = (h^i - g^i)(xa)$, $a \in R$. Let $\overline{Z}_i = t_i^{-1}(Y) \subseteq \overline{x}R$ and $\overline{W} = \overline{Z}_1 \cap \cdots , \cap \overline{Z}_n \subseteq \overline{x}R$. Since M is an essential extension of X, it follows from Lemma 1.20(1) that $\overline{x}R$ is an essential extension of each of the modules $\overline{Z}_i$. Therefore $\overline{x}R$ is an essential extension of $\overline{W}$. Then $0 \neq \overline{x}a \in \overline{W}$ for some $a \in R$. Therefore $h^i(xa) = t_i(xa) + g^i(xa) \in X$ for all $i = 1, \ldots, n$ and $h^{n+1}(xa) \in \sum_{i=0}^{n} h^i(xa)R \in X$. Then $h^{n+j}(xa) \in X$ for all $j \geq 1$. Therefore $h^k(xa) \in X$ for all $k \geq 1$, $xa \in Y$, $\overline{x}a = 0$. This is a contradiction.

(4) and (5). These statements follow from (3).

(6) This follows from (4) and Lemma 17.9. $\square$

Proposition 17.11 *If X is a skew-injective module which is a finite direct sum of completely integrally closed modules, then X is a completely integrally closed module.*

Proof By Proposition 17.6, it is sufficient to prove that $h(X) \subseteq X$ for any $h \in \Delta(\overline{X})$. It follows from induction that we can assume that $X = X_1 \oplus X_2$, where the modules X_1 and X_2 are completely integrally closed. By part (1) above, X is π-injective. Let $\overline{X}$ be the injective hull of X. Then $\overline{X} = \overline{X}_1 \oplus \overline{X}_2$, where $\overline{X}_1$ and $\overline{X}_2$ are the injective hulls of X_1 and X_2, respectively. Let $h \in \Delta(\overline{X})$, $u_1 \colon \overline{X}_1 \to \overline{X}$ and $u_2 \colon \overline{X}_2 \to \overline{X}$ be the natural embeddings, and let $\pi_1 \colon \overline{X} \to \overline{X}_1$ and $\pi_2 \colon \overline{X} \to \overline{X}_2$ be the natural projections. We set $h_{11} = \pi_1 h u_1 \in \mathrm{End}(\overline{X}_1)$, $h_{22} = \pi_2 h u_2 \in \mathrm{End}(\overline{X}_2)$, $h_{12} = \pi_2 h u_1 \in \mathrm{Hom}(\overline{X}_1, \overline{X}_2)$, and $h_{21} = \pi_1 h u_2 \in \mathrm{Hom}(\overline{X}_2, \overline{X}_1)$. Since X_1 and X_2 are completely integrally closed, it follows from Lemma 17.3(4) that $h_{11}(X_1) \subseteq X_1$ and $h_{22}(X_2) \subseteq X_2$. In addition, it follows from Lemma 1.19(9) that $h_{21}(X_2) \subseteq X_1$ and $h_{12}(X_1) \subseteq X_2$. Therefore

$$h(X) = h(X_1 \oplus X_2) = h_{11}(X_1) + h_{12}(X_1) + h_{22}(X_2) + h_{21}(X_2) \subseteq X. \qquad \square$$

Proposition 17.12 *For a nonsingular module X, the following conditions are equivalent:*

1. *X is a completely integrally closed module.*
2. *X is a skew-injective module.*

Proof The implication $(1) \Rightarrow (2)$ follows from Proposition 17.6.

$(2) \Rightarrow (1)$. Let M be an injective hull of X. By Proposition 17.6, it is sufficient to prove that $h(X) \subseteq X$ for every endomorphism h of M such that $\mathrm{Ker}(h)$ is an essential submodule of M. This is trivially true, since $h = 0$ by Lemma 1.20(3). $\qquad \square$

Lemma 17.13 ([75](19.34, 19.36)) *Let R be a ring, Q be the maximal right ring of quotients of R, and let M be the injective hull of the module R_R. If M is a rational extension of R_R, then Q_R is an injective hull of R_R, the ring Q is right self-injective and the mapping $\varphi \colon \mathrm{End}(Q_R) \to Q$ with $\varphi(f) = f(1)$ is a ring isomorphism.*

Proposition 17.14 *Let R be a ring and let Q be the maximal right ring of quotients of R. The following conditions are equivalent:*

1. *R_R is a completely integrally closed module.*
2. *R_R is a skew-injective module and $R_R = Y \oplus Z$, where the module Y is nonsingular and the module Z is quasi-injective.*
3. *R_R is completely integrally closed, Q_R is an injective hull of R_R, the ring Q is right self-injective, R is a right completely integrally closed subring in Q, and $R_R = Y \oplus Z$, where Y is nonsingular and Z is injective.*
4. *R is a right completely integrally closed subring of Q and Q_R is injective.*

Proof Let M be an injective hull of the module R_R.

The implications $(3) \Rightarrow (4)$ and $(3) \Rightarrow (1)$ are obvious.

$(1) \Rightarrow (2)$. By Proposition 17.6, R_R is a skew-injective module and $R_R = Y \oplus Z$, where $\Delta(Y) = 0$, the module Z is quasi-injective and Z is an essential extension of the module $\sum_{h \in \Delta(M)} h(R)$. Let $y \in Z(Y)$ and f be the endomorphism of the module R such that $f(a) = ya$ for all $a \in R$. Then f can be extended to an endomorphism h of the injective module M. Since $\mathrm{Ker}(f) = ann_r(x)$ is an essential right ideal in R, we have that $f \in \Delta(R_R)$ and $h \in \Delta(M)$. Then

$$ y \in f(R) = h(R) \subseteq \sum_{h \in \Delta(M)} h(R) \subseteq Z \cap Y = 0. $$

Therefore the module Y is nonsingular.

$(2) \Rightarrow (3)$. By Lemma 1.19(9), Z is Y-injective. In addition, Z is Z-injective. Thus Z is R_R-injective, that is, Z is injective. By Lemma 17.10(2), R_R is a completely integrally closed module and M is a rational extension of the module R_R. By Lemma 17.13, Q_R is a injective hull of R_R, the ring Q is right self-injective and the mapping $\varphi \colon \mathrm{End}(Q_R) \to Q$ with $\varphi(f) = f(1)$ is a ring isomorphism.

Let q be an element of the ring Q such that $qB \subseteq B$ for some essential right ideal B of R. Let f denote the endomorphism of Q_R given by $f(x) = qx$ for all $x \in Q$. Then f is an R_R-essential endomorphism of the injective hull Q_R of the completely integrally closed of the module R_R. By Proposition 17.6, $f(R) \subseteq R$. Therefore $q = f(1) \in R$ and R is a right completely integrally closed subring of Q.

$(4) \Rightarrow (1)$. The rational extension Q_R of R_R is an essential extension of R_R. In addition, Q_R is injective by assumption. Therefore Q_R is an injective hull of the module R_R. By Proposition 17.6, it is sufficient to prove that $\alpha(R) \subseteq R$ for every R_R-essential endomorphism α of Q_R. Since α is an R_R-essential endomorphism, $\alpha(B) \subseteq B$ for some essential right ideal B of the ring R. We set $q = \alpha(1) \in Q$. By Lemma 17.13, the mapping $\varphi \colon \mathrm{End}(Q_R) \to Q$ with $\varphi(f) = f(1)$ is a ring isomorphism. Since $qB \subseteq B$ and R is a right completely integrally closed subring of Q, we have $q \in R$. Therefore $\alpha(R) \subseteq R$. $\square$

Theorem 17.15 ([230]) *For a ring R with maximal right ring of quotients Q, the following conditions are equivalent:*

1. *R is a completely integrally closed right R-module.*
2. *R is a right completely integrally closed subring of Q and the module Q_R is injective.*

This follows from Proposition 17.14.

Corollary 17.16 ([230]) *If R is a right nonsingular ring and Q is the maximal right ring of quotients of R, then the following conditions are equivalent:*

1. *R_R is a completely integrally closed module.*
2. *R is a right completely integrally closed subring of Q.*
3. *R_R is a skew-injective module.*

Proof Since the ring R is right nonsingular, Q_R is an injective hull of R_R [75, 19.35]. Therefore the equivalence $(1) \Leftrightarrow (2)$ follows from Proposition 17.14. The equivalence $(1) \Leftrightarrow (3)$ follows from Proposition 17.12. $\qquad \square$

Remark 17.17 ([161](11.13, 13.15)) *A ring R is a semiprime right Goldie ring $\Leftrightarrow R$ is a right order in a semisimple artinian ring $Q \Leftrightarrow R$ is a semiprime right nonsingular right finite-dimensional ring $\Leftrightarrow$ in R, the set of all essential right ideals coincides with the set of all right ideals containing nonzero-divisors. In this case, Q is the maximal right ring of quotients of the ring R.*

Proposition 17.18 *Let R be a semiprime right Goldie ring and let Q be the classical right ring of quotients of R. The following conditions are equivalent:*

1. *R_R is a completely integrally closed module;*
2. *R is a right completely integrally closed subring of Q;*
3. *R_R is a skew-injective module;*
4. *R is a right classically completely integrally closed subring of Q.*

Proof The equivalence of conditions (1), (2), and (3) follows from Corollary 17.16 and Remark 17.17.

$(2) \Rightarrow (4)$. Let $q \in Q$ and let a be a non-zero-divisor in R such that $q^n a \in R$ for all $n \in \mathbb{N}$. Let B denote the right ideal $\sum_{n=0}^{\infty} q^n aR$. Then B contains a non-zero-divisor a so, by Remark 17.17, B is an essential right ideal. Since $qB \subseteq B$ and R is a right completely integrally closed subring of Q, we have $q \in R$.

$(4) \Rightarrow (2)$. Let B be an essential right ideal of the ring R, $q \in Q$, and let $qB \subseteq B$. By Remark 17.17, B contains some nonzero-divisor a. Then $q^n a \in qB \subseteq B \subseteq R$ for all $n \in \mathbb{N}$. Since R is a right classically completely integrally closed subring of Q, we have $q \in R$. $\qquad \square$

Remark 17.19 *A noetherian module X is skew-injective if and only if X is an integrally closed module.*

The above remark follows from the property that submodules of noetherian modules are finitely generated modules.

Proposition 17.20 *Let R be a right noetherian ring. Then*

1. *If X is a right R-module, then X is a completely integrally closed module $\Leftrightarrow X$ is a skew-injective module.*
2. *R_R is a completely integrally closed module $\Leftrightarrow R_R$ is a skew-injective module $\Leftrightarrow R_R$ is an integrally closed module.*

Proof

(1) The implication $\Rightarrow$ follows from Proposition 17.6. The implication $\Leftarrow$ follows from Lemma 17.10(6) and the property that every right module over a right noetherian ring is a locally noetherian module.

(2) The first equivalence follows from (1). The second equivalence follows from Remark 17.19. $\square$

Theorem 17.21 ([230]) *A right noetherian ring R is an integrally closed right R-module if and only if R is a completely integrally closed right R-module.*

This follows from Proposition 17.20(2).

Theorem 17.22 ([230]) *Let R be a semiprime right noetherian ring and let Q be the classical right ring of quotients of R. The following conditions are equivalent:*

1. *R_R is an integrally closed module.*
2. *R_R is a completely integrally closed module.*
3. *R is a right completely integrally closed subring of Q.*
4. *R is a right classically completely integrally closed subring of Q.*

This follows from Proposition 17.18 and Proposition 17.20(2).

Lemma 17.23 *Let R be a ring and let S be the set of all nonzero-divisors in R. We assume that every endomorphism of any principal right ideal of R can be extended to an endomorphism of R_R.*

1. *If a is a element of R, then $ann_r(a)$ is an ideal of R if and only if $aR \subseteq Ra$. In particular, if a is a left nonzero-divisor, then $aR \subseteq Ra$.*
2. *S is a left Ore set, R has a classical left ring of quotients Q and $sRs^{-1} \subseteq R$ for every $s \in S$. If R has a classical right ring of quotients, then Q is a two-sided classical ring of quotients of the ring R.*
3. *If $Rs \subseteq sR$ for every $s \in S$, then R has a classical two-sided ring of quotients Q and $R = sRs^{-1} = s^{-1}Rs$ for every $s \in S$.*

Proof

(1) We assume that $ann_r(a)$ is an ideal of R and $b \in R$. Then $ann_r(a) \subseteq ann_r(ab)$. Therefore $f(a) = ab$ induces an epimorphism $f \colon aR \to abR$. Then $f \in \text{End}(aR)$. By the assumption, f can be extended to some endomorphism f' of the module R_R. We set $c = f'(1)$. Then $ab = f(a) = ca$. Therefore Ra is an ideal.

We now assume that Ra is an ideal of R and $b \in R$. Then $ab = ca$ for some $c \in R$. Therefore $ab \cdot ann_r(a) = ca \cdot ann_r(a) = 0$, so $b \cdot ann_r(a) \subseteq ann_r(a)$ and so $ann_r(a)$ is an ideal.

(2) Let $s \in S$, $a \in R$. By (1), $sa = bs$ for some $b \in R$; therefore S is a left Ore set. Therefore R has a classical left ring of quotients. The relation $sRs^{-1} \subseteq R$ follows from (1). The second assertion follows from the property that if a ring has a classical left ring of quotients Q and a classical right ring of quotients, then Q is a two-sided classical ring of quotients [161, 10.14].

(3) Let $s \in S$. Since $Rs \subseteq sR$, we have that S is a right Ore set. Therefore R has a classical right ring of quotients Q. By (2), R has a classical two-sided ring of quotients Q. By (1), $sR \subseteq Rs$. In addition, $Rs \subseteq sR$ by assumption. Therefore $sR = Rs$, $sRs^{-1} = s^{-1}Rs = R$. $\qquad\square$

Proposition 17.24 *Let R be a right integrally closed ring, S be the set of all nonzero-divisors of R, Q be the classical left ring of quotients of the ring R.*

1. *If $q \in Q$ and $q^{n+1} = q^n a_n + \cdots + q_1 a + a_0$ for some $a_0, \ldots, a_n \in R$, then $q \in s^{-1}As$ for some $s \in S$.*
2. *If $Rs \subseteq sR$ for every $s \in S$, then Q is the classical two-sided ring of quotients of the ring R and R is a right classically integrally closed subring of Q.*

Proof

(1) Let M denote the finitely generated submodule $\sum_{i=0}^{n} q^i R$ of the module Q_R. Then $qM \subseteq M$. Since $\{q, q^2, \ldots, q^n\}$ is a finite subset of Q, there exist elements $s, a_1, \ldots, a_n \in R$ such that s is a nonzero-divisor and $sq^i \in R$ for $i = 1, \ldots, n$ [213, p. 61]. Therefore sM is a finitely generated right ideal of R and $s \in sM$. In addition,

$$qs^{-1} \cdot sM = qM \subseteq M = s^{-1} \cdot sM.$$

Therefore the relation $f(x) = sqs^{-1}x$ defines an endomorphism f of the finitely generated right ideal sM. By the assumption, f can be extended to an endomorphism g of the module R_R. We set $c = g(1)$. Then

$$sq = sqs^{-1} \cdot s = f(s) = cs, \quad q = s^{-1}cs \in s^{-1}Rs.$$

(2) By Lemma 17.23(3), Q is the two-sided classical ring of quotients of R and $R = sRs^{-1} = s^{-1}Rs$ for every $s \in S$. Let $q \in Q$ and $q^{n+1} = q^n a_n + \cdots + q_1 a + a_0$ for some $a_0, \ldots, a_n \in R$. By (1), $q \in s^{-1}Rs = R$ for some non-zero-divisor s in R. Therefore R is a right classically integrally closed subring of Q. $\qquad\square$

The following is wellknown.

Proposition 17.25 *If R is a commutative domain and Q is the field of quotients of R, then R_R is an integrally closed module if and only if R is a right classically integrally closed subring of Q.*

Proof The implication $\Rightarrow$ follows from Proposition 17.24(2). We prove the converse. Let $0 \neq B = \sum_{i=1}^{n} b_i R \subseteq R$ and let $0 \neq f \in \operatorname{End}(B_R)$. Since Q_R is injective, we can assume that f is an endomorphism of Q_R. We set $q = f(1) \in Q$. It is easily verified that $q \neq 0$ and $f(x) = qx$ for all $x \in Q$. Since $qB \subseteq B$, we have $qb_i = \sum_{j=1}^{n} b_j a_{ji}$, where $a_{ji} \in R$ and $1 \leq i, j \leq n$. Then $\{(q - a_{ii})b_i = \sum_{i \neq j} b_j a_{ji} = 0\}_{i,j=1}^{n}$ and the equation system $\{(q - a_{ii})x_i =$

$\sum_{i \neq j} x_j a_{ji} = 0\}_{i,j=1}^n$ has the nonzero solution $b_1, \ldots, b_n$. Therefore the determinant $D(q)$ of this system is zero, it is a polynomial in variable q with coefficients in R, and the leading coefficient of this polynomial is 1. Since R is classically integrally closed in Q, we have $q \in R$. Therefore the relation $h(a) = qa$ defines an endomorphism h of R_R which coincides with f on B. $\qquad\qquad\square$

17.26 Questions

1. Is each skew-injective module completely integrally closed?
2. Is each (right) skew-injective ring (right) completely integrally closed?
3. Is it true that each skew-injective module with essential socle is quasi-injective?
4. Which group rings (monoid rings) are right skew-injective (or right completely integrally closed)?

18 Rings each of whose cyclic modules is completely integrally closed

Over the ring of integers $\mathbb{Z}$, all cyclic modules are completely integrally closed, but the completely integrally closed cyclic $\mathbb{Z}$-module $\mathbb{Z}$ is not quasi-injective.

The main result of this chapter is that due to Tuganbaev that describes rings over which all (right or left) cyclic modules are completely integrally closed. For example, all commutative Dedekind domains and all artinian, principal one-sided ideal rings have the property that all cyclic modules over them are completely integrally closed. In particular, the ring of integers $\mathbb{Z}$ and each residue ring $\mathbb{Z}/n\mathbb{Z}$ have the property that all cyclic modules over them are completely integrally closed.

Lemma 18.1 *For a ring R, the following conditions are equivalent:*

1. *For any two elements $x, y \in R$ with $xy = 0$, we have $xRy = 0$.*
2. *All right annihilator ideals of R are two-sided ideals.*
3. *All left annihilator ideals of R are two-sided ideals.*

Lemma 18.2 *For a ring R, the following conditions are equivalent:*

1. *R is a left duo ring.*
2. *All right annihilator ideals of R are two-sided ideals and each endomorphism of each principal right ideal xR can be extended to an endomorphism of R_R.*

Consequently, if R is a right and left skew-injective ring and $xRy = 0$ for any two elements $x, y \in R$ with $xy = 0$, then R is a right and left duo ring.

Proof $(1) \Rightarrow (2)$. By Lemma 18.1, $ann_r(x)$ is an ideal of R. Let f be an endomorphism of the module xR_R. Then $f(x) = xb$ for some $b \in R$. Therefore $f(xc) = xbc$ for any element $c \in R$. Since R is left duo, $xb = dx$ for some element $d \in R$. Let g be the endomorphism of R_R given by $g(a) = da$. For each $c \in R$, we have $g(xc) = dxc = xbc = f(xc)$. Therefore g coincides with f at xR.

$(2) \Rightarrow (1)$. Let x and a be two elements of R. Since $ann_r(x)$ is an ideal of R, we have $ann_r(x) \subseteq ann_r(xa)$. Therefore there exists a homomorphism $f: xR \to xaR \subseteq xR$ with $f(x) = xa$. By assumption, f can be extended to an endomorphism g of R_R. We set $b = g(1)$. Then $xa = g(1 \cdot x) = g(1)x = bx$ and so R is left duo. $\qquad\square$

Lemma 18.3 *Let X be a skew-injective module, $\overline{X}$ be an injective hull of X, and let $\Delta(X)$ be the ideal of the ring $\mathrm{End}(X)$ consisting of all endomorphisms whose kernels are essential submodules in X.*

1. *If $\alpha(Z) \subseteq Z$ for any $\alpha \in \Delta(X)$ and every essential submodule Z in X, then X is a completely integrally closed module.*

2. *If each essential submodule Y in X is fully invariant in X, then X is a completely integrally closed module. In particular, if X is a uniserial module, then X is a completely integrally closed module.*

3. *If $\Delta(X) = 0$, then X is a completely integrally closed module.*

4. *If the module X is nonsingular, then X is a completely integrally closed module.*

5. *If X is a finite direct sum of uniserial modules, then X is a completely integrally closed module.*

6. *If X is a uniserial module and α is an automorphism of the module $\overline{X}$ with $\alpha(X) \not\subseteq X$, then $\alpha^{-1}(X) \subseteq X$.*

7. *If X is a uniserial module and $\Delta(X) \subseteq J(\mathrm{End}(X))$, then X is a completely integrally closed module.*

Proof

(1) Let $h \in \Delta(\overline{X})$. By Proposition 17.6, it suffices to prove that $h(X) \subseteq X$. Let Y denote the sum of all submodules Y' in X such that $h(Y') \subseteq Y'$. Since $X \cap \mathrm{Ker}(h) \subseteq Y$, we have that Y is an essential submodule of X. Since $h(Y) \subseteq Y$ and the module X is skew-injective, $(h - g)(Y) = 0$ for some endomorphism g of X. Since $g(X \cap \mathrm{Ker}(h)) = 0$, we have $g \in \Delta(X)$. Set $Z = \{x \in X \mid (h - g)(x) \in Y\}$ and $Z^* = \{x \in X \mid h(x) \in X\}$. Since $h(Z) \subseteq (h - g)(Z) + g(Z) \subseteq X$, we have $Z \subseteq Z^* \subseteq X$. In addition, $Y \subseteq Z$, Z is an essential submodule of X and $(h - g)(Z) \subseteq Y \subseteq Z$. This induces a homomorphism $\beta^* \colon Z^*/Y \to X$ with $\beta^*(z^* + Y) = (h - g)(z^*) \in X$. Then $Z/Y = \{z^* + Y \in Z^*/Y \mid \beta^*(z^* + Y) \in Y\}$. Since Y is an essential submodule of X, we have that Z/Y is an essential submodule of Z^*/Y. Since the module X is skew-injective and $(h - g)(Z) \subseteq Z$, we have $(h - g - f)(Z) = 0$ for some endomorphism f of the module X. Since $f(Y) = (h - g)(Y) = 0$, we have $f \in \Delta(X)$. Then $f + g \in \Delta(X)$. By assumption, $(g + f)(Z) \subseteq Z$. In addition, $(h - g - f)(Z) = 0$. Therefore $h(Z) = (g + f)(Z) \subseteq Z$. It follows from the definition of Y that $Z \subseteq Y \subseteq Z$ and $Z = Y$. In addition, Z/Y is an essential submodule of Z^*/Y. Therefore $Z^*/Y = 0$ and so $Z^* = Y$. By Lemma 1.19(5), $h(X) \subseteq X$.

(2) and (3). These statements follow from (1).

(4) This follows from (3) and the property that $\Delta(X) = 0$ for the nonsingular module X.

(5) This follows from (2) and Proposition 17.11(2).

(6) We set $N = \{x \in X \mid \alpha^{-1}(x) \in X\}$. Then $\alpha^{-1}(N) = \{x \in X \mid \alpha(x) \in X\}$. By Lemma 17.3(1) applied to α^{-1}, it is sufficient to prove that $\alpha^{-1}(N) \subseteq N$. We assume that $\alpha^{-1}(N) \nsubseteq N$. Since X is a uniserial module, $N \subseteq \alpha^{-1}(N)$. Then $\alpha(\alpha^{-1}(N)) \subseteq \alpha^{-1}(N)$. Since $\alpha^{-1}(N) = \{x \in X \mid \alpha(x) \in X\}$, it follows from Lemma 17.3(1) applied to α, that $\alpha(X) \subseteq X$. This is a contradiction.

(7) Let $h \in \Delta(\overline{X})$ and let $K = X \cap \mathrm{Ker}(h)$. By Proposition 17.6, it is sufficient to prove that $h(X) \subseteq X$. We have $\Delta(\overline{X}) = J(\mathrm{End}(\overline{X}))$. Therefore $1 - h$ is an automorphism of $\overline{X}$ which coincides with the identity mapping at the essential submodule K of X. Then the automorphism $(1 - h)^{-1}$ of $\overline{X}$ coincides with the identity mapping at the essential submodule K of X. If $(1 - h)(X) \subseteq X$, then $h(X) \subseteq X + (1 - h)(X) \subseteq X$.
Assume $(1 - h)(X) \nsubseteq X$. By (6), $(1 - h)^{-1}(X) \subseteq X$ and so there is an endomorphism f of X with $f(x) = (1 - h)^{-1}(x)$ for all $x \in X$. In addition, $(1 - f)(K) = 0$, since $(1 - h)^{-1}$ coincides with the identity mapping at K. Therefore $1 - f \in \Delta(X)$. In addition, $\Delta(X) \subseteq J(\mathrm{End}(X))$ by assumption. Therefore $f = 1 - (1 - f)$ is an automorphism of X. Since $f(x) = (1 - h)^{-1}(x)$ for all $x \in X$, we have $(1 - h)^{-1}(X) = X$. Therefore $X = (1 - h)(X)$. This is a contradiction, since $(1 - h)(X) \nsubseteq X$. $\qquad\square$

Proposition 18.4 *For a ring R, the following conditions are equivalent:*

1. *R is a right nonsingular ring without nontrivial idempotents and R_R is a completely integrally closed module.*

2. *R is a right nonsingular ring without nontrivial idempotents and R_R is a skew-injective module.*

3. *R is a right uniform domain and R_R is a completely integrally closed module.*

4. *R is a left duo domain, R has a classical two-sided division ring of fractions Q, R is a right completely integrally closed subring of Q, and R is a classically right completely integrally closed subring of Q.*

Proof The equivalence $(1) \Leftrightarrow (2)$ and the implication $(4) \Rightarrow (3)$ follow from Corollary 17.16.

The implication $(3) \Rightarrow (1)$ is obvious.

$(1) \Rightarrow (3)$. Since the ring R does not have nontrivial idempotents, R_R is a completely integrally closed indecomposable module. By Proposition 17.11(1), the ring R is right uniform. It is easily seen that the right nonsingular, right uniform ring R is a domain.

$(3) \Rightarrow (4)$. Since R is a right uniform domain, R has a classical right division ring of fractions Q. By Lemma 18.2, the domain R is left duo. Then the domain R is left uniform and R has a classical left division ring of fractions. Therefore the classical right division ring of fractions Q is a two-sided division ring of fractions of the domain R. By Corollary 17.16, R is a right completely integrally closed subring of Q.

Let $0 \neq q \in Q$, $0 \neq a \in R$, and let $q^i a \in R$ for all positive integers i. Let B denote the essential right ideal $\sum_{i=1}^{\infty} q^i a R$ of the uniform domain R. Since $qB \subseteq B$ and R is a right completely integrally closed subring of Q, we have $qR \subseteq R$ and $q \in R$. Thus R is a classically right completely integrally closed subring of Q. $\qquad\square$

Lemma 18.5 *For a module X, the following conditions are equivalent:*

1. *X is a completely integrally closed indecomposable module.*

2. *X is a uniform module and either X is a quasi-injective module with local endomorphism ring $\mathrm{End}(X)$ and $\Delta(X) = J(\mathrm{End}(X))$, or X is a skew-injective module and each nonzero endomorphism of X is a monomorphism.*

Proof Let M be the injective hull of the module X.

$(1) \Rightarrow (2)$. By Proposition 17.11(1), X is a skew-injective uniform module. If each nonzero endomorphism of X is a monomorphism, then there is nothing to prove. We assume that there exists a nonzero endomorphism $g \in \mathrm{End}(X)$ with nonzero kernel. Then g can be extended to a nonzero endomorphism $h \in \Delta(M)$. Let Y denote the nonzero fully invariant submodule $\sum_{f \in \Delta(M)} f(M)$ of the injective uniform module M. By Proposition 17.6, $Y \subseteq X$. By Lemma 1.20(3), X is a fully invariant submodule of M. By Lemma 1.19(4), X is quasi-injective. Since X is an indecomposable quasi-injective module, it follows that $\mathrm{End}(X)$ is a local ring and $\Delta(X) = J(\mathrm{End}(X))$.

$(2) \Rightarrow (1)$. Without loss of generality, we can assume that X is a skew-injective module and each nonzero endomorphism of X is a monomorphism. By Lemma 18.3(3), X is a completely integrally closed module. $\qquad\square$

Lemma 18.6 *Let R be a ring and let $N_\ell(R)$ be the set of all left zero-divisors in R. The following conditions are equivalent:*

1. *The ring R does not have nontrivial idempotents and R_R is a completely integrally closed module.*

2. *R is a right uniform ring and either R is a right self-injective, local ring and $J(R) = \Delta(R_R) = Z(R_R) = N_\ell(R)$, or R is a left duo domain and R_R is a skew-injective module.*

Proof The implication $(2) \Rightarrow (1)$ is obvious.

$(1) \Rightarrow (2)$. By Proposition 17.11(1), the ring R is right uniform. Therefore $Z(R_R) = N_\ell(R) = \Delta(R_R)$. The remaining assertions follow from Lemma 18.5 and Proposition 18.4. $\qquad\square$

Lemma 18.7 *Let R be a domain, a be a noninvertible element of R, and let the module R_R be skew-injective. Then*

1. *$\cap_{i=1}^{\infty} a^i R = 0$.*

2. *If the principal right ideal aR properly contains some prime ideal P of the domain R, then $P = 0$.*

Proof

(1) We can assume that $a \neq 0$. By Proposition 18.4, the domain R has a classical division ring of fractions Q which contains the element a^{-1}. Let X denote the right ideal $\cap_{i=1}^{\infty} a^i R$ of R. Consider the endomorphism f of X_R given by $f(x) = a^{-1}x$ for all $x \in X = \cap_{i=1}^{\infty} a^i R$. By assumption, f can be extended to some endomorphism g of R_R. We set $b = g(1) \in R$. Assume that $X \neq 0$. Choose a nonzero element $x \in X$. Then $a^{-1}x = f(x) = g(1 \cdot x) = bx$ and so $a^{-1} = b \in R$, that is, a is invertible in R. This contradiction gives $X = 0$.

(2) Let $h \colon R \to R/P$ be the natural ring epimorphism. Since $P \subsetneq aR$, we have that $h(a) \neq h(0)$, $P = aY \subseteq Y$ for some ideal Y of the domain R, and $h(a)h(Y) = h(0)$. By Lemma 18.2, R is left duo. Then $h(R)$ is a left duo, prime ring and so a domain. Then, since $h(a)h(Y) = h(0)$, we have $h(Y) = h(0)$ and so $Y \subseteq P \subseteq Y$ and $P = Y$. Therefore $P = aP$ and $P \subseteq \cap_{i=1}^{\infty} a^i R$. By (1), $P = 0$. $\qquad\square$

Lemma 18.8 *Let R be a right uniserial, right skew-injective domain. Then*

1. *R has at most two prime ideals.*
2. *R is a duo uniserial domain.*

Proof By Lemma 18.2, the domain R is left duo.

(1) We assume the contrary. Then the left duo, right uniserial domain R has a nonzero prime ideal P properly contained in $J(R)$. Let $a \in J(R) \setminus P$. This contradicts Lemma 18.7(2).

(2) Since R is a left duo, right uniserial domain, R is a left uniserial domain. It remains to prove that the domain R is right duo. It is sufficient to prove that $ba \in aR$ for any two nonzero elements $a \in J(R)$, $b \in R$. We assume that $ba \notin aR$. Since R is a right uniserial domain, $aR \subseteq baR$. Then $a = bac$ for some nonzero element $c \in R$, and the element c is not invertible in R, since $ba \notin aR$. Therefore $c \in J(R)$ and $a = b^i a c^i$ for all positive integers i. Then $0 \neq a \in \cap_{i=1}^{\infty} Rc^i$.

Since R is a domain and $0 \neq a \in J(R)$, we have $a \notin Ra^2$. Therefore the right ideal aR is not contained in $Ra^2 = Ra^2 R$. Let $h \colon R \to R/Ra^2$ be the natural ring epimorphism. Since $h(J(R))$ is the unique prime ideal of the ring $h(R)$, it is the prime radical of $h(R)$. Therefore $h(c)$ is a nilpotent element of $h(R)$. Then $c^n = xa^2$ for some $n \in \mathbb{N}$ and $x \in R$. Since $0 \neq a \in \cap_{i=1}^{\infty} Rc^i$, we have $0 \neq a \in Rxa^2$. Therefore $1 \in Ra$; this is a contradiction. $\qquad\square$

Lemma 18.9 *Let R be a duo uniserial domain and let Q be the classical division ring of fractions of R. If $q \in Q \setminus R$, then $q^{-1} \in R$. In addition, the following conditions are equivalent:*

1. *R is a classically right completely integrally closed subring of Q.*
2. *The module R_R is skew-injective.*

3. $\cap_{i=1}^{\infty} a^i R = 0$ *for any noninvertible element* a *of the domain* R.

4. R *has at most two prime ideals.*

Proof Let $q = ab^{-1} \in Q \setminus R$, where a and b are nonzero elements of the uniserial domain R. Since $ab^{-1} \notin R$, we have $a \notin Rb$. Therefore $b \in Ra$ and $q^{-1} = ba^{-1} \in R$. The implication $(1) \Rightarrow (2)$ follows from Proposition 18.4.

The implication $(2) \Rightarrow (3)$ follows from Lemma 18.7(1).

$(3) \Rightarrow (4)$. We assume the contrary. Then the duo uniserial domain R has a nonzero prime ideal X properly contained in $J(R)$. Since R is a duo ring, X is completely prime. Let $a \in J(R) \setminus X$. Since R is duo uniserial and X does not contain the ideal aR, the principal ideal aR properly contains the completely prime ideal X. Therefore $X = aX$. Then $X = a^i X$ for all i, giving $X \subseteq \cap_{i=1}^{\infty} a^i R = 0$ and so $X = 0$. This is a contradiction.

$(4) \Rightarrow (3)$. Let a be a non-invertible element of R, $X = \cap_{i=1}^{\infty} a^i R$, and let $h \colon R \to R/X$ be the natural epimorphism. Then a is an element of the prime ideal $J(R)$ of the domain R. We assume that $X \neq 0$. Since 0 is a prime ideal and R has at most two prime ideals, the ring $h(R)$ has exactly one prime ideal $h(J(R))$, which is the prime radical of the ring $h(R)$. Since the prime radical is a nil ideal, $h(a)$ is nilpotent. Therefore $a^n \in X = \cap_{i=1}^{\infty} a^i R$ for some positive integer n. Then $a^n = a^{n+1} b$ for some element b of R. Therefore $1 = ab$ and a is invertible. This is a contradiction.

$(3) \Rightarrow (1)$. Let $0 \neq q \in Q$, $0 \neq a \in R$, and let $q^i a \in R$ for all positive integers i. We assume that $q \notin R$. Then $q^{-1} = b$ is a noninvertible element of R. Let Y denote the nonzero ideal $\sum_{i=1}^{\infty} q^i a$ of R. Since $qY \subseteq Y$ and $q^{-1} = b \in R$, we have $Y \subseteq bY \subseteq Y$, $Y = bY$, and $Y = b^i Y$ for all i. Therefore $Y \subseteq \cap_{i=1}^{\infty} b^i R = 0$. This is a contradiction. $\qquad \square$

Lemma 18.10 *For a ring R, the following conditions are equivalent:*

1. R *is a right uniserial domain, and* R_R *is a completely integrally closed module.*

2. R *is a left uniserial domain, and* $_R R$ *is a completely integrally closed module.*

3. R *is a right uniserial domain, and the module* R_R *is skew-injective.*

4. R *is a left uniserial domain, and the module* $_R R$ *is skew-injective.*

5. R *is a duo uniserial domain with at most two prime ideals.*

Proof The equivalences $(1) \Leftrightarrow (3)$ and $(2) \Leftrightarrow (4)$ follow from Proposition 18.4.

The implication $(3) \Rightarrow (5)$ follows from Lemma 18.8.

The implication $(5) \Rightarrow (4)$ follows from Lemma 18.9.

$(4) \Rightarrow (3)$. Let a and b be two nonzero elements of the domain R. By Proposition 17.11(1), the domain R is right uniform; therefore $ac = bd \neq 0$ for some nonzero elements $c, d \in R$. Since R is a left uniserial domain, either $c = xd$ or $d = yc$ for some $x, y \in R$. Therefore either $axd = bd$ or $ac = byc$. Since R is a domain, either $b = ax$ or $a = by$. Therefore R is a right uniserial domain. $\qquad \square$

Lemma 18.11 *For a ring R, the following conditions are equivalent:*

1. *R is a right uniserial, right skew-injective ring.*
2. *Either R is a duo uniserial domain with at most two prime ideals, or R is a right self-injective, right and left uniserial ring.*

Proof The implication $(2) \Rightarrow (1)$ follows from Lemma 18.10.

$(1) \Rightarrow (2)$. If R is a domain, then it follows from Lemma 18.10 that R is a duo uniserial domain with at most two prime ideals. We assume that R is not a domain. Since R is a right uniserial ring and $R = \mathrm{End}(R_R)$, we have $\Delta(R_R) = J(R)$. By Lemma 18.3(7), R_R is a completely integrally closed module. By Lemma 18.6, R is a right self-injective ring.

Let $x, y \in R$. Since R is a right uniserial ring, we have that either $ann_r(y) \subseteq ann_r(x)$ or $ann_r(x) \subseteq ann_r(y)$. Therefore either there exists a homomorphism $f\colon yR \to xR$ with $f(y) = x$, or there exists a homomorphism $g\colon xR \to yR$ with $g(x) = y$. In addition, the ring R is right self-injective. Thus there is an element $a \in R$ with either $ay = x$ or $ax = y$. Therefore R is a left uniserial ring. $\qquad\square$

Lemma 18.12 *For a module M, the following conditions are equivalent:*

1. *M is skew-projective and all factor modules of M are skew-injective.*
2. *M is skew-injective and all submodules of M are skew-projective.*

Proof $(1) \Rightarrow (2)$. Let N be a submodule of M, $\overline{f}$ be an endomorphism of the factor module $\overline{N} = N/P$, and let $h\colon M \to M/P$ be the natural epimorphism. Since M/P is skew-injective, $\overline{f}$ can be extended to an endomorphism $\overline{g}$ of M/P. Since M is skew-projective, $\overline{g}h = hg$ for some endomorphism g of M. Therefore $g(N) \subseteq N$. Then g induces an endomorphism f of N and $\overline{f}h_N = h_N f$, where $h_N\colon N \to N/P$ is the natural epimorphism. Thus N is skew-projective.

$(2) \Rightarrow (1)$. Let M/N be a factor module of M, $\overline{X}$ be a submodule of M/N, and let $\overline{f}$ be an endomorphism of $\overline{X}$. There exists a submodule X of M such that $N \subseteq X$ and $\overline{X} = X/N$. Let $h\colon X \to X/N$ be the natural epimorphism. By assumption, the module X is skew-projective. Therefore $\overline{f}h = hf$ for some endomorphism f of X with $f(N) \subseteq N$. Since M is skew-injective, f can be extended to some endomorphism g of M. Since $g(N) = f(N) \subseteq N$, we have that g induces an endomorphism $\overline{g}$ of M/N. Since $\overline{g}$ coincides with $\overline{f}$ at $\overline{X}$, M/N is skew-injective. $\qquad\square$

Proposition 18.13 *For a ring R, the following conditions are equivalent:*

1. *All cyclic right R-modules are skew-injective.*
2. *R is right skew-injective and all right ideals of R are skew-projective right R-modules.*
3. *$R = \prod_{i=1}^{n} R_i$ where, for each $i \in \{1, \ldots, n\}$, R_i is either a simple artinian ring or a right uniform ring over which all cyclic right modules are skew-injective finite-dimensional modules.*

4. $R = \prod_{i=1}^{n} R_i$, *where, for each* $i \in \{1, \ldots, n\}$, R_i *is either a simple artinian ring or a right skew-injective right uniform ring in which all right ideals are skew-projective.*

Proof The equivalence $(1) \Leftrightarrow (2)$ follows from Lemma 18.12 and the projectivity of R_R.

The implication $(1) \Rightarrow (3)$ follows from Theorem 9.2 and Proposition 17.11(1) which shows that each skew-injective module is π-injective.

The equivalence $(3) \Leftrightarrow (4)$ follows from the equivalence $(1) \Leftrightarrow (2)$.

$(3) \Rightarrow (1)$. Since all modules over a simple artinian ring are injective, all cyclic right R_i-modules are skew-injective, $i = 1, \ldots, n$. Since $R = \prod_{i=1}^{n} R_i$, it is easily checked that all cyclic right R-modules are skew-injective. $\qquad \square$

Lemma 18.14 *For a local domain R, the following conditions are equivalent:*

1. *All cyclic right R-modules are completely integrally closed.*
2. *All cyclic left R-modules are completely integrally closed.*
3. *All cyclic right R-modules are skew-injective.*
4. *All cyclic left R-modules are skew-injective.*
5. *R is a duo uniserial domain and R has at most two prime ideals.*

Proof It is sufficient to prove the equivalence of conditions (1), (3), and (5).

The implication $(1) \Rightarrow (3)$ follows from Proposition 17.11(1).

$(3) \Rightarrow (5)$. By Lemma 9.1(2), R is a right uniserial ring. Therefore (5) follows from Lemma 18.10.

$(5) \Rightarrow (1)$. Let B be an ideal of the duo ring R. If $B = 0$, then it follows from Lemma 18.10 that R/B is a completely integrally closed right module over the ring R/B. If $B \neq 0$, then R/B has exactly one prime ideal, and it follows from Theorem 7.3 that each cyclic right (left) R/B-module is quasi-injective. Hence R/B is a completely integrally closed right module over the ring R/B. $\qquad \square$

Lemma 18.15 ([33], *Lemma 3.5, Theorem 3.6*) *Let R be a uniserial ring such that, for any element $x \in R$, either $xR \subseteq Rx$ or $Rx \subseteq xR$. Then:*

1. *Each prime ideal of R is completely prime.*
2. *If R has the minimum condition on prime ideals, then R is a duo ring.*

Lemma 18.16 *Let R be a local ring such that R is not a domain and all cyclic right R-modules are completely integrally closed. Then:*

1. *R is a right self-injective, (right and left) uniserial ring.*
2. *$Jx \subseteq xR$ for any $x \in R$.*
3. *$xR \subseteq Rx$ or $Rx \subseteq xR$ for any element $x \in R$.*
4. *R is a duo uniserial ring with nonzero prime radical P, R/P is a domain and R has at most two prime ideals.*

Proof

(1) By Lemma 9.1(2), R is a right uniserial ring. By Lemma 18.11, R is a right self-injective, left uniserial ring.

(2) We set $J = J(R)$. The natural isomorphism $R \to \mathrm{End}(R_R)$ maps the ideal J onto the ideal $\Delta(R_R)$. If $0 \neq x \in R$, then the injective uniserial module R_R is the injective hull of the cyclic completely integrally closed module xR, whence $JxR \subseteq xR$ by Proposition 17.6.

(3) We assume that $x \in R$ and $xa \notin Rx$ and $bx \notin xR$ for some $a, b \in R$. By (1), R is a uniserial ring. Therefore $x = cxa = bxd = cbxda$ for some $c, d \in R$. If the element c is invertible, then $xa = c^{-1}cxa = c^{-1}x \in Rx$; this is a contradiction. If the element d is invertible, then $bx = bxd\,d^{-1} = xd^{-1} \in xR$; again a contradiction. Therefore $c, d \in J$ and $x = cbxda \in JxJ$, and $Jx \subseteq xR$. Hence $x \in JxJ \subseteq xJ$ and $x = xy$ for some $y \in J$. Then the element $1 - y$ is invertible and $x(1 - y) = 0$, whence $x = 0$ and $xa \in Rx$; this is a contradiction.

(4) The prime radical P of the uniserial ring R is a prime ideal. By (3) and Lemma 18.15(1), P is completely prime. Then R/P is a uniserial domain, over which all cyclic right modules are completely integrally closed. By Lemma 18.14, R/P is a duo uniserial domain with at most two prime ideals. Since each prime ideal of R contains the prime radical P, R has at most two prime ideals. By Lemma 18.15(2), R is a duo ring. $\qquad\square$

Lemma 18.17 *For a ring R, the following conditions are equivalent:*

1. *R is a local ring and all cyclic right R-modules are completely integrally closed.*

2. *R is a duo uniserial ring with at most two prime ideals, and either R is a domain or, for any nilpotent nonprime ideal B of R, the factor ring R/B is right self-injective.*

Proof (1) $\Rightarrow$ (2). By Lemma 18.14, we can assume that R is not a domain. By Lemma 18.16(4), R is a duo uniserial ring with nonzero prime radical P, R/P is a domain and R has at most two prime ideals. Let B be a nilpotent nonprime ideal of R. Then B is properly contained in P. Therefore R/B is not a domain. Since R/B is a local ring and all cyclic right modules over R/B are completely integrally closed, it follows from Lemma 18.16(4) that the ring R/B is right self-injective.

(2) $\Rightarrow$ (1). Let B be an ideal of the duo ring R. It is sufficient to prove that the factor ring R/B is completely integrally closed right module over the ring R/B. The prime radical P of the duo uniserial ring R is a completely prime ideal. Then R/P is a duo uniserial domain and R/P has at most two prime ideals. If $P \subseteq B$, then it follows from Lemma 18.14 that R/B is a completely integrally closed right module over R/B. We assume that $B \subsetneq P$ and $p \in P \setminus B$. Since R is a duo uniserial ring, pR is a nilpotent ideal and $B \subset pR$. Then B is a nilpotent

nonprime ideal. By assumption, the ring R/B is right self-injective. Therefore R/B is a completely integrally closed right module over R/B. $\square$

Lemma 18.18 *For a ring R, the following conditions are equivalent:*

1. *All cyclic right R-modules are completely integrally closed.*
2. *$R = R_1 \times \cdots \times R_n$ and, for $i \in \{1, \ldots, n\}$, either R_i is a simple artinian ring or R_i is a right uniform ring over which all cyclic right modules are completely integrally closed, finite-dimensional modules.*
3. *$R = R_1 \times \cdots \times R_n$ and, for $i \in \{1, \ldots, n\}$, either (a) R_i is a simple artinian ring, or (b) R_i is a left duo, right uniform domain, over which all cyclic right modules are completely integrally closed, or (c) R_i is a duo uniserial ring with at most two prime ideals and, for any nilpotent nonprime ideal B of R_i, the factor ring R_i/B is right self-injective.*

Proof The implication $(3) \Rightarrow (2)$ follows from Lemma 18.17. The implication $(2) \Rightarrow (1)$ is easy to prove and $(1) \Rightarrow (2)$ follows from Proposition 17.11(1) and Theorem 9.2.

$(2) \Rightarrow (3)$. Without loss of generality, we can assume that $R = R_i$ is a right uniform ring. If R is a domain, then it follows from Lemma 18.2 that the domain R is left duo. It remains to consider the case where the right uniform ring R is not a domain. By Lemma 18.5, R is a right self-injective, local ring. Now we apply Lemma 18.17 to complete the proof. $\square$

Lemma 18.19 *Let R be a ring where all cyclic right R-modules are completely integrally closed. Then:*

1. *If R has an essential nilpotent left ideal, then R is a duo ring and R is a finite direct product of uniserial rings.*
2. *If R is a domain, then R is a left uniserial, right uniform, right completely integrally closed domain, and for any nonzero ideal B of R, the factor ring R/B is a finite direct product of duo uniserial rings.*

Proof

(1) Since R has an essential nilpotent left ideal, R does not have direct factors that are simple artinian rings or domains. Then it follows from Lemma 18.18 that R is a finite direct product of duo uniserial rings. In particular, R is a duo ring.

(2) By Lemma 18.18, R is a left duo, right uniform domain. Let B be a nonzero proper ideal in R containing a nonzero noninvertible element b. Let $h \colon R \to R/(Rb^2)$ be the natural ring epimorphism. Since R is a left duo domain and $Rb \neq R$, we have that Rb^2 is an ideal, $b \notin Rb^2$, and $h(Rb)$ is a nonzero nilpotent ideal of $h(R)$.

  We prove that $h(Rb)$ is an essential left ideal of $h(R)$. Let $a \in R$ with $h(a) \neq 0$. As R is a left duo ring, $h(ab) \in h(Ra)$. Suppose $h(ab) = 0$. Then

$ab = cb^2$ for some $c \in R$. This gives $a = cb$, $h(a) \in h(Rb)$ which proves that $h(Rb)$ is an essential left ideal of $h(R)$.

Since $h(R)$ has an essential nilpotent left ideal, it follows from (1) that $h(R)$ is a finite direct product of duo uniserial rings. Since $Rb^2 \subseteq B$, we have that R/B is a homomorphic image of the ring $h(R)$. Therefore R/B is a finite direct product of duo uniserial rings. $\square$

Lemma 18.20 *If R is a domain, then the following conditions are equivalent:*

1. *All cyclic right R-modules and all cyclic left R-modules are completely integrally closed.*

2. *R is a completely integrally closed, duo domain and, for any nonzero ideal B of R, the factor ring R/B is a finite direct product of duo uniserial rings $R_1, \ldots, R_n$, where each R_i has at most two prime ideals and either R_i is a domain or, for any nilpotent nonprime ideal B of R_i, the factor ring R_i/B is self-injective.*

Proof The implication $(1) \Rightarrow (2)$ follows from Lemma 18.19(2) and Lemma 18.17.

$(2) \Rightarrow (1)$. It is sufficient to prove that for any nonzero ideal B of the duo domain R, the ring R/B is a completely integrally closed right and left module over R/B. This follows from Lemma 18.17 and the property that any finite direct product of completely integrally closed rings is a completely integrally closed ring. $\square$

Theorem 18.21 *For a ring R, the following conditions are equivalent:*

1. *All cyclic right R-modules and all cyclic left R-modules are completely integrally closed modules.*

2. *$R = R_1 \times \cdots \times R_n$, where either R_i is a simple artinian ring or R_i is a duo uniserial ring with at most two prime ideals, and for any nilpotent ideal B of the ring R_i, the factor ring R_i/B is self-injective, or R_i is a duo completely integrally closed domain such that for any nonzero ideal B in R_i, the factor ring R_i/B is a finite direct product of duo uniserial rings $R_{i_1}, \ldots, R_{i_k}$, and either the ring R_{i_j} is self-injective, or R_{i_j} is a domain with at most two prime ideals $(i = 1, \ldots, n, \ j = 1, \ldots, k)$.*

The above theorem follows from Lemma 18.18 and Lemma 18.20.

Example Let K be the field $\mathbb{Z}/2\mathbb{Z}$, G be the quaternion group of order 8 with generators α, β and defining relations $\alpha^4 = 1$, $\alpha^2 = \beta^2$ and $\beta^{-1}\alpha\beta = \alpha^{-1}$. The group algebra $K[G]$ is a finite local self-injective ring. Note that $K[G]$ is not right uniserial, since it can be verified that the right ideals $(1 - \alpha)K[G]$ and $(1 - \beta)K[G]$ are incomparable. Thus it follows from the above theorem that cyclic right $K[G]$-modules are not necessarily completely integrally closed. $\square$

18.22 Questions

1. Let R be a ring such that all cyclic right R-modules are completely integrally closed. Is it true that all cyclic left R-modules are completely integrally closed?

2. Let R be a ring over which each cyclic right module is skew-injective. Is each cyclic right R-module completely integrally closed?

19 Rings characterized by their one-sided ideals

In this chapter we consider the dual notion, namely rings in which each one-sided ideal is quasi-injective.

19.1 Rings each of whose one-sided ideals is quasi-injective

A ring R is called a *right q-ring* if each right ideal of R is quasi-injective. The study of these rings was initiated by Jain, Mohamed, and Singh [141]. An obvious example of such a ring is a commutative self-injective ring which is, in fact, a two-sided q-ring.

We will start with some basic properties of right q-rings.

Theorem 19.1 *The following statements are equivalent for a ring R:*

1. *R is a right q-ring.*
2. *R is right self-injective and each right ideal of R is of the form eT, where e is an idempotent in R and T is a two-sided ideal of R.*
3. *R is right self-injective and each essential right ideal of R is two-sided.*

Proof $(1)\Rightarrow(2)$. Clearly, by assumption, R is right self-injective. Let A be any right ideal of R. Then the injective hull of A, $E(A) = eR$ for some idempotent e. Since A is quasi-injective, it is invariant under the endomorphisms of the injective hull $E(A)$. So $A = eReA = eRA = eT$, where $T = RA$ is a two-sided ideal of R.

$(2)\Rightarrow(3)$. Let E be any essential right ideal of R. Then by assumption, $E = eT$, where e is an idempotent in R and T is a two-sided ideal of R. Since $E \cap (1 - e) R = 0$, $(1 - e)R = 0$. This gives $e = 1$ and so $E = T$.

$(3)\Rightarrow(1)$. Let I be any right ideal of R. Let K be a complement of I in R. Then $I \oplus K$ is an essential right ideal of R. By assumption $I \oplus K$ is a two-sided ideal in R. Since R is right self-injective, $I \oplus K$ must be quasi-injective and hence I is quasi-injective. Thus R is a right q-ring. $\qquad\Box$

Lemma 19.2 *For $n > 1$, the matrix ring $\mathbb{M}_n(R)$ is a right q-ring if and only if R is semisimple artinian.*

Proof Suppose R is not semisimple artinian. We can find an essential right ideal A of R such that $A \neq R$. Let e_{ij}, $1 \leq i, j \leq n$ be the matrix units of $\mathbb{M}_n(R)$. Let $E = \{\Sigma a_{ij}e_{ij} : a_{ij} \in R \text{ with } a_{1j} \in A\}$. It may be checked that E is an essential right ideal of $\mathbb{M}_n(R)$. But E is not two-sided as $e_{nn} \in E$ and $e_{1n}e_{nn} = e_{1n} \notin E$.

Thus Theorem 19.1 yields that $\mathbb{M}_n(R)$ is not a right q-ring. The converse is obvious. $\qquad\square$

Lemma 19.3 *A simple ring is a right q-ring if and only if it is artinian.*

Proof Let R be a simple right q-ring. Let E be an essential right ideal of R. Then by Theorem 19.1, E is two-sided and hence $E = R$. This shows that R does not contain any proper essential right ideal. Hence R is artinian. $\qquad\square$

Lemma 19.4 *Let R be a right q-ring. Then the prime radical $N(R)$ is essential in the Jacobson radical $J(R)$ as a right R-module.*

Proof Since R is right self-injective, $J(R) = Z(R_R)$. Let $x \neq 0 \in Z(R_R)$. Then there exists an essential right ideal E of R such that $xE = 0$. Since R is a right q-ring, E is two-sided. Then, since $xE \subset P$ for every prime ideal P of R, we have either $x \in P$ or $E \subset P$. Let $\{P_i\}_{i \in I}$ be the set of all prime ideals of R such that $x \in P_i$ for every $i \in I$, and $\{P_j\}_{j \in J}$ be the set of all prime ideals of R such that $x \notin P_j$ for every $j \in J$. Let $X = \cap_{i \in I} P_i$ and $Y = \cap_{j \in J} P_j$. Note that $N(R) = X \cap Y$. We have $X \neq 0$, since $x\ (\neq 0) \in X$. On the other hand, $E \subset P_j$ for every $j \in J$. Thus $E \subset Y$. Hence Y is essential in R_R. As $X \neq 0$, we have $X \cap Y \neq 0$. Thus $N(R) \neq 0$. Moreover, there exists $a \in R$ such that $xa(\neq 0) \in Y$ and hence $xa\ (\neq 0) \in X \cap Y = N(R)$. This completes the proof. $\qquad\square$

Corollary 19.5 *A semiprime right q-ring is von Neumann regular.*

Proof Let R be a semiprime right q-ring. Then R is right self-injective and so $R/J(R)$ is von Neumann regular. By the above lemma, $J(R) = 0$. Thus R is von Neumann regular. $\qquad\square$

Lemma 19.6 *Let V be a vector space over a division ring D and let $R = \mathrm{End}_D(V)$. Then R is a right q-ring if and only if V is finite-dimensional over D.*

Proof Suppose V is infinite-dimensional over D. Let $X = \{v_1, v_2, \ldots\}$ be a countable linearly independent subset of V. Then X can be extended to a basis $X \cup Y$ of V. Let I be the ideal of R consisting of all elements of finite rank. Let $\sigma \in R$ be defined by $\sigma(v_{2i}) = v_{2i}$, $\sigma(v_{2i-1}) = 0$ for every i, and $\sigma(y) = 0$ for every $y \in Y$. Let $E = \sigma R + I$. Then $I \subset E$. Since I is a two-sided ideal of R, I is essential. Therefore E is an essential right ideal of R. Now we proceed to show that E is not a two-sided ideal. Let $\alpha_1, \alpha_2 \in R$ be defined by $\alpha_1(v_i) = v_{2i}$ for every i and $\alpha_1(y) = 0$ for every $y \in Y$; $\alpha_2(v_{2i}) = v_i$, $\alpha_2(v_{2i-1}) = 0$ for every i, and $\alpha_2(y) = 0$ for every $y \in Y$. Let $\alpha = \alpha_2 \sigma \alpha_1$. It can be shown that $\alpha \notin E$, however $\alpha \in R\sigma R \subset RE$. Hence E is not a two-sided ideal. Therefore R is not a right q-ring. The converse is obvious. $\qquad\square$

The structure of a semiprime right q-ring is given in the following result.

Theorem 19.7 *Let R be a semiprime right q-ring. Then $R = S \oplus T$ where S is semisimple artinian and T is a ring with zero socle.*

Proof Let R be a semiprime right q-ring. Then the prime radical $N(R) = 0$. Now, by the above lemma, $J(R) = 0$. Since R is right self-injective with $J(R) = 0$, R is von Neumann regular. Let $A = \mathrm{Soc}(R)$. Since R is right nonsingular, $E(A) = \{x \in R : xE \subset A$ for some essential right ideal E of $R\}$. Clearly $E(A)$ is a two-sided ideal of R. Now since R is right self-injective, $E(A) = eR$ for some idempotent $e \in R$. Then e is central since R is von Neumann regular. Let $S = eR$ and $T = (1 - e)R$. Hence $R = S \oplus T$. Clearly both S and T are right q-rings. Note that $\mathrm{Soc}(S) = A$, and A is essential in S. Since $\mathrm{Soc}(S)$ is essential in S, every nonzero right ideal of S contains a minimal right ideal. Hence, by [148, Theorem 3.1], S is a direct sum of rings S_i, where each S_i is the ring of all linear transformations of some vector space V_i over a division ring D_i. By Lemma 19.6, each S_i is simple artinian. Hence S is semisimple artinian and T is a ring with zero socle. $\qquad\square$

As a consequence, we have the following:

Corollary 19.8 *Let R be a prime right q-ring. Then R is simple artinian.*

If for a ring R, every matrix ring $\mathbb{M}_n(R)$ is directly finite then R is called a *stably-finite ring*.

Theorem 19.9 *Every right q-ring is stably-finite.*

Proof Let R be a right q-ring. Then R is a right self-injective ring in which each essential right ideal is two-sided. By Theorem 1.21, $R/J(R)$ is a von Neumann regular right self-injective ring. Let $E/J(R)$ be an essential right ideal of $R/J(R)$. Then E is an essential right ideal of R. Since R is a right q-ring, E is two-sided. Thus each essential right ideal of $R/J(R)$ is two-sided and hence $R/J(R)$ is a right q-ring (see Theorem 19.1). Since $R/J(R)$ is a von Neumann regular right q-ring, $R/J(R) = U \oplus T$ where U is semisimple artinian and T has zero socle. Now let M be a maximal right ideal of T. If M is a summand then the complement of M is a minimal right ideal which yields a contradiction as T has zero socle. Therefore M is essential. Since T is a right q-ring, M must be two-sided. Thus each maximal right ideal of T is two-sided. Hence T is a right quasi-duo ring. Consider $S = \mathbb{M}_n(R)$. Then $S/J(S) \cong \mathbb{M}_n(R/J(R)) = \mathbb{M}_n(U \oplus T)$. Thus $S/J(S) \cong \mathbb{M}_n(U) \oplus \mathbb{M}_n(T)$. Clearly, $\mathbb{M}_n(U)$ is directly finite. Now we proceed to show that $\mathbb{M}_n(T)$ is directly finite. Let $\{M_i\}$ be the set of maximal right ideals of the quasi-duo ring T. Then each M_i is a two-sided ideal and $J(T) = \cap M_i$. Clearly, each T/M_i is a division ring. Thus $\mathbb{M}_n(T)/\mathbb{M}_n(M_i) \cong \mathbb{M}_n(T/M_i)$ is a simple artinian ring which is clearly directly finite. Consider the natural ring homomorphism $\varphi : \mathbb{M}_n(T) \longrightarrow \prod_i \mathbb{M}_n(T/M_i)$. We have $\mathrm{Ker}(\varphi) = \mathbb{M}_n(J(T)) = J(\mathbb{M}_n(T))$. Since each $\mathbb{M}_n(T/M_i)$ is directly finite, $\prod_i \mathbb{M}_n(T/M_i)$ is directly finite and consequently, $\mathbb{M}_n(T)/J(\mathbb{M}_n(T))$ is directly finite being a subring of a directly finite ring. Hence $\mathbb{M}_n(T)$ is directly finite. Thus $S/J(S)$ is directly finite and therefore S is directly finite. Hence R is stably-finite. $\qquad\square$

Mohamed [177] studied various classes of q-rings and obtained the following results.

Theorem 19.10 *For a ring R, we have the following:*

1. *If R is a semilocal right q-ring and eRe is not a division ring for any primitive idempotent $e \in R$, then R is a finite direct sum of local right self-injective duo rings.*
2. *If R is a pseudo-Frobenius right q-ring, then $R = A \oplus B$ where A is a QF right q-ring and B is a finite direct sum of local right self-injective duo rings with nonzero socle.*
3. *If R is a right or left artinian right q-ring, then R is a left q-ring as well.*

Hill [109] gave a complete description of semiperfect right q-rings.

Theorem 19.11 *Let R be semiperfect right q-ring. Then R is a finite direct sum of a semisimple artinian ring and a right self-injective basic ring. Further, a right self-injective basic ring which is semiperfect need not be a right q-ring.*

The following is an easy observation.

Proposition 19.12 *A finite direct product of rings $\prod_{i=1}^{n} R_i$ is a right q-ring if and only if each R_i is a right q-ring.*

The above result does not extend to an infinite product of rings. The example below shows this.

Example Let R be a 2×2 matrix ring over a field F. Let $\{R_i : i \in \mathcal{I}\}$ be an infinite family of copies of R and let $S = \prod_{i \in \mathcal{I}} R_i$. Let E be the right ideal of S consisting of those elements $\{x_i\}$ of S such that all but finitely many x_i's are matrices with first row zero. Since $R_i \subset E$ for all $i \in \mathcal{I}$, E is an essential right ideal of S. To show that E is not two-sided, consider $\{x_i\} \in E$ where $x_i = \begin{bmatrix} 0 & 0 \\ 1 & 0 \end{bmatrix}$ for all $i \in \mathcal{I}$. Let $\{y_i\} \in S$ be such that $y_i = \begin{bmatrix} 0 & 1 \\ 0 & 0 \end{bmatrix}$ for all $i \in \mathcal{I}$. Then $\{y_i\}\{x_i\} = \{z_i\}$, where $z_i = \begin{bmatrix} 1 & 0 \\ 0 & 0 \end{bmatrix}$. But then $\{z_i\} \notin E$. This shows that E is not two-sided. Hence, by Theorem 19.1, $S = \prod_{i \in \mathcal{I}} R_i$ is not a right q-ring. $\square$

Byrd [35] and Ivanov ([127], [128]) gave structures of right q-rings but their characterizations were not complete. Finally, the structure of right q-rings was completely described in Beidar et al. [23].

Theorem 19.13 *A right q-ring R is isomorphic to a finite direct product of right q-rings of the following types:*

1. *Semisimple artinian ring.*
2. *$H(n; D; id_D)$ where id_D is the identity automorphism on division ring D.*
3. *$G(n; \Delta; P)$ where Δ is a right q-ring whose all idempotents are central.*

4. A right q-ring whose all idempotents are central.

Here

$$
H(n; D; \alpha) =
\begin{bmatrix}
D & V & 0 & . & . & . & 0 \\
0 & D & V & 0 & . & . & 0 \\
. & . & D & V & 0 & . & . \\
. & . & . & . & . & . & . \\
. & . & . & . & D & V & 0 \\
. & . & . & . & . & D & V \\
V(\alpha) & 0 & . & . & . & . & D
\end{bmatrix},
$$

where V is one-dimensional both as a left D-space and a right D-space, $V(\alpha)$ is also a one-dimensional left D-space as well as a right D-space with right scalar multiplication twisted by an automorphism α of D, that is, $vd = v \cdot \alpha(d)$ for all $v \in V$, $d \in D$,

and

$$
G(n; \Delta; P) =
\begin{bmatrix}
D & V & 0 & . & . & . & 0 \\
0 & D & V & 0 & . & . & 0 \\
. & . & D & V & 0 & . & . \\
. & . & . & . & . & . & . \\
. & . & . & . & D & V & 0 \\
. & . & . & . & . & D & V \\
0 & 0 & . & . & . & . & \Delta
\end{bmatrix},
$$

where V is as above and Δ is a right q-ring with maximal essential right ideal P and hence $D = \Delta/P$ is a division ring.

Moreover, R is a left q-ring if and only if it is left self-injective.

The relationship between q-rings, q^*-rings, and qc-rings was given by Koehler in the following theorem.

Theorem 19.14 *If R is a right qc-ring, then R is a right q-ring and a right q^*-ring.*

Proof It follows easily from Theorem 7.3. $\qquad\qquad\qquad\qquad\qquad\qquad\qquad\square$

Mohamed studied rings where each proper homomorphic image is a right q-ring [178] and proved the following.

Theorem 19.15 *Let R be a nonprime right noetherian ring. Then every proper homomorphic image of R is a right q-ring if and only if:*

1. *$R = S \oplus T$, where S is semisimple artinian and T is a principal ideal duo ring with DCC; or*
2. *$R/J(R)$ is artinian and $J(R)$ is a minimal ideal; or*

3. *R is a local ring whose maximal right ideal M satisfies $M^2 = 0$ and every proper homomorphic image of R contains at most one proper right (left) ideal.*

Theorem 19.16 *Let R be a prime right noetherian ring with the property that every proper homomorphic image of R is a right q-ring. Then*

1. *Every ideal of R is a product of prime ideals; and*
2. *for every nonzero prime ideal P of R, R/P is a division ring.*

19.2   Rings each of whose one-sided ideals is a direct sum of quasi-injectives

A ring R is called a *right Σ-q ring* if each right ideal is a finite direct sum of quasi-injective right ideals. Jain, Singh, and Srivastava [146] initiated the study of such rings. These rings may also be viewed as a generalization of artinian serial rings. Nakayama showed that every module over an artinian serial ring is a direct sum of uniserial modules [182]. Fuller proved that every indecomposable module over an artinian serial ring is quasi-injective [85]. Therefore every right ideal in an artinian serial ring is a finite direct sum of quasi-injective right ideals.

We will start with a basic lemma.

Lemma 19.17 *Let R be a right self-injective, right Σ-q ring. Then each essential right ideal E of R is of the form $E = e_1T_1 \oplus e_2T_2 \oplus \cdots \oplus e_nT_n$ where $T_1, \ldots, T_n$ are two-sided ideals of R and $e_1, \ldots, e_n$ are orthogonal idempotents with $e_1 + e_2 + \cdots + e_n = 1$.*

Proof By hypothesis, $E = E_1 \oplus E_2 \oplus \cdots \oplus E_n$, where each E_i is a quasi-injective right ideal. This gives $R = \widehat{E}_1 \oplus \cdots \oplus \widehat{E}_n$, where $\widehat{E}_i$ denotes the injective hull of E_i. Write each $\widehat{E}_i$ as e_iR with $e_1 + e_2 + \cdots + e_n = 1$. Since each E_i is quasi-injective, we have $e_iR \ e_iE_i = E_i$. This gives $e_iRE_i = E_i$. Denote each two-sided ideal RE_i as T_i. So we have $E = e_1T_1 \oplus e_2T_2 \oplus \cdots \oplus e_nT_n$ where $T_1, \ldots, T_n$ are two-sided ideals of R and $e_1, \ldots, e_n$ are orthogonal idempotents with $e_1 + e_2 + \cdots + e_n = 1$. $\square$

Lemma 19.18 *Let R be a right Σ-q ring with no nontrivial idempotents. Then R is a right q-ring and hence a duo ring.*

Proof It follows easily from Lemma 19.17 that if R is a right Σ-q ring with no nontrivial idempotents then R is a right q-ring. Since R has no nontrivial idempotents, it must be a right duo by Theorem 19.1. Now we will show that a right self-injective right duo ring must be a duo ring. To prove that R is a duo ring, it suffices to show that Ra is an ideal of R for any $a \in R$. We claim that $ann_l(ann_r(Ra)) = Ra$. Clearly, $Ra \subseteq ann_l(ann_r(Ra))$. Let $b \in ann_l(ann_r(Ra))$. Then $bx = 0$ for all $x \in R$ with $ax = 0$. Therefore the map $f : aR \to bR$, given by $ax \mapsto bx$ is a well-defined map and so, since R is right self-injective, there exists an element $c \in R$ such that $f(y) = cy$ for all $y \in Ra$. In particular, $b = f(a) = ac$

which proves our claim. Finally, $ann_r(Ra)$ is a right ideal and so it is an ideal of R. Therefore $Ra = ann_l(ann_r(Ra))$ is an ideal of R as well. Hence R is a duo ring. $\qquad\square$

Corollary 19.19 *Let R be a local ring. Then R is a right Σ-q ring if and only if R is a right q-ring, if and only if R is a right self-injective duo ring.*

Proof This follows from the above lemma and Theorem 19.1. $\qquad\square$

Proposition 19.20 *If R is a right Σ-q ring and e is an idempotent in R such that $ReR = R$ then eRe is also a right Σ-q ring.*

Proof We know that if $ReR = R$, then *mod-R* and *mod-eRe* are Morita equivalent under the functors given by $\mathcal{F} : mod\text{-}R \longrightarrow mod\text{-}eRe$, $\mathcal{G} : mod\text{-}eRe \longrightarrow mod\text{-}R$ such that for any M_R, $\mathcal{F}(M) = Me$ and for any module T over eRe, $\mathcal{G}(T) = T \otimes_{eRe} eR$.

Suppose R is a right Σ-q ring. Let A be any right ideal of eRe. Then $AeR \cong A \otimes_{eRe} eR$ and AeR is a right ideal of R. Therefore $AeR = A_1 \oplus \cdots \oplus A_n$ where the A_i's are quasi-injective right ideals in R. By Morita equivalence we get that each A_ie is quasi-injective as an eRe-module. Then $A = AeRe = A_1e \oplus \cdots \oplus A_ne$ is a direct sum of quasi-injective right ideals of eRe. Hence eRe is a right Σ-q ring. $\qquad\square$

Lemma 19.21 *If $\mathbb{M}_n(R)$ is a right Σ-q ring, then R is also a right Σ-q ring.*

Proof We have $R \cong e_{11}\mathbb{M}_n(R)e_{11}$ and $\mathbb{M}_n(R)e_{11}\mathbb{M}_n(R) = \mathbb{M}_n(R)$, where e_{11} is the usual matrix unit. Therefore the result follows from the above proposition. $\qquad\square$

Later, we will show that if R is a right Σ-q ring then $\mathbb{M}_n(R)$ need not be a right Σ-q ring.

First, we consider a prime, right self-injective, right Σ-q ring.

Lemma 19.22 *A prime, right self-injective, right Σ-q ring must be von Neumann regular.*

Proof We prove $Z(R_R) = 0$. If possible, let x be a nonzero element in $Z(R_R)$. Then there exists an essential right ideal E of R such that $xE = 0$. Now, by Lemma 19.17, $E = e_1T_1 \oplus e_2T_2 \oplus \cdots \oplus e_nT_n$ where $T_1, \ldots, T_n$ are two-sided ideals of R and $e_1, \ldots, e_n$ are orthogonal idempotents with $e_1 + e_2 + \cdots + e_n = 1$. Thus $xe_iT_i = 0$. But since R is prime, $xe_i = 0$, for all i. Therefore $x = 0$, a contradiction. Hence R is right nonsingular and, therefore, R must be von Neumann regular. $\qquad\square$

Theorem 19.23 *A prime, right self-injective ring R is a right Σ-q ring if and only if R is artinian.*

Proof Let R be a prime, right self-injective, right Σ-q ring. By Lemma 19.22, R is von Neumann regular. Since the two-sided ideals of R are well-ordered (see

[97], Proposition 8.5), R has a unique maximal two-sided ideal, say A. We claim R/A is simple artinian. If each maximal right ideal of R/A is a summand then it may be deduced that each right ideal of R/A is a summand and then we are done. Else, let N/A be a maximal essential right ideal of R/A. Then N is an essential maximal right ideal of R. By Lemma 19.17, $N = e_1 T_1 \oplus e_2 T_2 \oplus \cdots \oplus e_n T_n$, where $T_1, \ldots, T_n$ are two-sided ideals of R and $e_1, \ldots, e_n$ are orthogonal idempotents with $e_1 + e_2 + \cdots + e_n = 1$. Since R/N is a simple module, we have $R/N \cong e_i R/e_i T_i$. Now, since $e_i R/e_i T_i$ is a direct summand of R/T_i, it is projective as an R/T_i-module. As R/N is an R/T_i-module and $T_i \subset A \subset N$, it follows that R/N is a simple, projective R/A-module. Therefore R/A is a simple ring with nonzero socle and hence artinian. Thus R has bounded index of nilpotence (see [97], p. 79) and so R is artinian (see [97], Theorem 7.9).

The converse is obvious. $\qquad\square$

We remark that if in the above theorem R is not a prime ring then it need not be artinian. The following is an example of a nonprime von Neumann regular, right self-injective, right Σ-q ring which is not artinian.

Example Let S be an infinite boolean ring and suppose $R = Q^r_{\max}(S)$. Clearly R is commutative, von Neumann regular, self-injective, and hence a Σ-q ring. But R is not artinian. $\qquad\square$

The following result is a consequence of Theorem 19.23.

Corollary 19.24 *The ring of linear transformations, $R = \mathrm{End}_D(V)$ of a vector space V over a division ring D is a right Σ-q ring if and only if the vector space V is finite-dimensional.*

Next, we have the following:

Corollary 19.25 *A simple ring R is a right Σ-q ring if and only if R is artinian.*

Proof Let R be a simple right Σ-q ring. Then $R = e_1 R \oplus \cdots \oplus e_n R$, where each $e_i R$ is a quasi-injective right ideal. As R is a simple ring, $R e_1 R = R$. Therefore $R \cong (e_1 R)^k / L \cong S$ where S is a direct summand of $(e_1 R)^k$ and L is a submodule of $(e_1 R)^k$. Because $e_1 R$ is quasi-injective, S_R and hence R_R is quasi-injective. Thus R is right self-injective. Now, by Theorem 19.23, R is artinian.

The converse is trivial. $\qquad\square$

Next, we give an example of a simple right self-injective ring which is not a right Σ-q ring.

Example Let S be an integral domain which is not a right Ore domain. Consider $R = Q^r_{\max}(S)$. We know that R is a simple, right self-injective, von Neumann regular ring. However, if R is a right Σ-q ring then by the above corollary, R is artinian and hence S is right Ore, a contradiction. $\qquad\square$

Lemma 19.26 *Let R be a von Neumann regular ring. Suppose every ring homomorphic image of R is a right self-injective, right Σ-q ring. Then R is semisimple artinian.*

Proof Let P be a prime ideal of R. Then R/P is a prime, von Neumann regular, right self-injective, right Σ-q ring. By Theorem 19.23, R/P is artinian. Since R is a von Neumann regular, right self-injective ring and each primitive factor ring of R is artinian, we have $R \cong \Pi_{i=1}^{k}\mathbb{M}_{n_i}(S_i)$ where each S_i is an abelian regular right self-injective ring (see [97], Theorem 7.20). Thus each S_i is a right self-injective duo ring.

Now let C_i be any right ideal of S_i. Since S_i is a duo ring, C_i is a two-sided ideal and we have $\mathbb{M}_{n_i}(S_i)/\mathbb{M}_{n_i}(C_i) \cong \mathbb{M}_{n_i}(S_i/C_i)$. Since each $\mathbb{M}_{n_i}(S_i)$ satisfies the property that homomorphic images are right self-injective, $\mathbb{M}_{n_i}(S_i/C_i)$ is right self-injective and hence S_i/C_i is right self-injective. So each cyclic right S_i-module is quasi-injective. Hence by Theorem 7.3, S_i is a finite direct sum of rings each of which is semisimple artinian or a rank zero duo right linearly compact ring. But, since S_i is a von Neumann regular ring, we conclude that it must be semisimple artinian. Hence R is semisimple artinian. $\square$

Next, we show:

Proposition 19.27 *Let R be a right Σ-q ring and suppose every ring homomorphic image of $R/J(R)$ is right self-injective, right Σ-q ring. Then R is semiperfect.*

Proof By the above lemma, $R/J(R)$ is semisimple artinian. Now let the composition length of $R/J(R)$ be n. Then R_R cannot be a direct sum of more than n submodules. Thus R_R is a finite direct sum of indecomposable right ideals, each of which is quasi-injective. Let $A = eR$ be any indecomposable summand of R_R. As $A = eR$ is quasi-injective, its ring of endomorphisms eRe is a local ring (see [161, p. 244, Exercise 6.22]). Hence R is semiperfect (see [12, Theorem 27.6]). $\square$

Clearly, each right q-ring is a right Σ-q ring. However, there are plenty of examples of right Σ-q rings that are not right q-rings, as the following shows.

Recall that an artinian serial ring is a right Σ-q ring but such rings need not be q-rings. Furthermore, a left Σ-q ring need not be a right Σ-q ring. We give an example of an incidence ring which is a left Σ-q ring but not a right Σ-q ring.

Example Let $X = \{1, 2, 3, 4\}$ be a partially ordered set with $1 < 2 < 3$ and $1 < 2 < 4$. Let F be a field. The incidence ring is given as

$$R = I(X, F) = \begin{bmatrix} F & F & F & F \\ 0 & F & F & F \\ 0 & 0 & F & 0 \\ 0 & 0 & 0 & F \end{bmatrix}.$$

R is a left artinian, left serial ring. Note that $\mathrm{Soc}(_R R) = Fe_{11} + Fe_{12} + Fe_{13} + Fe_{14}$, where the e_{ij} are matrix units. Now we will show that $_R R$ is nonsingular.

Suppose $Z(_RR) \neq 0$. Then there exists z $(\neq 0) \in \text{Soc}(Z(_RR))$. Clearly, $z = a_{11}e_{11} + a_{12}e_{12} + a_{13}e_{13} + a_{14}e_{14}$ where $a_{ij} \in F$. Then $e_{11}z = z$, which implies $e_{11} \notin ann_l(z)$ and therefore, $Fe_{11} \cap ann_l(z) = 0$. Hence $ann_l(z)$ is not essential in $_RR$, a contradiction. So $_RR$ is nonsingular.

Re_{11} is simple, so quasi-injective. As Re_{11} is nonsingular and quasi-injective, $\text{End}(Re_{11}) \cong \text{End}(E(Re_{11}))$. Hence any endomorphism of $E = E(Re_{11})$ is given as multiplication by some element of F. Next, we show that Re_{22}, Re_{33}, and Re_{44} are quasi-injective.

The ring R is left serial, so Re_{33} is uniserial and hence uniform. We have $\text{Soc}(Re_{33}) = Fe_{13} \cong Fe_{11}$ as left R-modules under the mapping $ae_{13} \mapsto ae_{11}$. Hence Re_{33} being an essential extension of Fe_{13} embeds in the injective hull E of Re_{11}.

So $\sigma(Re_{33}) \subseteq Re_{33}$, for all $\sigma \in \text{End}(E)$. Therefore Re_{33} is quasi-injective. Similarly, Re_{44} is quasi-injective. For every left ideal $C \subseteq Re_{33}$ and for all $\sigma \in \text{End}(E)$, we have $\sigma(C) \subseteq C$ as Re_{33} is uniserial. Hence C is quasi-injective. Now $Re_{22} \cong Re_{23} \subseteq Re_{33}$. Therefore Re_{22} is quasi-injective. So $_RR$ is a direct sum of quasi-injectives.

Next, we show that any indecomposable left ideal is uniserial and quasi-injective. Suppose there exists a left ideal $A \neq 0$ which is indecomposable and not uniform. Choose such an A of smallest composition length. Let $\pi_i :_R R \longrightarrow Re_{ii}$ be the projection map. If for some i, $\pi_i(A) = Re_{ii}$, we get $A = A' \oplus B'$ for some B', $A' \cong Re_{ii}$. This gives $B' = 0$ and $A \cong Re_{ii}$, a contradiction as A is not uniform. Therefore $\pi_i(A) \subseteq J(R)e_{ii}$ for all i, which gives $A \subseteq \begin{bmatrix} 0 & F & F & F \\ 0 & 0 & F & F \\ 0 & 0 & 0 & 0 \\ 0 & 0 & 0 & 0 \end{bmatrix} = B_2 \oplus B_3 \oplus B_4.$

Let $\pi_i' : J(R) \longrightarrow B_i$. Suppose $\pi_2'(A) \neq 0$. Now B_2 is minimal and $B_2 \cong Re_{11}$ which is projective. So $A \cong B_2$ which is uniform, a contradiction. Hence $\pi_2'(A) = 0$. This gives $A \subseteq \begin{bmatrix} 0 & 0 & F & F \\ 0 & 0 & F & F \\ 0 & 0 & 0 & 0 \\ 0 & 0 & 0 & 0 \end{bmatrix} = L$. Clearly, $ann_l(L) = \begin{bmatrix} 0 & 0 & F & F \\ 0 & 0 & F & F \\ 0 & 0 & F & 0 \\ 0 & 0 & 0 & F \end{bmatrix}$. Let us denote $ann_l(L)$ by U. Then L is a left R/U-module. Now $R/U \cong \begin{bmatrix} F & F \\ 0 & F \end{bmatrix}$, which is artinian serial.

As $A \subseteq L$, we note that A is a module over $\begin{bmatrix} F & F \\ 0 & F \end{bmatrix}$. This implies that A is uniserial and hence uniform. Let B be any indecomposable left ideal in R. Then B is uniform. For some i, $\pi_i(\text{Soc}(B)) \neq 0$ and so π_i is one-one on $\text{Soc}(B)$. As B is uniform, $\pi_i|_B : B \longrightarrow Re_{ii}$ is one-one. We have $B \cong \pi_i(B) \subseteq Re_{ii}$. This gives that B is uniserial and quasi-injective. Now let I be any left ideal of R. Then I is a finite direct sum of indecomposable left ideals. Since we have already proved above that any indecomposable left ideal is quasi-injective, I is a finite direct sum of quasi-injective left ideals. Thus R is a left Σ-q ring.

Now we show that R is not a right Σ-q ring. We have $e_{11}R = Fe_{11} + Fe_{12} + Fe_{13} + Fe_{14}$. Note that Fe_{13}, Fe_{14} are minimal right ideals. Clearly, $e_{11}R$ is not uniform and hence not quasi-injective. Therefore R is not a right Σ-q ring. $\quad\square$

Here is another example analogous to the above example that might be of interest to the reader.

Example Let $X = \{1, 2, 3, 4\}$ be a partially ordered set with $1 < 3 < 4$ and $2 < 3 < 4$. Let F be a field. Then the incidence ring is given as

$$R = I(X, F) = \begin{bmatrix} F & 0 & F & F \\ 0 & F & F & F \\ 0 & 0 & F & F \\ 0 & 0 & 0 & F \end{bmatrix}.$$

By similar arguments as above R is a right Σ-q ring, but not a left Σ-q ring. $\quad\square$

Let R be any right Σ-q ring and $S = \mathbb{M}_n(R)$. Now $R \cong e_{11}\mathbb{M}_n(R)e_{11}$ is right Σ-q ring. We have the equivalence functor $\mathcal{F} : mod\text{-}\mathbb{M}_n(R) \longrightarrow mod\text{-}e_{11}\mathbb{M}_n(R)e_{11}$, where $\mathcal{F}(M) = Me_{11}$. In particular, $\mathcal{F}(e_{11}\mathbb{M}_n(R)) = e_{11}\mathbb{M}_n(R)e_{11}$ gives that any right ideal contained in $e_{11}\mathbb{M}_n(R)$ is a direct sum of finitely many quasi-injective modules. However, the ring $\mathbb{M}_n(R)$ need not be a right Σ-q ring, as we next show.

In fact, the following example shows that a matrix ring over even a q-ring need not be right Σ-q. Therefore being right Σ-q is not a Morita-invariant property.

Example Let F be any field. Consider the ring $R = F[x, y]$ where $x^2 = 0$ and $y^2 = 0$. Then $R = F + Fx + Fy + Fxy$ is a commutative, local, artinian ring. Here, $J(R) = Fx + Fy + Fxy$. Let $z = a + bx + cy + dxy \in ann(J)$. This implies that $(a + bx + cy + dxy)x = 0$, which gives $ax + cxy = 0$, and so, $a = 0$ and $c = 0$. Similarly, $(a + bx + cy + dxy)y = 0$, gives $a = 0$ and $b = 0$. So we have $z = dxy$. Therefore $\text{Soc}(R) = Fxy$. Thus R is a commutative local artinian ring with simple socle and hence a self-injective ring (see [74, p. 217, Exercise 5]). Since a commutative self-injective ring must be a q-ring, R is a q-ring.

Let $K = \text{Soc}(R)$. Then R/K is of length 3 and $J(R/K) = A \oplus B$ where A and B are simple. Let $T = \mathbb{M}_2(R)$. Then $T = e_{11}T \oplus e_{22}T$, where $e_{11}T \cong e_{22}T$.

Note that $e_{11}\mathbb{M}_2(R/K)/e_{11}\mathbb{M}_2(J(R/K))$ is simple. Therefore $e_{11}\mathbb{M}_2(R/K)$ is local as a $\mathbb{M}_2(R/K)$ ($\cong \mathbb{M}_2(R)/\mathbb{M}_2(K)$)-module, and hence as a $T = \mathbb{M}_2(R)$-module. Observe that $e_{11}\mathbb{M}_2(R/K)$ has two minimal right ideals $e_{11}A + e_{12}A$ and $e_{11}B + e_{12}B$. By suitably factoring, we then obtain a local T-module, say, N_T with $l(\text{Soc}(N)) = 2$. If S_T is a simple module, then N is embeddable in $E(N) \cong E(S) \oplus E(S) \cong T_T$. Thus N is a local, indecomposable, nonuniform module embeddable in T and so T contains a right ideal which is not quasi-injective. Therefore T is not a right Σ-q ring. $\quad\square$

The following lemma is well known (see [161, p. 359, Theorem 13.1]).

Lemma 19.28 *Let R be right artinian and let eR be a right nonsingular indecomposable quasi-injective right ideal, where e is an idempotent in R. Then eRe is a division ring.*

Lemma 19.29 *Let R be a right artinian, right nonsingular, right Σ-q ring. Let e, f be any two indecomposable idempotents in R such that $eRf \neq 0$. Let $D = eRe$ and $D' = fRf$.*

1. *Then eRf is a one-dimensional left vector space over D.*
2. *For any $0 \neq z \in eRf$, there exists an embedding $\sigma : D' \longrightarrow D$ such that for $fbf \in D'$, $\sigma(fbf)z = zfbf$.*
3. *If R is also right serial, then σ is an isomorphism.*

Proof

(1) Consider any two nonzero elements erf and esf in eRf. Define $\phi : fR \to erfR$ by $\phi(x) = erx$ for all $x \in fR$. Then ϕ is clearly a well-defined R-epimorphism. Moreover, since fR is indecomposable and quasi-injective, fR is uniform. Thus, if $\ker(\phi) \neq 0$, then $\ker(\phi)$ is essential in fR. This implies that $erfR \cong fR/\ker(\phi)$ is singular, a contradiction since R is nonsingular. Thus $\ker(\phi) = 0$ and so $fR \cong erfR$. Similarly $fR \cong esfR$. Thus we get an R-isomorphism $\sigma : erfR \longrightarrow esfR$ such that $\sigma(erf) = esf$. Since eR is quasi-injective, we extend σ to $\eta : eR \longrightarrow eR$. Now $\eta(e) = eue$. Then $\sigma(erf) = euerf$. Therefore $esf = euerf$. Hence eRf is a one-dimensional left vector space over D.

(2) Now $eRf = Dz$, for some $z \in eRf$. Therefore, given any $fbf \in fRf$, $zfbf = uz$, for some $u \in D$. This defines a monomorphism $\sigma : D' \longrightarrow D$ such that $\sigma(fbf) = u$. Also, we have $\sigma(fbf)z = uz = zfbf$.

(3) Consider any two nonzero elements erf and esf in eRf. As eR is uniserial, we may suppose $esfR \subseteq erfR$. Then $esf = erfuf$ for some $fuf \in fRf$. Since fRf is a division ring, it follows that $esfRf = erfRf$. Hence eRf is one-dimensional over D'. Therefore $eRf = Dz = zD'$. It is now immediate that σ is an isomorphism. $\qquad\square$

Proposition 19.30 *Let R be a right artinian, right nonsingular, right Σ-q ring.*

1. *If e, f are two indecomposable idempotents in R such that $eRf \neq 0$, then for any $0 \neq z \in eRf$, $eRez = zfRf$.*
2. *If R is an indecomposable ring, then $eRe \cong fRf$ for any two indecomposable idempotents e and f.*

Proof

(1) As eR is quasi-injective and indecomposable, it is uniform. So $\operatorname{Soc}(eR)$ is simple. For some indecomposable idempotent $g \in R$, $\operatorname{Soc}(eR)$ is a homomorphic image of gR under a mapping, say h. We have $\operatorname{Ker} h = gJ$. If $gJ \neq 0$, then as gJ is essential in gR, $\operatorname{Soc}(eR)$ is singular. This gives

a contradiction. Hence $\text{Soc}(eR) \cong gR$. Since $g = g^2$ and $h(g) \in eR$, we have $h(g) = ewg$ for some $w \in R$. Hence $\text{Soc}(eR) = ewgR$ for some indecomposable idempotent $g \in R$, and $w = ewg$. Let $\mu : gR \longrightarrow wgR$ be the nonzero R-homomorphism given by $\mu(gr) = wgr$. It is monic because gR is uniform. Then $gR \cong wgR$. Thus $\text{Soc}(eR) = eRgR$. By Lemma 19.28, $eRew = eRg$. This yields, $\text{Soc}(eR) = eRewR = eRgR$. Consider any nonzero erg, $esg \in eRg$. As $\text{Soc}(eR)$ is simple, we get $eRgR = ergR = esgR$, and so $eRgRg = ergRg = esgRg$, whence it is easy to verify that, $eRg = eRew = wgRg$ for any nonzero $w \in eRg$. So the induced monomorphism $\eta : gRg \longrightarrow eRe$ given by $\eta(grg)w = wgrg$ is an isomorphism.

Consider the embedding $\lambda : fR \longrightarrow eR$, where $\lambda(fr) = zfr$. Then $\lambda(\text{Soc}(fR)) = zfRgR$. Let $v \in fRg$ be such that $zv \neq 0$. Then $w = zv \in eRg$. Now v and w induce isomorphisms $\sigma_1 : gRg \longrightarrow fRf$, and $\sigma_2 : gRg \longrightarrow eRe$, respectively. Furthermore, $\sigma : fRf \longrightarrow eRe$ is a monomorphism induced by z. Because $\sigma_2 = \sigma\sigma_1$, σ is also an isomorphism.

(2) Let $S = \{e_1, \ldots, e_m\}$ be a basic orthogonal set of indecomposable idempotents in R. For any two distinct members $e, f \in S$, set $e \leq f$ if $eRf \neq 0$, equivalently, if fR embeds in eR. This is a partial ordering on S. As R is indecomposable, S is connected. We may take $e, f \in S$. There exists a path $e = e_1, e_2, \ldots, e_k = f$. By definition, for any $i < k$, $e_i Re_{i+1} \neq 0$. By (i), $e_i Re_i \cong e_{i+1} Re_{i+1}$. Hence $eRe \cong fRf$. $\qquad\square$

If we assume that R is right serial in addition to being right artinian and right nonsingular, we have the following equivalence. First, recall that by Warfield [234], such a ring is right hereditary.

Theorem 19.31 *Let R be a right artinian, right nonsingular, right serial ring. Then the following are equivalent;*

1. *R is a right Σ-q ring.*
2. *For any two indecomposable idempotents $e, f \in R$, if $eRf \neq 0$ then eRf is one-dimensional left vector space over eRe and one-dimensional right vector space over fRf.*

Proof (1)$\Rightarrow$(2) follows by Proposition 19.30.

Conversely, suppose (2) holds. Let e be any indecomposable idempotent in R. Let A be a nonzero right ideal contained in eR. As R is right nonsingular and right serial, $A \cong fR$ for some indecomposable idempotent $f \in R$. Let $\sigma : A \longrightarrow eR$ be a nonzero R-homomorphism. Then σ is a monomorphism, and $\sigma(A) \subseteq A$. For some indecomposable idempotent $g \in R$, $\text{Soc}(eR) = eugR$ for some $u \in R$, $\sigma(eug) = eugvg$ for some $v \in R$. By (2), $eugvg = eweug$, for some nonzero ewe. Let η be the endomorphism of eR given by left multiplication by ewe. If $\lambda = \eta|_A$, then $\sigma - \lambda$, being zero on $\text{Soc}(eR)$, is zero. Hence η extends σ. So eR is quasi-injective. This also proves that A is quasi-injective. In a right artinian right nonsingular right serial ring, any right ideal is a finite direct sum of uniserial right ideals. As shown above any uniserial right ideal is quasi-injective. Hence R is a right Σ-q ring. $\qquad\square$

Theorem 19.32 *Let R be an indecomposable right artinian, right nonsingular, right Σ-q ring. Then*

$$R \cong \begin{bmatrix} \mathbb{M}_{n_1}(e_1 R e_1) & \mathbb{M}_{n_1 \times n_2}(e_1 R e_2) & \cdot & \cdot\;\cdot & \mathbb{M}_{n_1 \times n_k}(e_1 R e_k) \\ 0 & \mathbb{M}_{n_2}(e_2 R e_2) & \cdot & \cdot\;\cdot & \mathbb{M}_{n_2 \times n_k}(e_2 R e_k) \\ 0 & 0 & \mathbb{M}_{n_3}(e_3 R e_3) & \cdot\;\cdot & \mathbb{M}_{n_3 \times n_k}(e_3 R e_k) \\ \cdot & \cdot & \cdot & \cdot\;\cdot & \cdot \\ \cdot & \cdot & \cdot & \cdot\;\cdot & \cdot \\ 0 & 0 & \cdot & \cdot\;\cdot & \mathbb{M}_{n_k}(e_k R e_k) \end{bmatrix}$$

where $e_i R e_i$ is a division ring, $e_i R e_i \cong e_j R e_j$ for each $1 \leq i,j \leq k$ and $n_1, \ldots, n_k$ are any positive integers. Furthermore, if $e_i R e_j \neq 0$, then it is a one-dimensional left vector space over $e_i R e_i$ and a one-dimensional right vector space over $e_j R e_j$.

Proof Let R be an indecomposable right artinian, right *nonsingular*, right Σ-q ring. There exists an independent family $\mathcal{F} = \{e_i R : 1 \leq i \leq n\}$ of indecomposable right ideals such that $R = \oplus_{i=1}^{n} e_i R$. After renumbering, we may write $R = [e_1 R] \oplus [e_2 R] \oplus \cdots \oplus [e_k R]$, where for $1 \leq i \leq k$, $[e_i R]$ denotes the direct sum of those $e_j R$ that are isomorphic to $e_i R$. Let $[e_i R]$ be a direct sum of n_i copies of $e_i R$. Consider $1 \leq i < j \leq k$. We arrange the summands $[e_i R]$ in such a way that $l(e_j R) \leq l(e_i R)$. Suppose $e_j R e_i \neq 0$. Then we have an embedding of $e_i R$ into $e_j R$, hence $l(e_i R) \leq l(e_j R)$. But by assumption $l(e_j R) \leq l(e_i R)$, so $l(e_i R) = l(e_j R)$, we get $e_j R \cong e_i R$, which is a contradiction. Hence $e_j R e_i = 0$ for $j > i$.

Thus we have

$$R \cong \begin{bmatrix} \mathbb{M}_{n_1}(e_1 R e_1) & \mathbb{M}_{n_1 \times n_2}(e_1 R e_2) & \cdot & \cdot\;\cdot & \mathbb{M}_{n_1 \times n_k}(e_1 R e_k) \\ 0 & \mathbb{M}_{n_2}(e_2 R e_2) & \cdot & \cdot\;\cdot & \mathbb{M}_{n_2 \times n_k}(e_2 R e_k) \\ 0 & 0 & \mathbb{M}_{n_3}(e_3 R e_3) & \cdot\;\cdot & \mathbb{M}_{n_3 \times n_k}(e_3 R e_k) \\ \cdot & \cdot & \cdot & \cdot\;\cdot & \cdot \\ \cdot & \cdot & \cdot & \cdot\;\cdot & \cdot \\ 0 & 0 & \cdot & \cdot\;\cdot & \mathbb{M}_{n_k}(e_k R e_k) \end{bmatrix}$$

We have already seen earlier that each $e_i R e_i$ is a division ring, $e_i R e_i \cong e_j R e_j$ for each $1 \leq i,j \leq k$, and if $e_i R e_j \neq 0$ then it is a one-dimensional left vector space over $e_i R e_i$ as well as a one-dimensional right vector space over $e_j R e_j$ (see Lemma 19.28 and Proposition 19.30). $\qquad\square$

Let us consider the following condition:

($*$) : *For $1 \leq i,j \leq k$ with $i \neq j$ and primitive orthogonal idempotents e_i, e_j, either $e_i R e_j \neq 0$ or $e_j R e_i \neq 0$. In other words, for a right nonsingular ring either $e_i R$ is embeddable in $e_j R$ or $e_j R$ is embeddable in $e_i R$.*

We remark that ($*$) holds if R is an indecomposable right nonsingular serial ring.

Under the hypothesis (*) we have the following:

Theorem 19.33 *Let R be an indecomposable right artinian, right nonsingular ring with condition (*). Then R is a right Σ-q ring if and only if*

$$R \cong \begin{bmatrix} \mathbb{M}_{n_1}(D) & \mathbb{M}_{n_1 \times n_2}(D) & \cdot & \cdot \ \cdot & \mathbb{M}_{n_1 \times n_k}(D) \\ 0 & \mathbb{M}_{n_2}(D) & \cdot & \cdot \ \cdot & \mathbb{M}_{n_2 \times n_k}(D) \\ 0 & 0 & \mathbb{M}_{n_3}(D) & \cdot \ \cdot & \mathbb{M}_{n_3 \times n_k}(D) \\ \cdot & \cdot & \cdot & \cdot \ \cdot & \cdot \\ \cdot & \cdot & \cdot & \cdot \ \cdot & \cdot \\ 0 & 0 & \cdot & \cdot \ \cdot & \mathbb{M}_{n_k}(D) \end{bmatrix}$$

where D is a division ring and $n_1, \ldots, n_k$ are any positive integers.

Proof Let R be an indecomposable right artinian, right nonsingular, right Σ-q ring. By the above theorem,

$$R \cong \begin{bmatrix} \mathbb{M}_{n_1}(e_1 R e_1) & \mathbb{M}_{n_1 \times n_2}(e_1 R e_2) & \cdot & \cdot \ \cdot & \mathbb{M}_{n_1 \times n_k}(e_1 R e_k) \\ 0 & \mathbb{M}_{n_2}(e_2 R e_2) & \cdot & \cdot \ \cdot & \mathbb{M}_{n_2 \times n_k}(e_2 R e_k) \\ 0 & 0 & \mathbb{M}_{n_3}(e_3 R e_3) & \cdot \ \cdot & \mathbb{M}_{n_3 \times n_k}(e_3 R e_k) \\ \cdot & \cdot & \cdot & \cdot \ \cdot & \cdot \\ \cdot & \cdot & \cdot & \cdot \ \cdot & \cdot \\ 0 & 0 & \cdot & \cdot \ \cdot & \mathbb{M}_{n_k}(e_k R e_k) \end{bmatrix}$$

where each $e_i R e_i$ is a division ring and $e_i R e_i \cong e_j R e_j$ for each $1 \leq i, j \leq k$. Furthermore, by condition (*), we have $e_i R e_j \neq 0$ for each $1 \leq i < j \leq k$. Therefore each $e_i R e_j$ is a one-dimensional left vector space over $e_i R e_i$ as well as a one-dimensional right vector space over $e_j R e_j$. Let us denote the division ring $e_i R e_i$ by D. Then we have:

$$R \cong \begin{bmatrix} \mathbb{M}_{n_1}(D) & \mathbb{M}_{n_1 \times n_2}(D) & \cdot & \cdot \ \cdot & \mathbb{M}_{n_1 \times n_k}(D) \\ 0 & \mathbb{M}_{n_2}(D) & \cdot & \cdot \ \cdot & \mathbb{M}_{n_2 \times n_k}(D) \\ 0 & 0 & \mathbb{M}_{n_3}(D) & \cdot \ \cdot & \mathbb{M}_{n_3 \times n_k}(D) \\ \cdot & \cdot & \cdot & \cdot \ \cdot & \cdot \\ \cdot & \cdot & \cdot & \cdot \ \cdot & \cdot \\ 0 & 0 & \cdot & \cdot \ \cdot & \mathbb{M}_{n_k}(D) \end{bmatrix}$$

Conversely, suppose that

$$R \cong \begin{bmatrix} \mathbb{M}_{n_1}(D) & \mathbb{M}_{n_1 \times n_2}(D) & \cdot & \cdot \ \cdot & \mathbb{M}_{n_1 \times n_k}(D) \\ 0 & \mathbb{M}_{n_2}(D) & \cdot & \cdot \ \cdot & \mathbb{M}_{n_2 \times n_k}(D) \\ 0 & 0 & \mathbb{M}_{n_3}(D) & \cdot \ \cdot & \mathbb{M}_{n_3 \times n_k}(D) \\ \cdot & \cdot & \cdot & \cdot \ \cdot & \cdot \\ \cdot & \cdot & \cdot & \cdot \ \cdot & \cdot \\ 0 & 0 & \cdot & \cdot \ \cdot & \mathbb{M}_{n_k}(D) \end{bmatrix}$$

where D is a division ring and $n_1, \ldots, n_k$ are positive integers.

Clearly, R is an indecomposable, right nonsingular ring. By Eisenbud and Griffith [67], we know that R is an artinian serial ring. Therefore R is a right Σ-q ring. $\square$

19.3 Rings each of whose one-sided ideals is π-injective

A ring R is called a *right π-ring* if each right ideal of R is π-injective [138]. Clearly, a right uniform ring is a right π-ring.

Jain, López-Permouth, and Syed [138] gave the structure of right π-rings. Recall that a module M is said to be a *square* if $M \cong X^2$ for some module X. A module N is called a *square root* in M if N^2 is embeddable in M. If M contains no square roots then M is called *square free*. If every submodule of M contains a square root in M then M is called *square full*.

Theorem 19.34 *A right π-ring R is the direct sum of a square full semisimple artinian ring and a right square free right π-ring.*

Beidar and Jain [24] described the structure of right continuous right π-rings in the following theorem.

Theorem 19.35 *A ring R is a right continuous right π-ring if and only if R is the direct sum of finitely many rings of the following types:*

1. $G_n(D_1, \ldots, D_n, \Delta, V_1, \ldots, V_n)$.
2. An indecomposable nonlocal right continuous right π-ring.
3. A right continuous right π-ring all of whose idempotents are central.

Here $G_n(D_1, \ldots, D_n, \Delta, V_1, \ldots, V_n)$ denotes the ring of $(n+1) \times (n+1)$ matrices of the form

$$\begin{bmatrix} D_1 & V_1 & 0 & . & . & . & 0 \\ 0 & D_2 & V_2 & 0 & . & . & 0 \\ . & . & D_3 & V_3 & 0 & . & . \\ . & . & . & . & . & . & . \\ . & . & . & . & D_{n-1} & V_{n-1} & 0 \\ . & . & . & . & . & D_n & V_n \\ 0 & 0 & . & . & . & . & \Delta \end{bmatrix},$$

where Δ is a right continuous right π-ring all of whose idempotents are central with essential ideal P such that $D = \Delta/P$ is a division ring.

19.4 Rings each of whose one-sided ideals is a direct sum of π-injective right ideals

A ring R is called a right Σ-π *ring* if each right ideal of R is a direct sum of π-injective right ideals. Clark and Huynh have studied simple right Σ-π rings [50]. They proved the following.

Theorem 19.36 *If R is a simple right Σ-π ring. Then*

1. *R is either artinian; or*
2. *R is a nonselfinjective right Goldie ring in which every right ideal is a direct sum of uniform right ideals.*

Proof We first show that if A is any nonzero right ideal of any simple ring R, then R_R is a direct summand of A^n for some $n \in \mathbb{N}$. Since R is a simple ring and RA is a nonzero two-sided ideal of R, we have $R = RA$. Then, for some $n \in \mathbb{N}$, there exist $x_j \in R$ and $a_j \in A_i$ such that $1 = x_1 a_1 + \cdots + x_n a_n$. Thus it follows that $R = x_1 A + x_2 A + \cdots + x_n A$. Consider the epimorphism $\varphi : A^n \to R$ defined by $\varphi(a_1, \ldots, a_n) = x_1 a_1 + \cdots + x_n a_n$ for all $a_1, \ldots, a_n \in A$. Since R_R is projective, φ splits and so R_R is isomorphic to a direct summand of A^n, as desired.

We next show that R contains a uniform right ideal. Suppose to the contrary that R has no uniform right ideal. Then we must have $\mathrm{Soc}(R_R) = 0$. If A is a nonzero right ideal of R, then by assumption, $A = \oplus_{i \in \mathcal{I}} A_i$ where each A_i is π-injective. Since we have assumed that R does not contain a uniform right ideal, each A_i must be decomposable, say $A_i = C_i \oplus D_i$ where C_i and D_i are nonzero right ideals. Then C_i and D_i are mutually injective. Consequently, C_i is D_i^n-injective and D_i is C_i^m-injective for all $n, m \in \mathbb{N}$. Since, by above, R_R is isomorphic to a direct summand of C^k for some $k \in \mathbb{N}$, it follows that D_i is R-injective, that is, D_i is injective. Similarly, C_i is also injective. Thus A_i is injective for each $i \in \mathcal{I}$ and R_R is also injective. Now we choose A to be a maximal right ideal of R. Suppose that the index set $\mathcal{I}$ is finite. Then A_R is injective and hence a direct summand of R_R. Thus any complement of A_R in R_R is a minimal right ideal of R. This contradicts the fact that $\mathrm{Soc}(R_R) = 0$. Thus $\mathcal{I}$ must be infinite. In this case, let $\mathcal{H}, \mathcal{K}$ be two infinite disjoint subsets of $\mathcal{I}$ with $\mathcal{H} \cup \mathcal{K} = \mathcal{I}$. Further set $U = \oplus_{i \in \mathcal{H}} A_i$ and $V = \oplus_{i \in \mathcal{K}} A_i$. Then, since $\mathcal{H}$ and $\mathcal{K}$ are infinite, U and V cannot be direct summands of R_R. However their injective hulls $\widehat{U}$ and $\widehat{V}$, respectively, are summands of R_R (since R_R is injective). It follows that $\widehat{U} \neq U$ and $\widehat{V} \neq V$. Now write $R_R = \widehat{U} \oplus \widehat{V} \oplus W$ for some right ideal W of R. Then, since $A = U \oplus V$ is a maximal right ideal, W must be zero and this gives $R/A \cong (\widehat{U}/U) \oplus (\widehat{V}/V)$ where both $\widehat{U}/U$ and $\widehat{V}/V$ are nonzero. This contradicts the maximality of A. This contradiction shows that R has a uniform right ideal, as claimed.

Then, by ([107], Theorem 2), R is right Goldie. Once again, let A be a nonzero right ideal of R and write $A = \oplus_{i \in \mathcal{I}} A_i$ where each A_i is a nonzero π-injective right ideal. Then, since R is right Goldie, the index set $\mathcal{I}$ must be finite and we may assume that each A_i is indecomposable and so uniform. Now if there are distinct indices $i, j \in \mathcal{I}$ for which $A_i \oplus A_j$ is not π-injective, it follows that A_i and A_j are not mutually injective. It is then straightforward to see that R_R is not injective.

We now show that R is also not left self-injective. Assume to the contrary that $_R R$ is injective and let n be the (finite) Goldie dimension of R_R. If $_R R$ has infinite Goldie dimension, then by the left injectivity of R we can produce a set of $n+1$ nonzero orthogonal idempotents e_i in R. This produces the right ideal $e_1 R \oplus \cdots \oplus e_n R \oplus e_{n+1} R$ in R, which is impossible since the Goldie dimension of R_R is n. Hence $_R R$ has finite Goldie dimension and so R is left Goldie. This is a contradiction because R is not right self-injective. Thus R is not left self-injective. Hence, in this case, R is a ring satisfying (2).

On the other hand, if there is a right ideal A for which there are $i \neq j$ such that $A_i \oplus A_j$ is π-injective, then A_i and A_j are mutually injective. By our first paragraph, there are $m, n \in \mathbb{N}$ for which R_R is isomorphic to a direct summand of A_i^n and to a direct summand of A_j^m. Then, by ([12], Propositions 16.10 and 16.13), A_i^n and A_j^m are also mutually injective. Then, applying ([12], Propositions 16.13 and 16.10) once more, it follows that A_i^n is R_R-injective and so R_R is R_R-injective, that is, R is right self-injective. Thus it follows that R is right artinian (see [161, Corollary 13.4]). $\qquad\qquad\qquad\qquad\qquad\qquad\qquad\qquad\qquad\qquad\quad\square$

19.5 Rings each of whose one-sided ideals is weakly injective

López-Permouth, Rizvi, and Yousif [168] studied rings where each one-sided ideal is weakly injective and provided a characterization for such rings. We start with some definitions and observations. Let M, N be right R-modules. Then M is called an *N-tight module* if for every submodule K of N, if N/K embeds in $E(M)$ then N/K embeds in M. It is easy to observe that any weakly N-injective module is N-tight. If a module M is N-tight for every finitely generated module N, then we say that M is *tight*.

Theorem 19.37 *The following are equivalent for a ring R:*

1. *Each right ideal of R is weakly injective.*
2. *R is a semiprime Goldie ring.*
3. *Each left ideal of R is weakly injective.*

Proof $(1) \Rightarrow (2)$. We first claim that R has finite right Goldie dimension. Assume to the contrary that R has infinite right Goldie dimension. Then there exists an essential right ideal A of R such that $A = \oplus_{i=1}^{\infty} x_i R$. By assumption, A is weakly injective. So there exists a submodule X of $E(A) = E(R_R)$, such that $R_R \subset X \cong A$. We may then write $X = \oplus_{i=1}^{\infty} x_i' R$, where $x_i' R \cong x_i R$. Since R_R is finitely generated, $R_R \subset \oplus_{i=1}^{k} x_i' R$, for some $k \in \mathbb{N}$. Thus $R \cap x_i' R = 0$ for all $i > k$, which contradicts the essentiality of R in $E(R_R)$. Thus R has finite Goldie dimension.

Now we will show that R is semiprime. Let I be any nonzero right ideal of R. Let I' be a complement of I in R. Then $I \oplus I' \subset_e R$. As $I \oplus I'$ is weakly injective, there exists an embedding $f : R \to I \oplus I'$. Let $x = f(1) = a + b, a \in I, b \in I'$. For any nonzero element $y \in I, f(y) = f(1)y = xy = (a+b)y = ay$, since $by \in I \cap I' = 0$. Therefore $ay \in I^2$ and $ay \neq 0$. This shows that R has no nonzero nilpotent right ideal. Thus R is semiprime.

Next, we claim that R is right nonsingular. Assume $Z_r(R) \neq 0$. Write $E(Z_r(R)) \oplus K = E(R_R)$ and consider $1 \in R \subset E(R)$ as a sum $1 = a + b$, where $a \in E(Z_r(R))$ and $b \in K$. As $E(Z_r(R))$ is singular, there exists an essential right ideal I of R such that $aI = 0$. But then $bI = I$ and hence $I \subset K$ is nonsingular. Since $0 = Z(I) = I \cap Z_r(R)$ and I is essential in R_R, we conclude that $Z_r(R) = 0$, as claimed.

This shows that R is semiprime right Goldie. Thus the injective hull of R_R is its complete ring of right quotients Q. Let $q \in Q$. Since R is right weakly injective there exists $q' \in Q$ such that the right annihilator of q' is zero and $1, q \in q'R$. It follows that $q' = r^{-1}$ for some $r \in R$ and there exists $s \in R$ such that $q = r^{-1}s$. Therefore Q is also a left ring of quotients for R and hence R is left Goldie as well. Thus R is a semiprime Goldie ring.

$(2) \Rightarrow (1)$. We will show that every nonsingular right R-module is weakly injective. Let M be an arbitrary nonsingular right R-module. Since R is semiprime Goldie there exists $M' = \oplus_{i \in I} U_i$, a direct sum of (possibly infinitely many) nonzero uniform submodules of M, which is essential in M. Let N be a finitely generated submodule of $E(M) = E(M')$. It is not difficult to see that every nonsingular module over a semiprime Goldie ring is tight. Thus M' is tight and so there exists an embedding $\varphi : N \to M'$. Note that $\varphi(N)$ is finitely generated. Thus there exists a finite subset $J \subset I$ such that $\varphi(N) \subset_{i \in J} U_i$. It follows that, for some K, $E(\varphi(N)) \oplus K = \oplus_{i \in J} E(U_i)$. By the Krull–Schmidt Theorem, we get that $E(\varphi(N)) \cong \oplus_{i \in I'} E(U_i)$ for some $I' \subset J$. So there exists a new embedding $\varphi_1 : N \to E(M)$ such that $\varphi_1(N) \subset_e \oplus_{i \in I'} E(U_i)$. Now $\oplus_{i \in I'} U_i$ is finite dimensional and hence weakly injective. Thus there exists $X_1 \subset E(\oplus_{i \in I'} U_i)$ such that $X_1 = \oplus_{i \in I'} V_i$ and $U_i \cong V_i$, for all $i \in I'$ and $\varphi_1(N) \subset_e \oplus_{i \in I'} V_i$. Note that $\varphi_1(N) \subset \oplus_{i \in I'} V_i \oplus (\oplus_{i \in I - I'} U_i) \cong \oplus_{i \in I} U_i$. Thus we have that, for some $K_1 \subset E(M)$, $E(N) \oplus K_1 = E(M) = E(\varphi_1(N)) \oplus E(\oplus_{i \in I - I'} U_i) = E((\oplus_{i \in I'} E_i) \oplus (\oplus_{i \in I - I'} E_i))$ where $E_i = E(V_i)$ when $i \in I'$ and $E_i = E(U_i)$ when $i \in I - I'$.

On the other hand, there exists a collection $\{T_\lambda\}_{\lambda \in \Lambda}$ of indecomposable injectives such that $K_1 = E(K_1) = E(\oplus_{\lambda \in \Lambda} T_\lambda$. Since $E(N) \cong E(\varphi_1(N))$, we may write $E(N) = \oplus_{i \in I'} W_i$, with $W_i \cong E_i$ for all $i \in I'$. Thus $E(M) = E(N) \oplus K_1 = E(\oplus_{i \in I'} W_i \oplus \oplus_{\lambda \in \Lambda} T_\lambda)$. It follows then that $(\oplus_{i \in I'} W_i) \oplus (\oplus_{\lambda \in \Lambda} T_\lambda) \cong (\oplus_{i \in I'} E_i) \oplus (\oplus_{i \in I - I'} E_i)$. Then, using the Azumaya–Krull–Schmidt Theorem, we get a bijection $\sigma : I - I' \to \Lambda$ such that, for every $i \in I - I'$, $E_i \cong V_{\sigma(i)}$. Without loss of generality, let us assume that $\Lambda = I - I'$ and that σ is the identity map. Consider the obvious isomorphism $f : \varphi_1(N) \oplus (\oplus_{i \in I - I'} U_i) \to E(M)$. Since the domain of f embeds as an essential submodule of M, the injectivity of $E(M)$ yields the existence of a monomorphism $\widehat{f} : M \to E(M)$ extending f for which $N \subset \widehat{f}(M)$. Thus M is weakly injective, as claimed. Now, since R is right nonsingular, it follows that each right ideal of R is weakly injective.

The equivalence of (2) and (3) follows similarly. $\qquad\qquad \square$

19.6   Rings each of whose one-sided ideals is quasi-projective

A ring R is called a *left qp-ring* if each left ideal of R is quasi-projective as a left R-module. Jain and Singh [145] proved the following for perfect left qp-rings.

Theorem 19.38 *Let R be a local perfect ring. Then R is a left qp-ring if and only if:*

1. *$J(R)^2 = 0$; or*
2. *R is a left artinian left principal ideal ring.*

In the next result Jain and Singh [145] have shown that perfect left qp-rings may be presented as triangular matrix rings.

Theorem 19.39 *Let R be a perfect left qp-ring. Then:*

1. *R is semiprimary.*

2. *R is an upper triangular matrix ring of the form $\begin{pmatrix} S & M \\ 0 & T \end{pmatrix}$ where S is a hereditary semiprimary ring, T is a finite direct sum of local left qp-rings, and M is an $(S;T)$-bimodule such that M is projective as a left S-module.*

Theorem 19.40 *Let R be an indecomposable perfect ring which has a faithful projective injective left R-module M. Then R is a left qp-ring if and only if:*

1. *R is a local principal left ideal ring; or*
2. *R is a left artinian left hereditary ring.*

Theorem 19.41 *Let R be a perfect left qp-ring and A be a left ideal of R. Then the projective dimension of A as a left R-module is 0 or ∞.*

Theorem 19.42 *Let R be a perfect left qp-ring. Then the left global dimension of R is either $0, 1$, or ∞.*

Jain and Singh [145] have also given the following example of a local primary ring which is a left qp-ring but not a right qp-ring.

Example Let F be a field which has an isomorphism $a \mapsto \bar{a}$ that is not an automorphism and let $\overline{F}$ be the subfield of images $\bar{a}$, where $a \in F$. Take x to be an indeterminate over F. Let $F[x]$ be the ring of polynomials of the form $a_0 + a_1 x + a_2 x^2$, $a_i \in F$ with multiplication defined by the rule $xa = \bar{a}x$, $x^3 = 0$ together with the distributive law. Its radical $J(R) = \{a_1 x + a_2 x^2 : a_i \in F\}$ is such that $J(R)^2 \neq 0$ and $J(R)^3 = 0$ and is a maximal left ideal of R. Thus R is a local perfect ring. Also, $J(R)$ is not principal as a right ideal. Therefore R is a left qp-ring but not a right qp-ring. $\square$

Goel and Jain [88] continued the study of left qp-rings and proved the following.

Theorem 19.43 *Let R be a semiprime left noetherian left qp-ring. Then R is left hereditary.*

The following example of Goel and Jain [88] shows that the left qp-property is not Morita-invariant.

Example Let $R = \mathbb{Z}/(p^2)$ where p is a prime number. Then R is a qp-ring but for $n > 1$ the matrix ring $M_n(R)$ is not a left or right qp-ring. $\qquad\square$

19.44 Questions

1. Is every right Σ-q ring directly finite?
2. Let R be a von Neumann regular right self-injective right Σ-q ring. Must R have a bounded index of nilpotence?
3. Do the lower nil radical and upper nil radical coincide for a right q ring? Do these coincide for a right Σ-q ring?
4. Characterize when group rings $R[G]$ are q-rings.
5. Characterize when group rings $R[G]$ are Σ-q rings.

References

[1] A. N. Abyzov, Weakly regular models over normal rings, *Siberian Mathematical Journal*, 49, 4 (2008), 576–586.

[2] A. N. Abyzov, Generalized SV-modules, *Siberian Mathematical Journal*, 50, 3 (2009), 379–384.

[3] A. N. Abyzov, A. A. Tuganbaev, Rings over which all modules are I_0-modules. II, *J. Math. Sci. (New-York)*, 162, 5 (2009), 587–593.

[4] A. N. Abyzov, A. A. Tuganbaev, Submodules and direct summands, *J. Math. Sci. (New-York)*, 164, 1 (2010), 1–20.

[5] J. Abuhlail, S. K. Jain, A. Laradji, On almost perfect rings, preprint.

[6] J. Ahsan, Rings all of whose cyclic modules are quasi-injective, *Proc. London Math. Soc.*, 27 (1973), 425–439.

[7] Y. Akizuki, Teilerkettensatz und Vielfachenkettensatz, *Proceedings of the Physico-Mathematical Society of Japan*, 17, 3 (1935), 337–345.

[8] A. N. Alahmadi, S. K. Jain, Pramod Kanwar, J. B. Srivastava, Group algebras in which the complements are direct summands, (Russian) *Fundamental and Applied Math*, 11, 3 (2005), 3–11.

[9] A. N. Alahmadi, S. K. Jain, A. Leroy, *ADS* modules, *J. Algebra*, 352, 1, 215–222.

[10] A. Al-Huzali, S. K. Jain, S. R. López-Permouth, On the weak relative-injectivity of rings and modules, *Noncommutative Ring Theory, Lecture Notes in Mathematics*, no. 1448, Springer-Verlag, Heidelberg Berlin - New York 1990, 93–98.

[11] A. Al-Huzali, S. K. Jain, S. R. López-Permouth, Rings whose cyclics have finite Goldie dimension, *J. Algebra*, 153 (1992), 37–40.

[12] F. W. Anderson, K. R. Fuller, *Rings and Categories of Modules*. Second edition, Springer, New York, 1992.

[13] E. Armendariz and K. Hummel, Restricted semiprimary rings, Ring theory (*Proc. Conf., Park City, Utah*, 1971), Academic Press, New York, 1972, pp. 1–8.

[14] E. Armendariz, Rings with DCC on essential left ideals, *Comm. Alg.*, 8 (1980), 299–308.

[15] K. Asano, Uber verallgemeinerte abelsche gruppe mit hyperkomplexem operatorenring und ihre Anwendungen, (German) *Jap. J. Math.*, 15 (1939), 231–253.

[16] M. Auslander, On the dimension of modules and algebras III. Global dimension, *Nagoya Math. J.*, 9 (1955), 67–77.

[17] S. Barthwal, S. Jhingan, P. Kanwar, A simple ring over which proper cyclics are continuous is a PCI-domain, *Can. Math. Bull.*, 41 (1998), 261–266.

[18] G. Baccella, Von Neumann regularity of V-rings with artinian primitive factor rings, *Proc. Amer. Math. Soc.*, 103, 3 (1988), 747–749.

[19] G. Baccella, Semiartinian V-rings and Semiartinian von Neumann Regular Rings, *J. Algebra*, 173 (1995), 587–612.

[20] H. Bass, Finitistic dimension and a homological generalization of semi-primary rings, *Trans. Amer. Math. Soc.*, 95 (1960), 466–488.

[21] S. Bazzoni, L. Salce, Strongly flat covers, *J. London Math. Soc.*, 2, 66 (2002), 276–294.

[22] M. Behboodi, A. Ghorbani, A. Moradzadeh-Dehkordi, S. H. Shojaee, On left Koethe rings and an analogue of the Cohen-Kaplansky theorem, *Proc. Amer. Math. Soc.*, to appear.

[23] K. I. Beidar, Y. Fong, W.-F. Ke, and S. K. Jain, An example of a right q-ring, *Israel Journal of Mathematics*, 127 (2002), 303–316.

[24] K. I. Beidar, S. K. Jain, The structure of right continuous right π-rings, *Comm. Alg.*, 32, 1 (2004), 315–332.

[25] L. Bican, R. El Bashir, E. E. Enochs, All modules have flat covers, *Bull. London Math. Soc.*, 33 (2001), 385–390.

[26] J. E. Björk, Rings satisfying a minimum condition on principal ideals, *J. Reine Angew. Math.*, 236 (1969), 112–119.

[27] J. E. Björk, Radical properties of perfect modules, *J. Reine Angew. Math.*, 253 (1972), 78–86.

[28] N. Bourbaki, *Elements de Mathematique, Fasc. XXVII Algebre Commutative*, Actualites Sci. Indust., no. 1290, Hermann, Paris, 1961.

[29] A. K. Boyle, Ph.D. thesis, Rutgers, The State University, New Brunswick, New Jersey, 1971.

[30] A. K. Boyle, Hereditary QI rings, *Trans. Amer. Math. Soc.*, 192 (1974), 115–120.

[31] A. K. Boyle, K. R. Goodearl, Rings over which certain modules are injective, *Pacific J. Math.*, 58, 1 (1975), 43–53.

[32] B. Brown, N. H. McCoy, The maximal regular ideal of a ring, *Proc. Amer. Math. Soc.*, (1950), 165–171.

[33] H. H. Brungs, G. Törner, Chain rings and prime ideals, *Arch. Math.*, Vol. 27 (1976), 253–260.

[34] R. T. Bumby, Modules which are isomorphic to submodules of each other, *Arch. Math.*, 16 (1965), 184–185.

[35] K. A. Byrd, Right self-injective rings whose essential ideals are two-sided, *Pacific J. Math.*, 82 (1979), 23–41.

[36] W. Caldwell, Ph.D. thesis, Rutgers, The State University, New Brunswick, New Jersey, 1966.

[37] W. Caldwell, Hypercyclic rings, *Pacific J. Math.*, 24 (1968), 29–44.

[38] V. Camillo, G. Krause: Problem 12. In: R. Gordon, (ed.) *Ring Theory, Proceedings of a Conference on Ring Theory, Park City, Utah*, Academic Press, New York, 1972, p. 377.

[39] V. Camillo, Commutative rings whose quotients are Goldie, *Glasgow Math. J.*, 16 (1975), 32–33.

[40] V. Camillo, Distributive modules, *J. Algebra*, 36, 1 (1975), 16–25.

[41] H. Cartan, S. Eilenberg, *Homological Algebra*. The Princeton University Press, Princeton, 1956.

[42] S. Chairat, D. V. Huynh, C. Somsup, On rings over which the injective hull of each cyclic module is Σ-extending, *J. Alg. Appl.*, to appear.

[43] S. U. Chase, Direct products of modules, *Trans. Amer. Math. Soc.*, 97 (1960), 457–473.

[44] A. W. Chatters, The restricted minimum condition in noetherian hereditary rings, *J. London Math. Soc.*, 2, 4 (1971), 83–87.

[45] A. W. Chatters, A characterization of right noetherian rings, *Quart. J. Math. Oxford*, 33, 2 (1982), 65–69.

[46] A. W. Chatters and C. R. Hajarnavis, Rings in which every complement right ideal is a direct summand, *Quart. J. Math.*, 28 (1977), 61–80.

[47] A. W. Chatters, C. R. Hajarnavis, *Rings with Chain Conditions*. Pitman, London 1980.

[48] J. Chen, W. Li, On artinness of right CF rings, *Comm. Alg.*, 32, 11 (2004), 4485–4494.

[49] J. Clark, C. Lomp, N. Vanaja and R. Wisbauer, *Lifting Modules: Supplements and Projectivity in Module Theory*, Frontiers in Mathematics, Birkhäuser Verlag, Basel, 2006.

[50] J. Clark, D. V. Huynh, Simple rings with injectivity conditions on one-sided ideals, *Bull. Aust. Math. Soc.*, 76 (2007), 315–320.

[51] I. S. Cohen, Commutative rings with restricted minimum condition, *Duke Math. J.*, 17, 1 (1950), 27–42.

[52] I. S. Cohen, I. Kaplansky, Rings for which every module is a direct sum of cyclic modules, *Math. Z.*, 54 (1951), 97–101.

[53] P. M. Cohn, Free associative algebras, *Bull. London Math. Soc.*, 1 (1969), 1–39.

[54] J. Cozzens, Ph.D. thesis, Rutgers, The State University, New Brunswick, New Jersey, 1969.

[55] J. Cozzens and C. Faith, *Simple Noetherian Rings*, Cambridge Tracts in Math. and Math. Phys., Cambridge University Press, Cambridge, 1975.

[56] R. F. Damiano, A right *PCI* ring is right noetherian, *Proc. Amer. Math. Soc.*, 77, 1 (1979), 11–14.

[57] J. Dauns, *Modules and Rings*, Cambridge University Press, Cambridge, 1994.

[58] J. Dauns, L. Fuchs, Infinite Goldie dimensions, *J. Algebra*, 115, 2 (1988), 297–302.

[59] H. Q. Dinh, P. A. Guil Asensio, S. R. López-Permouth, On the Goldie dimension of rings and modules, *J. Algebra*, 305 (2006), 937–948.

[60] Yu. A. Drozd, On generalized uniserial rings, *Mat. Zametki*, 18, 5 (1975), 707–710.

[61] N. I. Dubrovin, Rational closures of group rings of left-ordered groups, *Russian Acad. Sci. Sb. Math.*, 79, 2 (1994), 231–263.

[62] N. V. Dung, Some conditions for a self-injective ring to be quasi-Frobenius. *Studia Sci. Math. Hungar.*, 24, 2 (1989), 349–354.

[63] N. V. Dung, D. V. Huynh, R. Wisbauer, On modules with finite uniform and Krull dimension, *Arch. Math.*, 57 (1991), 122–132.

[64] N. V. Dung, P. F. Smith, On semi-artinian *V*-modules, *J. Pure Appl. Algebra*, 82, 1 (1992), 27–37.

[65] N. V. Dung, D. V. Huynh, P. F. Smith, R. Wisbauer, *Extending Modules*, Pitman, London, 1994.

[66] S. Eilenberg, T. Nakayama, On the dimension of modules and algebras II, *Nagoya J. Math.*, 9 (1955), 1–16.

[67] D. Eisenbud, P. Griffith, The structure of serial rings, *Pacific J. Math.*, 36, 1 (1971), 109–121.

[68] N. Er, Some remarks on a question of Faith, *Contemp. Math.*, 480 (2009), 133–137.

[69] A. Facchini, C. Parolin, Rings whose proper factors are right perfect, *Colloq. Math.*, 122 (2011), 191–202.

[70] C. Faith, Rings with ascending chain condition on annihilators, *Nagoya Math. J.*, 27 (1966), 179–191.

[71] C. Faith, On Kothe rings, *Mathematische Annalen*, 164 (1966), 207–212.

[72] C. Faith, Modules finite over endomorphism ring, *Proceedings of the Tulane University Symposium in Ring Theory*, Lecture Notes in Math., 246 (1971), 145–189.

[73] C. Faith, When are proper cyclics injective?, *Pacific J. Math.*, 45 (1973), 97–112.

[74] C. Faith, *Algebra: Rings, Modules, and Categories I*, Springer, Berlin, 1973.

[75] C. Faith, *Algebra II*, Springer, Berlin, 1976.

[76] C. Faith, On hereditary rings and Boyle's conjecture, *Arch. Math.*, 27 (1976), 113–119.

[77] C. Faith, *Embedding Modules in Projectives. A Report on a Problem*, Lecture Notes in Math., Vol. 951, Springer-Verlag, 1982, 21–40.

[78] C. Faith, *Rings and Things and a Fine Array of Twentieth Century Associative Algebra*, Mathematical Surveys and Monographs 65, AMS, Providence, RI, 1999.

[79] C. Faith, Indecomposable injective modules and a theorem of Kaplansky, *Comm. Alg.*, 30, 12, (2002), 5875–5889.

[80] C. Faith, When cyclic modules have Σ-injective hulls, *Comm. Alg.*, 31, 9 (2003), 4161–4173.

[81] D. R. Farkas, R. L. Snider, Group algebras whose simple modules are injective, *Trans. Amer. Math. Soc.*, 194 (1974), 241–248.

[82] J. W. Fisher, *Von Neumann Regular Rings Versus V-rings*, Lecture Notes in Pure and Appl. Math., Vol. 7, Dekker, New York, 1974, 101–119.

[83] J. W. Fisher, R. L. Snider, On the von Neumann regularity of rings with regular prime factor rings, *Pacific J. Math.*, 54, 1 (1974), 135–144.

[84] L. Fuchs, *Infinite Abelian Groups, Pure and Applied Mathematics*, A Series of Monographs and Textbooks, Vol. 36, No. 1, Academic Press, New York, San Francisco, London, 1970.

[85] K. R. Fuller, On indecomposable injectives over Artinian rings, *Pacific J. Math.*, 22, 1 (1969), 115–135.

[86] J. L. Garcia Hernández, J. L. Gómez Pardo, V-rings relative to Gabriel topologies, *Comm. Alg.*, 13, 1 (1985), 59–83.

[87] S. C. Goel, S. K. Jain, S. Singh, Rings whose cyclic modules are injective or projective, *Proc. Amer. Math. Soc.*, 53 (1975), 16–18.

[88] S. C. Goel, S. K. Jain, Semiperfect rings with quasi-projective left ideals, *Math. J. Okayama*, 19 (1976), 39–43.

[89] V. K. Goel, S. K. Jain, π-injective modules and rings whose cyclic are π-injective, *Comm. Alg.*, 6 (1978), 59–72.

[90] J. L. Gómez Pardo, N. V. Dung, R. Wisbauer, Complete pure-injectivity and endomorphism rings, *Proc. Amer. Math. Soc.*, 118, 4 (1993), 1029–1034.

[91] J. L. Gómez Pardo, P. A. Guil Asensio, Essential embedding of cyclic modules in projectives, *Trans. Amer. Math. Soc.*, 349, 11 (1997), 4343–4353.

[92] J. L. Gómez Pardo, P. A. Guil Asensio, Rings with finite essential socle, *Proc. Amer. Math. Soc.*, 125, 4 (1997), 971–977.

[93] J. L. Gómez Pardo, P. A. Guil Asensio, Indecomposable decompositions of finitely presented pure-injective modules, *J. Algebra*, 192 (1997), 200–208.

[94] J. L. Gómez Pardo, P. A. Guil Asensio, On the Goldie dimension of injective modules, *Proc. Edinb. Math. Soc.*, 2, 41 (1998), 265–275.

[95] J. L. Gómez Pardo, P. A. Guil Asensio, *Chain Conditions on Direct Summands and Pure Quotient Modules, Interactions Between Ring Theory and Representations of Algebras* (Murcia), Lecture Notes in Pure and Appl. Math., 210, Dekker, New York, 2000, 195–203.

[96] K. R. Goodearl, Singular torsion and the splitting properties, *Mem. Amer. Math. Soc.*, 124, 1972.

[97] K. R. Goodearl, *Von Neumann Regular Rings*, Pitman, London, 1979.

[98] K. R. Goodearl, Artinian and noetherian modules over regular rings, *Comm. Alg.*, 8 (1980), 477–504.

[99] K. R. Goodearl, R. B. Warfield, *An Introduction to Noncommutative Noetherian Rings*, London Math. Soc. 61, Cambridge University Press, Cambridge, 2004.

[100] R. Gordon, J. C. Robson, Krull dimensions, *Memoirs of the Amer. Math. Soc.*, 1978.

[101] J. M. Goursaud, Sur les V-anneaux réguliers, *Séminaire d'Algébre Noncommutative*, U. E. R. Math., Univ. Paris XI, 1976.

[102] J. M. Goursaud and J. Valette, Sur l'enveloppe injective des anneaux de groupes réguliers, *Bulletin of Math. Soc. France*, 103 (1975), 91–102.

[103] P. A. Guil-Asensio, S. K. Jain, A. K. Srivastava, Direct sums of injective and projective modules, *J. Algebra*, 24 (2010), 1429–1434.

[104] R. N. Gupta, On f-injective modules and semi-hereditary rings, *Proc. Nat. Inst. Sci. India Part A*, 35 (1969), 323–328.

[105] C. R. Hajarnavis, Noncommutative rings whose homomorphic images are self-injective, *Bull. London Math. Soc.*, 5 (1973), 70–74.

[106] C. R. Hajarnavis and T. H. Lenagan, Localisation in Asano Orders, *J. Algebra*, 21 (1972), 441–449.

[107] R. Hart, Simple rings with uniform right ideals, *J. London Math. Soc. (Ser. 1)*, 42 (1967), 614–617.

[108] C. Hopkins, Rings with minimal conditions for left ideals, *Ann. of Math.*, 40 (1939), 712–730.

[109] D. A. Hill, Semi-perfect q-rings, *Mathematische Annalen*, 200 (1973), 113–121.

[110] C. J. Holston, S. K. Jain, A. Leroy, Rings over which cyclics are direct sums of projective and CS or noetherian, *Glasgow Math. J.*, 52 (2010), 103–110.

[111] C. Holston, D. V. Huynh, Some results on V-rings and WV-rings, preprint.

[112] D. V. Huynh, Some characterizations of hereditarily Artinian rings, *Glasgow Math. J.*, 28 (1986), 21–23.

[113] D. V. Huynh, Rings with ACC on essential right ideals, *Math. Japonica*, 35 (1990), 707–712.

[114] D. V. Huynh, A characterization of noetherian rings by cyclic modules, *Proc. Edinburgh Math. Soc.*, 39 (1996), 253–262.

[115] D. V. Huynh, Structure of some noetherian SI rings, *J. Algebra*, 254 (2002), 362–374.

[116] D. V. Huynh, P. Dan, On rings with restricted minimum condition, *Arch. Math.*, 51 (1988), 313–326.

[117] D. V. Huynh, N. V. Dung, A characterization of Artinian rings, *Glasgow Math. J.*, 30 (1988), 67–73.

[118] D. V. Huynh, N. V. Dung, P. F. Smith, Rings characterized by their right ideals or cyclic modules, *Proc. Edinburgh Math. Soc.*, 32 (1989), 356–362.

[119] D. V. Huynh, S. K. Jain, S. R. López-Permouth, When is a simple ring noetherian?, *J. Algebra*, 184 (1996), 784–794.

[120] D. V. Huynh, S. K. Jain, S. R. López-Permouth, When cyclic singular modules over a simple ring are injective, *J. Algebra*, 263 (2003), 185–192.

[121] D. V. Huynh, N. V. Dung and R. Wisbauer, On quasi-injective modules with ACC or DCC on essential submodules, *Arch. Math.*, 53 (1989), 252–255.

[122] D. V. Huynh and P. F. Smith, Some rings characterised by their modules, *Comm. Alg.*, 18, 6 (1990), 1971–1988.

[123] D. V. Huynh, S. T. Rizvi, An affirmative answer to a question on noetherian rings, *J. Alg. Appl.*, 7, 1 (2008), 47–59.

[124] D. V. Huynh, R. Wisbauer, Self-projective modules with π-injective factor modules, *J. Algebra*, 153 (1992), 13–21.

[125] M. Ikeda, Some generalisations of quasi-Frobenius rings, *Osaka J. Math.*, 3 (1951), 228–239.

[126] G. Ivanov, Ph.D. thesis, Australian National University, Canberra, 1972.

[127] G. Ivavov, Non-local rings whose ideals are quasi-injective, *Bulletin of the Australian Mathematical Society*, 6 (1972), 45–52.

[128] G. Ivavov, Non-local rings whose ideals are quasi-injective: Addendum, *Bulletin of the Australian Mathematical Society*, 12 (1975), 159–160.

[129] G. Ivanov, Decomposition of modules over serial rings, *Comm. Alg.*, 3 (1975), 1031–1036.

[130] S. K. Jain, Rings whose cyclic modules have certain properties and the duals, *Ring Theory: Lecture Notes Series in Pure and Applied Math.*, Marcel Dekker, 25 (1977), 143–160.

[131] S. K. Jain, S. Jain, Restricted regular rings, *Math. Z.*, 121 (1971), 51–54.

[132] S. K. Jain, P. Kanwar, S. Malik, J. B. Srivastava, KD_∞ is a CS algebra, *Proc. Amer. Math. Soc.*, 128, 2 (2000), 397–400.

[133] S. K. Jain, T. Y. Lam, A. Leroy, Ore extensions and V-domains, *Rings, Modules and Representations, Contemp. Math. Series AMS*, 480 (2009), 263–288.

[134] S. K. Jain, S. R. López-Permouth, A generalization of the Wedderburn–Artin theorem, *Proc. Amer. Math Soc.*, 106 (1989) 10–23.

[135] S. K. Jain, S. R. López-Permouth, Rings whose cyclic are essentially embeddable in projective modules, *J. Algebra*, 128 (1990), 257–269.

[136] S. K. Jain, S. R. López-Permouth, A survey of theory of weakly injective modules, *Computational Algebra*, Marcel Dekker, N.Y., 1994, 205–232.

[137] S. K. Jain, S. R. López-Permouth, S. Singh, On a class of QI-rings, *Glasgow Math. J.*, 34 (1992), 75–81.

[138] S. K. Jain, S. R. López-Permouth, S. R. Syed, Rings with π-injective right ideals, *Glasgow Math. J.*, 41 (1999), 167–181.

[139] S. K. Jain, D. S. Malik, q-Hypercyclic Rings, *Canadian J. Math.*, 37 (1985), 452–466.

[140] S. K. Jain, S. Mohamed, Rings whose cyclic modules are continuous, *Journal Indian Math. Soc.*, 42 (1978), 197–202.

[141] S. K. Jain, S. H. Mohamed, S. Singh, Rings in which each right ideal is quasi-injective, *Pacific J. Math.*, 31 (1969), 73–79.

[142] S. K. Jain, B. Mueller, Semiperfect rings whose proper cyclic modules are continuous, *Archiv der Math.*, 37 (1981), 140–143.

[143] S. K. Jain, H. Saleh, Rings with finitely generated injective (quasi-injective) hulls of cyclic modules, *Comm. Alg.*, 15 (1987), 1679–1687.

[144] S. K. Jain, H. Saleh, Rings whose (proper) cyclic modules have cyclic π-injective hulls, *Archiv der Math.*, 48 (1987), 109–115.

[145] S. K. Jain, S. Singh, Rings with quasi-projective left ideals, *Pacific J. Math.*, 60 (1975), 169–181.

[146] S. K. Jain, S. Singh, A. K. Srivastava, On Σ-q rings, *J. Pure and Appl. Alg.*, 213 (2009), 969–976.

[147] S. K. Jain, S. Singh, R. G. Symonds, Rings whose proper cyclic modules are quasi-injective, *Pacific J. Math.*, 67 (1976), 461–472.

[148] R. E. Johnson, Quotient rings of rings with zero singular ideal, *Pacific J. Math.*, 11, 4 (1961), 1385–1392.

[149] R. E. Johnson, E. T. Wong, Quasi-injective modules and irreducible rings, *J. London Math. Soc.*, 36 (1961), 260–268.

[150] I. Kaplansky, Elementary divisors and modules, *Trans. Amer. Math. Soc.*, 66 (1949), 464–491.

[151] I. Kaplansky, Modules over Dedekind rings and valuation rings, *Trans. Amer. Math. Soc.*, 72 (1952), 327–340.

[152] I. Kaplansky, *Algebraic and Analytic Aspects of Operator Algebras*, Amer. Math. Soc., Providence, R. L., 1970.

[153] F. Kasch, *Modules and Rings*, Academic Press, London, England 1982.

[154] G. Klatt, L. Levy, Pre self-injective rings, *Trans. Amer. Math. Soc.*, 137 (1969), 407–419.

[155] A. Koehler, Ph.D. thesis, Indiana University, Bloomington, Indiana, 1968.

[156] A. Koehler, Rings for which every cyclic module is quasi-projective, *Math. Ann.*, 189 (1970), 311–316.

[157] A. Koehler, Rings with quasi-injective cyclic modules, *Quart. J. Math.*, Oxford, 25 (1974), 51–55.

[158] G. Koethe, Verallgemeinerte Abelsche Gruppen mit hyperkomplexem Operatorenring, *Math. Z.*, 39 (1935), 31–44.

[159] R. P. Kurshan, Rings whose cyclic modules have finitely generated socles, *J. Algebra*, 15 (1970), 376–386.

[160] T. Y. Lam, *A First Course in Noncommutative Rings*, Second Edition, Springer-Verlag, New York, 2001.

[161] T. Y. Lam, *Lectures on Modules and Rings*, Springer-Verlag, New York, 1999.

[162] J. Lambek, *Lectures on Rings and Modules*, Blaisdell, Waltham, Mass., 1966.

[163] T. H. Lenagan, Bounded Asano orders are hereditary, *Bull. London Math. Soc.*, 3 (1971), 67–69.

[164] T. H. Lenagan, Nil radical of rings with Krull dimension, *Bull. London. Math. Soc.*, 5 (1973), 306–311.

[165] J. Levitzki, On rings which satisfy the minimum condition for right-hand ideals, *Compositio Math.*, 7 (1939), 214–222.

[166] L. S. Levy, Torsion-free and divisible modules over non-integral domains, *Canad. J. Math.*, 15 (1963), 132–151.

[167] L. S. Levy, Commutative rings whose homomorphic images are self-injective, *Pacific J. Math.*, 18, 1 (1966), 149–153.

[168] S. R. López-Permouth, S. T. Rizvi, M. F. Yousif, Some characterizations of semiprime Goldie rings, *Glasgow Math J.*, 35, (1993), 357–365.

[169] D. Malik, Ph.D. thesis, Ohio University, Athens, Ohio, 1985.

[170] J. C. McConnell, J. C. Robson, *Noncommutative Noetherian Rings*, GSM, Vol. 30, Amer. Math. Soc., 2001.

[171] P. Menal, On the endomorphism ring of a free module, *Publ. Mat. Univ. Autonoma Barcelona*, 27 (1983), 141–154.

[172] W. Menzel, Über den Untergruppenverband einer Abelschen Operatorgruppe. Teil II. Distributive und M-Verbande von Untergruppen einer Abelschen Operatorgruppe, *Math. Z.*, 74, 1 (1960), 52–65.

[173] W. Menzel, Ein Kriterium für Distributivität des Untergruppenverbands einer Abelschen Operatorgruppe, *Math. Z.*, 75, 3 (1961), 271–276.

[174] G. Michler, Prime right ideals and right noetherian rings. In: R. Gordon, (ed.) *Ring Theory*, Academic Press, New York, 1972.

[175] G. O. Michler and O. E. Villamayor, On rings whose simple modules are injective, *J. Algebra*, 25 (1973), 185–201.

[176] Y. Miyashita, On quasi-injective modules, *J. Fac. Sci. Hokkaido Univ.*, 18 (1965), 158–187.

[177] S. H. Mohamed, q-rings with chain conditions, *J. London Math. Soc.*, 2 (1972), 455–460.

[178] S. H. Mohamed, Rings whose homomorphic images are q-rings, *Pacific J. Math.*, 35 (1970), 727–735.

[179] S. H. Mohamed, B. J. Mueller, *Continuous and Discrete Modules*, Cambridge University Press, Cambridge, 1990.

[180] B. J. Mueller, S. T. Rizvi, On injective and π-injective modules, *J. Pure Appl. Algebra*, 28 (1983), 197–210.

[181] T. Nakayama, Note on uni-serial and generalized uni-serial rings, *Proc. Imp. Acad. Tokyo*, 16 (1940), 285–289.

[182] T. Nakayama, On Frobeniusean algebras II, *Ann. of Math.*, 42 (1941), 1–21.

[183] W. K. Nicholson, Semiregular modules and rings, *Can. J. Math.*, 28 (1976), 1105–1120.

[184] A. J. Ornstein, Rings with restricted minimum condition, *Proc. Amer. Math. Soc.*, 19 (1968), 1145–1150.

[185] B. L. Osofsky, Ph.D. thesis, Rutgers, The State University, New Brunswick, New Jersey, 1964.

[186] B. L. Osofsky, Rings all of whose finitely generated modules are injective, *Pacific J. Math.*, 14 (1964), 645–650.

[187] B. L. Osofsky, A counter-example to a lemma of Skornjakov, *Pacific J. Math.*, 15 (1965), 985–987.

[188] B. L. Osofsky, A generalization of quasi-Frobenius rings, *J. Algebra*, 4 (1966), 373–387; errata, 9 (1968), 120.

[189] B. L. Osofsky, Noninjective cyclic modules, *Proc. Amer. Math. Soc.*, 19 (1968), 1383–1384.

[190] B. L. Osofsky, Noncommutative rings whose cyclic modules have cyclic injective hull, *Pacific J. Math.*, 25 (1968), 331–340.

[191] B. L. Osofsky, On twisted polynomial rings, *J. Algebra*, 18 (1971), 597–607.

[192] B. L. Osofsky, Injective modules over twisted polynomial rings, *Nagoya Math. J.*, 119 (1990), 107–114.

[193] B. L. Osofsky, P. F. Smith, Cyclic modules whose quotients have complements direct summands, *J. Algebra*, 139 (1991), 342–354.

[194] S. Plubtieng and H. Tansee, Conditions for a ring to be noetherian or artinian, *Comm. Alg.*, 30, 2 (2002), 783–786.

[195] E. C. Posner, Prime rings satisfying a polynomial identity, *Proc. Amer. Math. Soc.*, 11 (1960), 180–183.

[196] G. Puninski, Some model theory over nearly simple uniserial domain and decomposition of serial modules, *J. Pure Appl. Algebra*, 21 (2001), 319–337.

[197] J. Rada, M. Saorin, On two open problems about embedding of modules in free modules, *Comm. Alg.*, 27, 1 (1999), 105–118.

[198] C. M. Ringel, Infinite length modules. Some examples as introduction, *Infinite Length Modules*, Trends Math., Birkhauser, Basel, 2000, 1–73.

[199] A. Rosenberg, D. Zelinsky, Finiteness of the injective hull, *Math. Z.*, 70 (1959), 372–380.

[200] L. Salce, Almost perfect domains and their modules, *Commutative Algebra*, Springer, (2011), 363–386.

[201] A. Shamsuddin, Rings with Krull dimension one, *Comm. Alg.*, 26, 7, (1998), 2147–2158.

[202] D. W. Sharpe and P. Vamos, *Injective Modules*, Cambridge University Press, London, 1972.

[203] R. C. Shock, Dual generalizations of the artinian and noetherian conditions, *Pacific J. Math.*, 54 (1974), 227–235.

[204] L. A. Skornyakov, Rings with injective cyclic modules, *Dokl. Akad. Nauk SSSR*, 148 (1963), 40–43.

[205] L. A. Skornyakov, When are all modules serial?, *Mat. Zametki*, 5, 2 (1969), 173–182.

[206] L. W. Small, An example in Noetherian rings, *Proc. Nat. Acad. Sci. U.S.A.*, 54 (1965), 1035–1036.

[207] L. W. Small, Semi-hereditary rings, *Bull. Amer. Math. Soc.*, 73 (1967), 656–658.

[208] P. F. Smith, Some rings which are characterized by their finitely generated modules, *Quart. J. Math. Oxford*, 29 (1978), 101–109.

[209] P. F. Smith, Rings which are characterized by their cyclic modules, *Canad. J. Math.*, 24 (1979), 93–111.

[210] C. Somsup, N. V. Sanh, P. Dan, On serial Noetherian rings. *Comm. Alg.*, 34, 10 (2006), 3701–3703.

[211] A. K. Srivastava, Ph.D. thesis, Ohio University, Athens, Ohio, 2007.

[212] A. K. Srivastava, On Σ-V rings, *Comm. Alg.*, 39, 7 (2011), 2430–2436.

[213] B. Stenström, *Rings of Quotients: An Introduction to Methods of Ring Theory*, Springer, Berlin, 1975.

[214] W. Stephenson, Modules whose lattice of submodules is distributive, *Proc. London Math. Soc.*, 28, 2 (1974), 291–310.

[215] S. Singh, Serial right noetherian rings. *Canad. J. Math.*, 36 (1984), 22–37.

[216] R. Symonds, Ph.D. thesis, Ohio University, Athens, Ohio, 1975.

[217] T. S. Tolskaya, When are all cyclic modules essentially embedded in free modules, *Mat. Issled.*, 5 (1970), 187–192.

[218] A. A. Tuganbaev, Rings all modules over which are direct sums of distributive modules, *Vestnik Mosk. Gos. Univer. Mat., Mekh.*, 1 (1980), 61–64.

[219] A. A. Tuganbaev, Rings over which all cyclic modules are poorly injective, *Trudy Sem. Petrovsk.*, 6 (1981), 257–262. See English translation in *J. Math. Sci. (New-York)*, 33, 4 (1986), 1153–1157.

[220] A. A. Tuganbaev, Distributive rings of series, *Mat. Zametki*, 39, 4 (1986), 518–528.

[221] A. A. Tuganbaev, Distributive rings and modules, *Trudy Mosk. Mat. Obshch.*, 51 (1988), 95–113.

[222] A. A. Tuganbaev, *Semidistributive Rings and Modules*, Kluwer Academic Publishers, Dordrecht-Boston-London, 1998.

[223] A. A. Tuganbaev, Skew-injective modules, *J. Math. Sci. (New-York)*, 95, 4 (1999), 2328–2420.

[224] A. A. Tuganbaev, *Rings Close to Regular*, Kluwer Academic Publishers, Dordrecht-Boston-London, 2002.

[225] A. A. Tuganbaev, Max rings and V-rings, *Handbook of Algebra*, Vol. 3, North-Holland, Amsterdam, 2003, 565–584.

[226] A. A. Tuganbaev, Rings over which all modules are semiregular, *J. Math. Sci. (New-York)*, 154, 2 (2008), 249–255.

[227] A. A. Tuganbaev, Modules with many direct summands, *J. Math. Sci. (New-York)*, 152, 2 (2008), 3928–3933.

[228] A. A. Tuganbaev, Rings over which all finitely generated modules are $\aleph_0$-injective, *Discrete Math. Appl.*, 19, 6 (2009), 619–624.

[229] A. A. Tuganbaev, Rings without infinite sets of noncentral orthogonal idempotents, *J. Math. Sci. (New-York)*, 162, 5 (2009), 730–739.

[230] A. A. Tuganbaev, Completely integrally closed modules and rings, *J. Math. Sci. (New-York)*, 171, 2 (2010), 296–306.

[231] A. A. Tuganbaev, Rings over which all cyclic modules are completely integrally closed, *Discrete Math. Appl.*, 21, 4 (2011), 477–497.

[232] D. V. Tyukavkin, Regular self-injective rings and V-rings, *Algebra i Logica*, 33, 5 (1994), 546–575.

[233] Y. Utumi, On continuous regular rings and semisimple self-injective rings, *Canad. J. Math.*, 12 (1960), 597–605.

[234] R. B. Warfield, Serial rings and finitely presented modules, *J. Algebra*, 37, 3 (1975), 187–222.

[235] R. Wisbauer, *Foundations of Module and Ring Theory*, Gordon and Breach, New York, 1991.

[236] L. E. T. Wu, J. P. Jans, On quasi-projectives, *Ill. J. Math.*, 11 (1967), 439–447.

Index*

*Disregard the prefix right or left, if any, of a term when searching for its location.